D0821719

The Road to Jaramillo

WILLIAM GLEN

The Road to Jaramillo

Critical Years of the Revolution in Earth Science

STANFORD UNIVERSITY PRESS, STANFORD, CALIFORNIA
1982

LS

Illustration credits appear on p. 437

Stanford University Press
Stanford, California

Printed in the United States of America
ISBN 0-8047-1119-4
LC 80-51647

TO THOSE WHO CARED

Preface

THIS WORK is an outgrowth of my realization, over a decade, that more than with a scientific body of knowledge, I was concerned with the ways in which science acquires its information, and with how scientists behave while doing science. I was attracted to the disparity between, on the one hand, the perfect symmetry and patness of professional identities, programs, and eventualities conveyed in some histories of science and, on the other hand, the real world of conflict, unrealized prospects, and halting progress that I had observed in the scientific community. The semblance of symmetry I had found so unsettling seemed somehow to accrue with increasing historical remoteness; in that surmise I turned to oral history and to an important, partly familiar, and conveniently available subject for study.*

During the completion of a textbook on continental drift and plate tectonics in 1974, I began to think about writing a history of the intellectual components, drawn from several branches of earth science, that had been synthesized to yield the recently emergent theory of plate tectonics. The almost endless character of such a project was mercifully clouded, for me, both by inexperience in doing history and by a zeal that deafened me to admonitions. Nonetheless, reconnaissance suggested that an assault on a more compact objective would more likely yield a successful first effort. After what then seemed an interminable search period, it became clear that within the broad spectrum of disciplines that had contributed

*I was emboldened in my surmise by Thomas Kuhn's view that "scientific education makes use of no equivalent for the art museum or the library of classics, and the result is a sometimes drastic distortion in the scientist's perception of his discipline's past. . . . He comes to see it as leading in a straight line to the discipline's present vantage. . . . He comes to see it as progress. No alternative is available to him while he remains in the field" (1970, p. 167).

to the eventual development and acceptance of plate theory, the paleomagnetists had been most outspoken and influential during the two decades prior to the theory's confirmation.

Surprisingly, many of those paleomagnetists, largely Englishmen, who worked assiduously during the 1950's and early 1960's to plot ancient magnetic-pole positions in order to demonstrate continental displacement, and who almost single-handedly resurrected the long-dormant question of continental drift, were not greatly concerned with reversals of the earth's magnetic field. Instead they left it to American and Australian colleagues, paleomagnetists and radiometrists, to demonstrate that the earth's magnetic field has repeatedly reversed itself—and by so demonstrating, to remove the earth's polarity from the roster of physical constants. It was the time scale that served as a calendar of those polarity shifts that serendipitously became the standard of reference by which seafloor magnetic-anomaly data were interpreted to virtually prove seafloor spreading (which had come to subsume continental-drift theory) in 1966. Almost simultaneously, polarity reversals found in deep-sea cores further validated the polarity-reversal time scale.

The research program in which the reversal scale was formulated was minimally staffed, and was pursued systematically in only two research centers; yet in less than a decade it had produced a scientific tool unanticipated in its almost magical usefulness.

As I reviewed the development and application of the reversal scale, I saw that it appeared to be a manageable and important subject for a detailed intellectual history. It was also a subject in which most of the major participants and informants were accessible in person or by telephone; thus was it eventually possible to secure a large quantity of oral data, much of it tape-recorded. In the end a major part of the study included above 500 hours of interviews with more than 100 respondents, of which 130 hours with more than 70 respondents are preserved on tape. These data were integrated with published and unpublished references, personal correspondence, administrative and laboratory records, memorabilia, and other primary historical materials.

My doctoral dissertation, researched and written mainly from November 1977 to December 1978, in large measure comprises Parts I and II of this book. Additional interviews, research, and writing from February 1979 through June 1981 were required to assemble the first draft of the complete manuscript. And in the end I had expanded my account, almost against my will, to far beyond its original scope: I had drawn in both the potassium-argon rock daters, whose new technology made possible the young-rock dating so crucial to the work of the paleomagnetists, and the seafloor investigators, to whose puzzling data the polarity-reversal scale was applied to finally confirm the inexorable spread of the seafloor.

The History of Science and Technology Program at the Bancroft Library, University of California at Berkeley, and the Center for History of Physics of the American Institute of Physics in New York City have both provided technical support and now serve as the repositories for the primary historical materials, but the Bancroft Library contains the complete archive as the "Project in Geomagnetic-Reversal Time Scale History" (see the Table of Interviewees, p. 433).

Arthur Norberg of the Bancroft Library provided early encouragement and served as a most helpful consultant during much of the reconnaissance and data-gathering stages, and Spencer Weart, Director of the Center for History of Physics, provided valuable criticism on certain parts of the manuscript and kind advice at several junctures. Charles Repenning of the United States Geological Survey at Menlo Park, California, has helped me in many ways, including arranging my stay at the Survey during the time in which my doctoral dissertation was written; my debt to him is profound. George E. Robinson, Assistant Director of the Survey, later arranged my appointment as a visiting research fellow and provided access to Survey facilities so that I might complete the book. The project has benefited from the cooperation of many members of the Survey; William Sanders and others of the library staff at Menlo Park provided extraordinary service. Telephone conversations with Survey historian Harold Burstyn were splendidly helpful; and Virginia Langenheim, with coffee and cheer, provided insights into Survey programs and introductions to many of the staff. Henry Frankel of the University of Missouri kindly commented on an early draft of a summary of the dissertation.

The final stages of the manuscript were completed during my tenure as a research fellow in the Office for History of Science and Technology, University of California at Berkeley, which was kindly arranged by Roger Hahn and John Heilbron.

The members of my doctoral committee, including Richard Lichtman, Peter Eckman, Reda Sobky, and Kathy Spangler of the Union Graduate School; Joseph Gregory of U.C., Berkeley; Batia

Sharon of U.C., Santa Cruz; and David Hopkins and Sherman Grommé of the U.S. Geological Survey, were supportive and helpful, each in unique and important ways.

Viktoria Langenheim and Neona Pubols were valuable secretarial aides and Isago Tanaka generously prepared many of the photographs.

I am also in the debt of those who contributed the most significant data by patiently answering questions; many spent considerable time searching their own files for historical materials, and others reviewed certain parts of the manuscript to provide important criticisms and additions. It is not possible to detail all of these many kindnesses, but among the most noteworthy contributors were James Balsley, Richard Blakely, Edward Bullard, Allan Cox, Garniss Curtis, Brent Dalrymple, Bruce Doe, Richard Doell, Ruth Doell, Walter Elsasser, Jack Evernden, Robert Folinsbee, John Foster, Sherman Grommé, Christopher Harrison, Bruce Heezen, James Heirtzler, David Hopkins, Edward Irving, Joseph Lipson, Harold Malde, Drummond Matthews, Ian McDougall, Jason Morgan, Lawrence Morley, Alfred Nier, Neil Opdyke, Walter Pitman, Arthur Raff, Charles Repenning, John Reynolds, Don Tarling, John Verhoogen, Fred Vine, Clyde Wahrhaftig, Gerald Wasserburg, and Tuzo Wilson.

Joseph Gregory first aroused my interest in the history of science by his course of lectures at Berkeley on the origins of paleontology; from the beginning of the project, his advice, criticism, and encouragement have been truly vital.

David Hopkins, a most valuable informant, critic, and friend, generously shared his broad knowledge of many aspects of this history—he lit my journey through much of the labyrinth.

William Carver, Senior Editor of Stanford University Press, was sublime in catering to the parent while directing editors Andrew Alden and Joy Dickinson through surgery on the child. I am thankful for having learned much from them during a delightful experience.

Aida Larson has been of invaluable service in almost all phases of the project; her tireless and intelligent effort, tact, initiative, and good humor have been peerless. The book would surely have been greatly delayed without her signal contribution.

During this undertaking my family suffered neglect; they returned care and love.

W.G.

Contents

Part II. Uncovering the Key: Geomagnetic Polarity-Reversal Scales

Part III. Turning the Key: Applying the Scale

Figures

Note on Citation

Asterisks in the text denote additional substantive commentary appearing as foot-of-page notes. *Arabic superscripts*, numbered consecutively in each chapter, signal notes given in the back of the book; these notes designate sources or offer substantive but less important excursuses. Most *published works* cited in the text or the notes are given simply by author's name and date (of publication), and are not marked by either asterisk or arabic superscript; full bibliographic information, however, for all works mentioned, is given in the References, p. 393. *References to tapes* include the interviewee's name, the cassette-tape number and side, and, when the interviews took place on two or more days, by the date of the interview; e.g. "Cox, t. 2, s. 1, v-4-78." The Table of Interviewees, p. 433, lists taped and untaped interviews, by date and duration. All *documents cited by number* are now part of the Archive in Geomagnetic Polarity-Reversal Time Scale History, History of Science and Technology Program, Bancroft Library, University of California at Berkeley.

The Road to Jaramillo

Introduction

"YOU OWE ME A MARTINI," read the telegram. It had reached Allan Cox in a remote Eskimo village on the Yukon, and he was "pretty sure it meant that Dick Doell had done the heating experiments and there was no indication of self-reversal." He was right: the rock samples had not somehow reversed their own polarity. He knew, too, what that meant: the Rock Magnetics group, back in Menlo Park, had confirmed that the earth's magnetic field reversed itself 900,000 years ago.

"Allan was a little doubtful about the Jaramillo event at the time," Doell recalled years later. "He thought we were going too far out, to be basing this event just upon a few rocks. 'But hell,' we argued with him, 'we had no more rocks than that when we hypothesized the whole polarity-reversal time scale in the beginning!'"

For more than four years, Cox, Doell, and Brent Dalrymple had been collecting rocks from around the world and carefully determining the age and magnetic polarity of each rock sample. From these data they had been assembling a polarity-reversal time scale—a calendar stretching back millions of years from the present, in which were identified intervals when the earth's magnetic field was of normal polarity, as it is today, alternated with others of reverse polarity. Throughout those four years, the geological community paid little heed to their efforts, or to the work of their competitors in Australia, Ian McDougall and Don Tarling. Both the importance and the validity of the research were questioned (after all, could the earth actually have reversed its own magnetic field, on several occasions?), and it was clear in any case that the time scale was incomplete.

Each time a new reversal, or "event," was discovered, the Americans or the Australians had published a new, more refined, version of the scale—to modest applause. But now a crucial datum, from rocks collected near Jaramillo Creek, in the Jemez Mountains of New Mexico, had fallen into place. That newly found short interval of normal polarity of the earth's field, recorded in the rocks almost a million years ago, was the basis for the eleventh version of the time scale—the version that found immediate corroboration in data from the seafloor and became the key to a revolution.

The discovery of the event was first revealed privately, at a meeting of the Geological Society of America in November 1965. Fred Vine, who would make the most of the discovery, vividly recalled the day: "The crucial thing at that meeting was that I met Brent Dalrymple for the first time. He told me in private discussion between sessions, 'We think we've sharpened up the polarity-reversal scale a bit, but in particular we've defined a new event—the Jaramillo event.' I realized immediately that with the new time scale, the Juan de Fuca Ridge [in the Pacific, southwest of Vancouver Island] could be interpreted in terms of a constant spreading rate. And that was fantastic, because we realized that the record was more clearly written than we had anticipated. Now we had evidence of constant seafloor spreading."

In February, Vine went to the Lamont-Doherty Geological Observatory of Columbia University to visit Neil Opdyke. "Neil was poring over papers on a light table. He must have been drawing up magnetic profiles. . . . In the same room, on the wall, Walter Pitman had pinned up all the *Eltanin* profiles from the South Pacific; he was working on interpretation of the magnetic anomalies. While talking to me about research in general, Neil said, 'We've discovered a new event, and we're just about to publish it. Look, we've got all the details of the time scale of Cox, Doell, and Dalrymple; and we've got a new event at 0.9 million years.' I said, 'Neil, I hate to tell you this, but Cox, Doell, and Dalrymple have already defined it and named it the Jaramillo event.' Neil just about fell off his chair—he said they'd given it another name."

The disappointment could not have troubled Opdyke long, for the discovery his group had just made went beyond confirming Jaramillo. They had been studying profiles showing variations in the magnetism of seafloor rock—diagrams of alternating polarity stretching across hundreds of kilometers of seafloor. One of the profiles had just been drawn up, from data they had recorded the previous fall aboard the vessel *Eltanin* in crossing the East Pacific

Rise on leg 19. Once they had interpreted that profile, the *Eltanin* 19—by using time scale number eleven as their key—there was no mistaking its importance. Walter Pitman confided later that after working all night running out magnified profiles, he "pinned up all the profiles of *Eltanin* 19, 20, and 21 on Neil Opdyke's door and went home for a bit of rest." When he came back, Opdyke (like Vine, an advocate of continental drift) had seen them and "was just beside himself! He knew that we'd proved seafloor spreading! It was the first time you could see the total similarity between the profiles—the correlation, anomaly by anomaly." (Pitman recalled also that Vine, whose theory of magnetization of the spreading seafloor had recently been published, had not felt comfortable with his own theory until he saw *Eltanin* 19.)

Simultaneously, the Jaramillo/*Eltanin* revelations triggered the development of a still greater idea, one that has transformed the earth sciences during the past two decades just as profoundly as the propositions of Copernicus, Darwin, and Einstein overturned their worlds and disciplines. For if seafloors could spread, continents could drift. This powerful new theory, which we know as plate tectonics, explained continental drift and seafloor spreading and went beyond both. It opened up new visions of how the planet works, with implications beyond our ken. During the tumultuous half-decade beginning with that confirmation of seafloor spreading in 1966, the centuries-old belief that continents and oceans were fixed in position was displaced, almost overnight, by the contradictory hypothesis that continent-size slabs or plates of rigid crustal shell move continuously about, opening and closing ocean basins, thrusting up mountains or forming deep-sea trenches where they collide, and forming and reforming the major features of the earth's surface.

This grand unifying theory has finally interconnected a great storehouse of isolated geological facts that had been accumulating for centuries. Plate tectonics holds that the lithosphere—the rigid outer shell of the earth, as much as 150 kilometers thick—is divided into about 25 huge plates of diverse expanse and shape. These plates ride or float on a plastic zone beneath them in the weak, upper part of the earth's mantle called the asthenosphere. The plates move about independently, and those that make up the seafloor are propelled outward from the axes of very long, mid-ocean submarine ridges: by the almost constant injection of lava from below the ridge, new crust is added along the two opposing plate edges as they pull steadily apart along the line of spreading; and far to each side, on the two plate edges thousands of kilometers away from the mid-ocean ridge, the plate boundaries dive into deep-sea trenches to be consumed in the earth's mantle. Moving from spreading axis to trench at the rate of only several centimeters per year, the entire seafloor is recycled every hundred million years or so.

Some of the other plates slide past each other laterally, along great transform faults. California's San Andreas Fault is such a fault, the land to its west moving steadily north with respect to the larger North American landmass. The continents, for so long thought immutable, are in fact only passengers riding the plates, and often comprise parts of several plates, as is the case in the lands about the Mediterranean. The major mountain belts, earthquake zones, and chains of volcanoes—including the great "Ring of Fire" about the Pacific—all tend to occur at plate boundaries.

Once these ideas were grasped, progress was rapid in learning how plates grow and are destroyed, and when and where they moved in the geologic past. What it is that drives the plates, however, is not well understood. Heat-diffusing convection currents moving at several centimeters per year in the plastic upper mantle, and acting somehow in concert with lesser forces, seem at present the best answer to a still difficult question. That problem notwithstanding, plate tectonics has been shown to be a theory of enormous breadth and predictive power, offering convincing answers to a host of centuries-old questions about how the earth works—and to still other questions no one had thought to ask. Almost two decades after its advent, the theory continues to be confirmed daily in the practice of earth science.

The several elements of the theory were long in the making, and many of them, as is often the case, formed parts of simpler predecessor theories. Plate theory grew from the hypothesis of seafloor spreading, which had earlier developed from a series of grafts onto the idea of continental drift.

Casual speculations about continental drift, inspired mainly by the jigsaw-puzzle fit of such continental edges as eastern South America and western Africa, had been advanced for centuries. But it was not until 1915 that Alfred Wegener drew sustained attention to the question by the publication of his monumental book *Die Entstehung der Kontinente und Ozeane*, which had been heralded by his lectures in Frankfurt and Marburg in 1912. Brilliantly marshaling the evidence then available, he spoke for consolidation of the existing continents into a single ancient landmass, which he called Pangaea. His panoply included intercontinental correlations of mountain ranges, rock formations, the great faults, fossil animals and plants, glacial features, and much more. There was little then available that he omitted. But what was not available was crucial, and the counterattack was predictably aimed at what he was unable to explain, and what is still problematic today: What force rips the continents apart and moves the pieces across the face of the earth?

The geophysicists, led by Harold Jeffreys of Cambridge, were Wegener's most outspoken opponents—their strictures did not permit a weak continental ship to sail through an unyielding oceanic crust. And with a firm institutional and intellectual grip on the earth sciences, they had no difficulty in attracting allies from other subdisciplines. Continental-drift theory found few supporters, outspoken or otherwise; orthodoxy would have its day.

Much of the criticism was in fact logically indefensible, but history is replete with examples of purely emotional attacks on bold theorists like Wegener. Although the main argument against drift—the absence of an acceptable mechanism—was sensible enough, many of the strongest critics displayed abysmal command of their own materials.

Chief among Wegener's supporters during the 1920's and 1930's were Alexander Du Toit of South Africa and Arthur Holmes of Edinburgh. It was Holmes who offered convection in the plastic mantle of the earth as a driving force; the idea was part of an early version of what was to become the theory of seafloor spreading. At about the same time Felix Vening-Meinesz's gravity studies over deep-sea trenches, near the great landmasses, suggested that an unknown force there was depressing the crust in defiance of flotation—a proposition that seemed to support Holmes's convection theory. And in the 1930's, K. Wadati, and later Bruno Benioff, showed that earthquake zones hundreds of kilometers deep slanted away from the submarine trenches at steep angles, far under the adjacent continents. But like other important observations, the Benioff-Wadati zone could not be fitted into a workable model explaining continental drift, and by and large the question of drift marked time for another decade.

At about the time of World War II, a new body of evidence about the earth began to emerge from studies of seafloor geology. By 1954

Jean Pierre Rothé, who had studied earthquake distribution, suggested that the mid-Atlantic ridge continued around South Africa to connect with the ridges of the Indian Ocean. Two years later, Maurice Ewing and Bruce Heezen remarked on the continuous, worldwide character of the mid-oceanic ridges, and observed that they were punctuated by earthquake foci and often split by great central valleys. Crustal mobility seemed to be supported by tensional structures and unusually high heat flow at the axes of the mid-ocean ridges; and the youthfulness of all seafloor rock demanded a device for disposing of the mysteriously absent *old* oceanic crust. These observations were later crucial in the formation of plate theory.

Much other evidence in favor of drift, some supporting Wegener's classical arguments, some based on a rash of newer kinds of studies derived largely from oceanographic surveys of the 1940's and 1950's, attracted increasing attention. But no program of research seemed as promising in its ability to prove drift as did that of fossil rock magnetism. Using recordings of the earth's magnetic field within certain kinds of rocks extending back hundreds of millions of years, English paleomagnetists showed, during the late 1950's and early 1960's, that the magnetism in rocks of some periods was not magnetically aligned with the earth's present field. They thus provided the first clearly numerical evidence for drift, and revived the continental-drift question from 20 years of virtual stagnation. But their arguments, drawn from an esoteric specialty with weak theoretical underpinnings, were not widely understood; only a minority was convinced, and paleomagnetism had not yet had its day.

By 1960, Harry Hess of Princeton University, enlarging upon the idea of Arthur Holmes, spoke of a mobile oceanic crust that grew at, and was carried away from, spreading centers along the axes of mid-oceanic ridges. On a slow, relentless journey, the crust traveled across an ocean basin to plunge finally into the mantle at the sites of deep-sea trenches. Hess combined several extant elements with two novel ideas: raising the mid-oceanic ridges by swelling the crust through the addition of water in a chemical reaction, and getting rid of old seafloor crust by flexing it down into the mantle along Benioff-Wadati earthquake zones at trench sites. Hess's theory, named seafloor spreading by Robert Dietz in 1961, stimulated wide interest; it embraced drift and largely explained the method of transport.

At about the same time, Ronald Mason and Arthur Raff were describing a unique pattern of parallel magnetic stripes on the ocean floor off the west coast of North America. Victor Vacquier believed the stripes were broken and offset hundreds of kilometers by great faults; he thus supported the idea of great lateral movement of the seafloor.

In 1963, that enigmatic zebra-striping of the oceanic crust was articulated with Hess's ideas of seafloor spreading by Fred Vine and Drummond Matthews, in England, and also independently by Lawrence Morley, in Canada. They suggested, in an outlandish hypothesis, that at the moment of its formation the new seafloor along the axis of a mid-oceanic ridge became magnetized as the rock cooled, capturing the direction and intensity of the earth's magnetic field. They also embraced the theory that the earth's magnetic field had repeatedly reversed its polarity in the past. Thus, as the seafloor departed from the spreading centers at the ridges, a mirror-image record of alternating normally and reversely magnetized stripes of seafloor crust would be produced. And because seafloor spreading encompassed the theory of continental drift, to prove the Vine-Matthews-Morley hypothesis would be to prove drift. But in subsuming the earlier, very questionable, hypotheses of seafloor spreading and reversals of the earth's magnetic field, Vine-Matthews-Morley was itself suspect, and it was poorly received. (As late as early 1966 it was rejected by geophysicists who examined it on the basis of seafloor magnetics data.) Thus lodged against a complex theoretical impasse, Vine-Matthews-Morley looked to an unpromising future.

The future seemed brighter in July 1965, when J. Tuzo Wilson postulated a new kind of fault—the transform fault—as the third boundary of a rigid crustal plate. Although his landmark paper contained the essence of plate tectonics theory, it too suffered inattention. Not for long, however: in less than a year, Jaramillo and *Eltanin* 19 would confirm Vine-Matthews-Morley and vindicate Wilson, Hess, Holmes, Wegener, and other mobilists.

But that is history, and several books now offer useful overviews of continental drift, seafloor spreading, and plate tectonics. In 1973 Allan Cox's anthology appeared, an important work laced with his own vivid historical commentary. In the same year Anthony Hallam published a terse but rich history informed by an insider's reflections on the revolution, and Ursula Marvin produced a lengthier work that examines more closely the period preceding theories of seafloor spreading. Walter Sullivan's 1974 history is comprehensive and delightfully readable. Two textbooks, my own brief one of 1975 and Peter Wyllie's more extensive one of the following year, are also historically oriented. Seiya Uyeda's excellent, up-to-date summary of 1978 examines recent developments in plate tectonics from the viewpoint of a front-line practitioner of geophysics. But there are no detailed, book-length histories treating any of the many research programs that contributed to the development of plate theory. Among others, there were programs that investigated large-scale deformational structures, earthquakes, gravity, heat flow, and magnetic polar wander. All were important and deserve full historical treatment.

In this work I have concentrated for the most part on three programs, those whose roles appear to have been decisive in launching the revolution: the rise of methods for *dating geologically young rocks* by the counting of potassium-argon isotopes; the use of these methods in dating magnetically determined rocks in order to formulate a *time scale of reversals of the earth's magnetic field*; and the *applications of that scale* in deciphering a large body of puzzling magnetic data from the rocks and sediment of the seafloor.

Initially, these programs had been undertaken with more modest prospects, and each with no thought of serving the others. Their story is in fact a suitable model against which to try the templates of philosophers, historians, and behavioral scientists interested in how science works. The geophysics community, for example, had known for more than half a century that the rocks in many parts of the earth's crust were magnetized with a polarity opposite that of the present-day magnetic field: magnetic north in these rocks points south. But in the late 1950's opinion was still divided about the cause of the phenomenon. Was the reverse magnetization due to reversals of the earth's magnetic field during the billions of years of geologic time? Or simply to certain minerals that could cause reversals of magnetization within the rock itself? The geomagnetic polarity-reversal time scale, a chronology based on identifying and dating the successive reversals of the earth's magnetic field, grew from efforts to decide that long-standing question—not to prove continental drift.

Although much evidence had been gathered for a reversing global magnetic field by the mid-1950's, proof was still wanting in the form of a conclusive demonstration that rocks of identical age have the same magnetic polarity, all over the earth. Means for accurately measuring fossil magnetism in rock had been available since the early 1950's; what was needed was a companion technology that would permit the precise *dating* of geologically very young rocks. By December 1954, at the University of California in Berkeley, John Reynolds had designed and built the first static-mode mass spectrometer. With this atom-counting machine he achieved an astounding refinement in the measurement of radiogenic argon in rock. Two Berkeley geologists, Garniss Curtis and Jack Evernden, learned the dating techniques, built another Reynolds machine, used the gas-extraction equipment designed and built by Joseph Lipson, and rapidly evolved techniques for dating even younger rocks. A program promising accurate potassium-argon dating of young rocks was thus established.

Among the minority who subscribed to the field-reversal hypothesis, only Allan Cox and Richard Doell, trained in paleomagnetism under John Verhoogen at Berkeley, actually conceived a research program aimed specifically at resolving the reversal question. And in 1959, at the United States Geological Survey in Menlo Park, California, they undertook the program, in short order inviting Brent Dalrymple, a young Berkeley-trained rock-dater, to collaborate in formulating a polarity-reversal time scale. During the early 1960's, they produced several increasingly refined versions of the scale through painstaking examination of countless rock samples. They found that the individual magnetic-polarity intervals appeared to vary greatly in duration, some only a few tens of thousands of years long, others persisting for hundreds of thousands of years. But the record remained incomplete, and their efforts were marked by a series of misadventures—they were conducting research of little interest to their community, and the difficulty of their experiments was matched by the strain of securing adequate instruments, technical support, and space.

While their research was getting under way, a young Australian petrologist, Ian McDougall, was being trained in young-rock dating with Dalrymple at Berkeley, as part of his 1960–61 postdoctoral year. That same year, John Jaeger, of the Australian National University in Canberra, commissioned Jack Evernden to assemble there a duplicate of the Berkeley dating laboratory. When McDougall then resumed his work in Canberra, with the new machine, he acquired great skill in dating young rocks and began a fruitful collaboration with Don H. Tarling, a paleomagnetist. Their success in the study of reversals gave the Menlo Park group a sense of urgency, and research in both centers profited from the competition. Although starting later than Cox, Doell, and Dalrymple, McDougall and Tarling benefited from their example and produced two of the crucial early scales.

The eleventh version of the polarity-reversal scale, containing the newly discovered Jaramillo event, was revealed at Menlo Park in 1966. That scale, developed from potassium-argon measurements of terrestrial rocks, became the standard by which to interpret not only the *Eltanin* 19 profile, which recorded the magnetism of great oceanic crustal blocks constituting a broad expanse of the seafloor, but also a third set of magnetic data. Neil Opdyke's group at the Lamont-Doherty Geological Observatory showed further that deep-sea cores from widely spaced localities held layers of alternately magnetized sediment in ratios identical to the ratios in the other records. Reversals were thus found to have been recorded by three different kinds of geological mechanisms.

This confluence of evidence instantly fired intense interest in the Vine-Matthews-Morley hypothesis, which had predicted what was in fact confirmed. Within two years of the Jaramillo discovery, seafloor spreading evolved into an extended model of plate tectonics through a number of ingenious contributions treating both geologic structures and kinematics. J. Tuzo Wilson's crucial, but dormant, transform-fault concept, which defined the tectonic plate, was suddenly thrust into prominence. And plate tectonics became a ruling paradigm perhaps more quickly than had any before in any discipline.

The road to the eleventh polarity-reversal scale was marked by important personal relationships, by institutional support (or the lack of it), and by rivalries and frictions between research groups. The road was broad, spanning research programs in the United States, Australia, Canada, England, Japan, Iceland, Germany, France, Italy, the Netherlands, the Soviet Union, and beyond. The road was also crooked: the uneven patterns of its progress repeatedly defied the expectations of those who formulated the research programs. And only toward the end of the road was a common goal recognized. Until then, successive discoveries, theories, and false starts marked the course of seemingly blind progress toward a titanic revolution.

PART I

Building the Hourglass:

Young-Rock Potassium-Argon Dating

Chapter 1

Developing the Potassium-Argon Method

THE MEASUREMENT of geologic time is the proper domain of geochronology; time is the essential framework in which most geologic data are ordered. Classical geology emerged during the eighteenth and nineteenth centuries from field studies often conducted in connection with mining, engineering, antiquarian pursuits, or "natural philosophy." The earliest widely applicable generalizations regarding the dating of rock bodies and geologic structures were based on the time value of the vertical succession of sedimentary rock layers, and in turn on the diagnostic value of certain time-restricted fossil organisms that were embedded in those rock layers and widely distributed geographically. Rocks from widely separated localities, even from different continents, can be assumed to be of roughly similar age if they contain the same complement of fossil organisms. But until the twentieth century, all such dating was relative, in the sense that although a rock's origin could be placed within a known sequence of geologic periods, no actual age, expressed in years, could be assigned to the rock. Those relative time intervals that were formally named—Precambrian, Devonian, Pleistocene, and so on—became part of the modern geologic time scale. Although not all rocks are arranged in layers and not all contain fossils, stratigraphy and paleontology are still the primary tools used by geologists to assign rocks their place in the scale.

Today we bracket the subdivisions of that time scale by dates expressed as spans of years before present (see Fig. 1.1). It was through the efforts of isotopic daters that earth history came to be placed within this numerically expressed temporal framework. All

Subdivisions in use by the U. S. Geological Survey (and their map symbols)					Age estimates[1] of boundaries in million years (m.y.)	
Phanerozoic Eon or Eonothem	Cenozoic Era or Erathem (Cz)	Quaternary Period or System (Q)		Holocene Epoch or Series	0.010	
				Pleistocene Epoch or Series	2	(1.7-2.2)
		Tertiary Period or System (T)	Neogene Subperiod or Subsystem (N)	Pliocene Epoch or Series	5	(4.9-5.3)
				Miocene Epoch or Series	24	(23-26)
			Paleogene Subperiod or Subsystem (Pε)	Oligocene Epoch or Series	38	(34-38)
				Eocene Epoch or Series	55	(54-56)
				Paleocene Epoch or Series	63	(63-66)
	Mesozoic Era or Erathem (Mz)	Cretaceous Period or System (K)		Late Cretaceous Epoch or Upper Cretaceous Series	96	(95-97)
				Early Cretaceous Epoch or Lower Cretaceous Series	138	(135-141)
		Jurassic Period or System (J)			205	(200-215)
		Triassic Period or System (Ћ)			~240	
	Paleozoic Era or Erathem (Pz)	Permian Period or System (P)			290	(290-305)
		Carboniferous Periods or Systems (C)	Pennsylvanian Period or System (ℙ)		~330	
			Mississippian Period or System (M)		360	(360-365)
		Devonian Period or System (D)			410	(405-415)
		Silurian Period or System (S)			435	(435-440)
		Ordovician Period or System (O)			500	(495-510)
		Cambrian Period or System (€)			~570[2]	
Proterozoic Eon or Eonothem (P̱)	Proterozoic Z (Z)[3]				800	
	Proterozoic Y (Y)[3]				1,600	
	Proterozoic X (X)[3]				2,500	
Archean Eon or Eonothem (A)				Oldest known rocks in U. S.	3,600	

[1] Ranges reflect uncertainties of isotopic and biostratigraphic age assignments. Age of boundaries not closely bracketed by existing data shown by ~. Decay constants and isotope ratios employed are cited in Steiger and Jager (1977).

[2] Rocks older than 570 m.y. also called Precambrian (p€), a time term without specific rank.

[3] Time terms without specific rank.

Fig. 1.1. The modern geologic time scale, 1980 edition (courtesy of the Geologic Names Committee, United States Geological Survey).

of the many earlier attempts at reckoning the age of the earth in years were based on physical changes, and all were defective, because the changes they examined are commonly both cyclical and nonprogressive and do not occur uniformly; they included rates for the deposition of sediments, the accumulation of salt in the sea, and the cooling of a once-incandescent earth. In contrast, isotopic dating is based on the natural radioactive decay of certain elements, an extremely constant and uniformly progressive process that is consistently predictable. The process is not reversible, and a constant rate of decay of parent to daughter atoms is maintained under all environmental conditions. By means of painstaking techniques and sophisticated equipment, the atoms in a piece of rock can be separated and counted, and on the basis of that count a true or absolute age can be calculated with great accuracy. Potassium-argon radiometry is one of several isotopic dating techniques in use today—and for the dating of geologically young rocks (as old as about 5 million years), it is the most effective.*

The extent to which the potassium-argon dating of young rocks contributed to the development of the time scale of reversals of the earth's magnetic field has only been touched upon in the few brief attempts to document that history that have been written to date.[1] Although active researchers are seldom concerned with systematically documenting a history of their studies, their neglect in this case is curious in view of the central role that geochronologists played in the research programs that produced the geomagnetic polarity-reversal scale. Brent Dalrymple, for example, in a 1972 paper entitled "Potassium-Argon Dating of Geomagnetic Reversals and North American Glaciations," included a six-page section on early "history" but made no mention of the development of potassium-argon dating, in which he himself had attained international stature. He did, however, devote almost a full page to early rock magnetics, which is another important part of this history. The polarity-reversal time scale grew from an attempt to decide whether reversely magnetized rocks were the result of reversals of the earth's magnetic field or reversals of magnetization within the rock itself owing to its mineralogical properties. Dalrymple was quite aware that "it was not until the 1960's that advances in radiometric dating techniques made it possible to design an experiment using both paleomagnetism and potassium-argon dates that offered a reasonable chance of deciding this question."[2]

*It is the only technique that works consistently in the range 10^5 to 5×10^7 years.

In contrast, the methods and instrumentation necessary to perform magnetic determinations of rock samples were in hand by the 1950's,[3] and several magnetostratigraphic and rock-magnetic studies had demonstrated the need for geochronologic control far more refined than that afforded by the fossil record.[4] How and why the growth of the study of paleomagnetism has so thoroughly overshadowed that of potassium-argon dating in discussions of the development of the polarity scale is a complex question, but it is noteworthy that the young-rock geochronologists themselves have contributed to the imbalance. When geology's most coveted honor, the Vetleson Prize, was given to Allan Cox and Richard Doell in April 1971 for their efforts in producing the polarity-reversal scale, the award was presented by Maurice Ewing, who had himself received the first Vetleson Prize; Ewing's long and complicated recounting of the development of paleomagnetism was in sharp contrast to his brief mention of the great significance of radiometry.[5]

Because there are no historical studies of the role of radiometry in the formation of the polarity-reversal scale, this and the following chapters will therefore examine that history in some detail.

The Nature and Rise of Isotopic Dating

Atoms of the same element that have the same number of protons but different numbers of neutrons, and are all but perfectly identical chemically, are called isotopes. Oxygen's three naturally occurring isotopes, for example, are designated oxygen 16, 17, and 18, according to their atomic weights (actually protons plus neutrons). In isotopic dating, radioactive isotopes in particular are dealt with; when one of these—the parent isotope (atom or element)—decays, the decay products, or daughter isotopes (atoms), are commonly isotopes of a different element.* Among the parent-daughter pairs currently in common use in isotopic dating are uranium 238, which decays to form lead 206; uranium 235, to lead 207; potassium 40, to argon 40; thorium 232, to lead 208; rubidium 87, to strontium 87; and carbon 14, to nitrogen 14. Because decay proceeds at a different rate for each isotope, each of the dating methods based on such isotope pairs lends itself best to the dating of rocks of a certain age range. After 4.5 billion years the lead atoms produced by the decay of uranium 238 would be equal in number to the remaining uranium-238 atoms; that time interval is called the

*Except for isomeric transitions (not really decay), this is always true.

half-life of uranium 238. By contrast, carbon 14, with a half-life of only 5,730 years, converts two-thirds of its original number of atoms to nitrogen 14 in 10,000 years, and is thus useful in dating geologically very young materials; whereas, after such a geologically brief interval, rubidium 87, which has an extremely long half-life—48.8 billion years—would produce an immeasurably small amount of strontium 87.

Potassium-argon isotopic dating, based on the decay of potassium 40 to argon 40, has emerged as perhaps the most generally useful method.* Potassium is the seventh most abundant element in the crust of the earth and occurs in a great variety of commonly available rock types. Argon is an inert gas that constitutes about 1 percent of the earth's atmosphere; it provides a great advantage in analysis, since, being chemically inert, it is almost never incorporated in any mineral at the time of original formation. Thus the argon 40 found in a sample is derived only by decay from potassium 40 or by contamination from argon 40 in the atmosphere introduced during sample analysis. (As much cannot be said for elements such as lead and strontium, whose isotopes are present in minerals at the time of their formation and thus complicate the dating analysis.) The half-life of potassium 40—1.25 billion years—is an additional advantage, for it means that measurable quantities of argon 40 are present in rocks extending in age from a few tens of thousands of years up to more than 4 billion years. The greatest difficulty with the method, as we shall see, is the contamination of samples by atmospheric argon 40 during analysis.

All dating methods rely on the assumption that when a mineral forms, each element incorporated in it is composed of fixed proportions of that element's isotopes. Most methods also assume that the proportion of isotopes for each element has been virtually constant during all of the earth's 4.7-billion-year history and, further, that at the time the earth's atmosphere was formed, it contained a

*"Decay" signifies any spontaneous change in the makeup of an atomic nucleus. In alpha decay, two protons and two neutrons are emitted as a lump of atomic number 2 and weight number 4; this particle is the same as a nucleus of helium. In beta decay, a neutron disintegrates into a proton and an electron. The proton stays in the nucleus, raising its atomic number but not its weight; the electron, ejected at high speed, is called a beta particle. In electron capture, a proton combines with an orbital electron to form a neutron, thus lowering the atomic number but not the weight. Lead decays by alpha emission. Uranium decays to lead through a series of transformations that involves all three types of decay. Potassium 40 yields calcium 40 by beta decay, or argon 40 by electron capture. The potassium 40–calcium 40 dating method has been unsuccessful because rock-forming minerals incorporate much calcium 40 at the time of their formation; that primary calcium 40 encumbers the measurement of the much smaller quantity of radiogenic calcium 40.

certain proportion of argon isotopes of weight numbers 36, 38, and 40, as inferred from their present abundances in the sun. Only argon 40 is being produced by decay; argon 36 and 38 are constant. It can also be determined that the potassium incorporated in a mineral at its origin included a precisely known fraction of potassium 40 (natural potassium now contains only 0.01167 percent potassium 40, 93.08 percent potassium 39, and 6.91 percent potassium 41). In order to determine age, the concentration of potassium and the amount of radiogenic argon 40—the amount produced by decay—must be measured. Measuring the argon 36 in the sample provides a standard by which to measure the proportion between radiogenic argon 40 and argon 40 from atmospheric contamination. The final calculation of age requires knowledge of the amount of potassium 40 now present, the present quantity of argon 40 produced by its decay, and the rate of decay from parent to daughter isotope. Piecing this information together was one of the premier scientific accomplishments of this century.

The earliest discoveries bearing directly on the growth of isotopic dating may have begun in 1855 with Heinrich Geissler's transmitting of electrical charges through vacuums by means of metal electrodes in evacuated glass vessels; this invention resulted in the discovery of cathode rays, which Joseph J. Thomson later identified as streams of electrons, thereby winning the Nobel Prize in physics in 1906. Wilhelm K. Roentgen was one of several physicists who by chance observed luminescence in the walls of cathode-ray tubes, but he alone discovered that the cathode-ray tube, even when enclosed by black cardboard, caused luminescence in a sheet of paper coated with barium platinocyanide. In 1895 he named the unknown form of radiation that could pass through black paper, calling it x-rays. Henri Becquerel, who around 1880 had experimented with luminescence induced by the exposure of uranyl double-sulfate crystals to ultraviolet light, was moved by Roentgen's work to determine that such crystals, when exposed to sunlight, produced radiation that could penetrate black paper and fog a photographic plate. Shortly afterward, in 1896, he determined that uranium-bearing minerals and salts produce such radiation even in the absence of sunlight.[6] Marya Sklodowska (Marie Curie), excited by Becquerel's early work, pursued the question of radioactivity in her doctoral dissertation at the Sorbonne and discovered not only that thorium was also radioactive, but that natural uranium and thorium minerals were more active than their respective salts.

Thus she and her physicist husband Pierre Curie, in their attempt to learn what it was in those minerals that produced such powerful rays, discovered the elements polonium and radium and coined the term radioactivity in 1898.

The work of the Curies led the New Zealander Ernest Rutherford, while at Cambridge University in 1898 and later at McGill University in Montreal, to determine that radioactive substances emit three distinguishably different components, which he termed alpha, beta, and gamma. The gamma rays were similar to x-rays, the beta proved to be electrons, and the alpha were identified as helium nuclei. Rutherford was joined at Montreal by Frederick Soddy, trained at Oxford. Together they formulated, in 1902, the principles underlying the radioactive transformation of unstable elements, in the course of learning that uranium and thorium disintegrate slowly through a series of reactions that eventually yield lead and helium.[7] They suggested for the first time that the atoms of radioactive elements are unstable and spontaneously disintegrate at a steady rate to yield a fixed proportion of atoms of other elements. Rutherford was also quick to point out, in 1905, that the progressive accumulation of stable new daughter atoms, such as helium in uranium-bearing minerals, might be used in the measurement of geologic time.[8] The following year, in the course of the Silliman lectures at Yale, he explicated the necessary procedures for calculating the age of uranium-bearing minerals. Bertrand Boltwood of Yale had earlier surmised that lead was the final decay product of the uranium decay series, as had Rutherford; then, acting upon a suggestion of Rutherford's concerning methodology, Boltwood calculated and published the first radiometric lead ages, in 1907.

After a number of disappointing attempts at helium dating of uranium minerals, John William Strutt (Lord Rayleigh) began work in the lead method in 1910 at Imperial College, London, with his most gifted student, Arthur Holmes. The publication of more than four dozen papers treating radiometry, from 1911 to 1962, would establish Holmes as the founder of the radiometric time scale. A geologist, Holmes recognized early the important ways in which radiometry ought to be applied to geological problems and centered much of his work on the dating of the Precambrian interval, the most ancient part of earth history.[9] Early work in radiometry produced dates one hundred times, or two orders of magnitude, older than the prevailing authoritative estimates of a 20- to 40-

million-year-old earth, which Lord Kelvin had based on a simple cooling model of a nonradioactive earth.[10]

Despite these dramatic findings, the radically new and generally little-understood radiometry research program pursued by Holmes and others was not well received. One problem was the very limited availability of minerals suitable for dating. A more substantial objection concerned the uncertain character of the decay process and the possibility of movement of the parent and daughter atoms to or from the minerals in question. Such an unrecognized gain or loss would, of course, yield discrepant ages when a mineral is dated by more than one method. (To some extent this problem continues unresolved today.) Owing to deficiencies of knowledge and technique, discordant results were not uncommon during the early years; one of the more notable, in 1908, was George Becker's much too great uranium-lead age of more than 10 billion years for granitic rock from Texas, which prompted his derision of radiometric dating. Thus, most geologists remained skeptical, but the dating contributions that accrued clarified a number of poorly understood points and won increasing support.

In 1913, at the age of 23, Arthur Holmes published *The Age of the Earth*, a comprehensive review of the state of radiometry in this period. Here he presented the first geological time scale, which he based on helium and lead in conjunction with the thicknesses of sedimentary rock sequences. The major question, of course, remained: How was it possible to be certain that radioactive decay had gone on uniformly through all of earth history? This question was answered only by the subsequent accumulation of a great number of dates, derived by means of different isotopic methods; the dates began to fall regularly into an orderly chronologic sequence in consistent agreement with the earlier, well-established, but only relative time scale based on fossils. This process was to occupy decades, until the 1950's, when potassium-argon dating got well under way.

Although radiometrists were beginning to produce an orderly time sequence, their goal of establishing absolute dates continued to be thwarted by their inability to measure quantities of isotopes accurately. Then, during the late years of World War I, Francis W. Aston of Cambridge University and Arthur J. Dempster of the University of Chicago developed the mass spectrograph, a great advance in radiometric instrumentation.[11] Their design was based on Thomson's "positive ray apparatus" at the Cavendish Laboratory. A mass spectrograph separates charged atoms and molecules, ac-

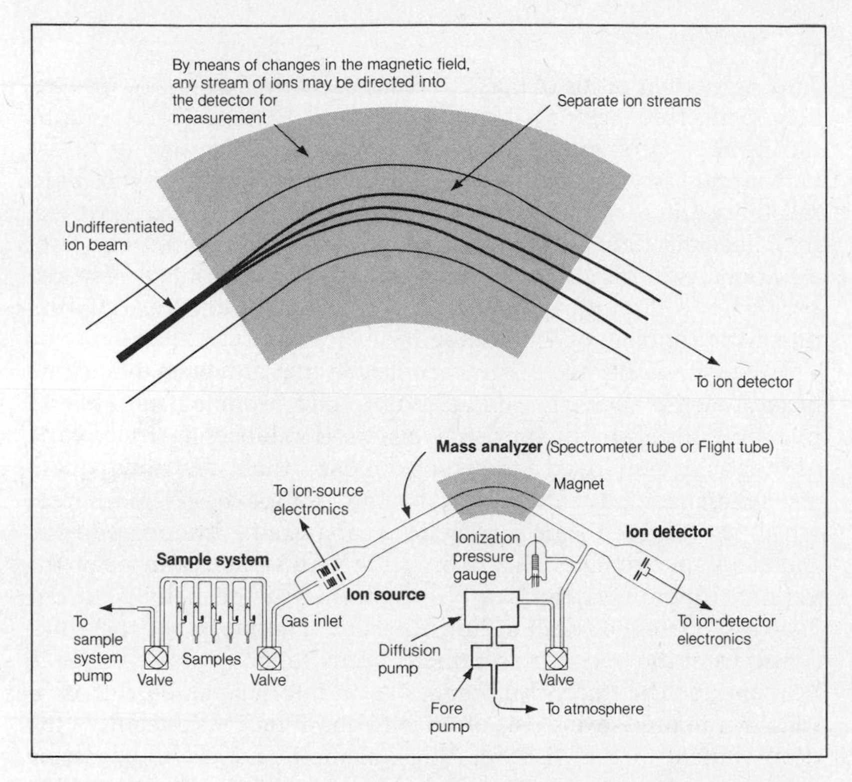

Fig. 1.2. Modern mass spectrometer for argon analysis (after Dalrymple and Lanphere, 1969 [© 1969 W. H. Freeman & Co.]). The mass spectrometer, descendant of the spectrograph developed by Aston and Dempster in the late 1910's, is used to sort ions (charged atoms, molecules, or fragments of molecules) by their motion in electric and magnetic fields: the ions behave according to their mass-to-electric-charge ratios. This instrument may be likened to an optical spectroscope in which a beam of white light, which contains various colors, is passed through a prism to separate the light into the spectrum of colors actually present. A typical mass spectrometer consists of several parts. The *mass analyzer* (also called the spectrometer tube or flight tube) is a curved, continuously pumped (except during analysis in the static mode) vacuum chamber into which a gas sample is admitted; the equilibrium pressure in the tube approximates 10^{-6} mm of mercury. The pumping system for maintaining the vacuum includes a *fore pump* and either a mercury or oil *diffusion pump*, and a *coldtrap* for trapping contaminants and preventing mercury or oil from the diffusion pump from getting into the mass spectrometer. (By the 1960's titanium devices had replaced most such diffusion pumps.) The *sample system* is the apparatus for introducing the gas sample into the area in which ions are generated. The *ion source* bombards the gas samples with electrons, producing positive ions. These ions are then accelerated and focused electrically into a narrow beam that passes through the mass analyzer at great speed. In the curved part of the analyzer, the beam passes through a magnetic field in which the ions are deflected into separate paths according to their respective ratios of mass to electric charge. The paths of ions of lower mass and/or higher electrical charge are bent more than those of higher mass and/or lower charge. The beam of ions is thus separated into streams of ions, each stream with ions of different mass/charge ratio. The *ion detector*, or receiver, records the number of ions in each of the separated mass streams (like the sectors of the spectrum of light) and transmits these data to the ion-detector electronics circuits. There the electrical signal of the separated ion beams is converted to data from which the relative isotope abundances can be calculated. Alfred Nier was the first to replace the photographic plate in the older mass spectrograph with fast-reading, electronic ion detectors in his spectrometer of 1937; that improvement resulted in a great increase in speed, sensitivity, and accuracy.

cording to their ratios of mass to charge, by their behavior in magnetic fields (see Fig. 1.2). The improvement in mass spectrographs during the 1930's was to culminate by 1937 in the design of Alfred O. Nier (at Harvard, later at Minnesota), whose machine attained a reliability and accuracy that set the standards for mass spectrometry. Nier's machine permitted rapid growth in the measurement of elemental isotopic compositions across a broad front of research problems, including isotopic dating of rocks. By the 1940's, absolute geochronology was well begun.[12]

By 1945, another technique, applicable to many elements useful in radiometric analysis, was opened to geochronologists. This remarkably simple method of analysis, used in biochemistry as early as the 1930's, employed a tracer, or diluent—that is, a known quantity of an isotope, different from the isotope being measured, which is mixed with the sample to be analyzed.[13] The record from the mass spectrometer shows only the relative abundances of the various isotopes in the sample, in terms of the heights of the "spikes" (peaks) corresponding to them. The height of the spike (peak) from the tracer corresponds exactly to the quantity mixed in the sample. The tracer thus serves as an internal standard that facilitates the precise analysis of minute quantities. As a result of the development of this "isotope dilution method,"[14] isotopic dating could be extended to a great variety of minerals not formerly datable because of their extremely low concentrations of those elements required for radiometry, most notably minerals in young rocks, in which the concentrations are vanishingly small.*

The advent of electronic circuits in the 1930's contributed greatly to the development of the modern mass spectrometer. Whereas in the older mass spectrograph the intensity distributions of different isotopes were recorded on photographic plates, in the mass spectrometer these distributions are recorded automatically, continuously, more precisely, and on a far greater scale by the electronic circuits. The mass spectrometer would come to play a crucial role in the dating of young rocks.

Early Measurements and Old Rocks

The roots of potassium-argon radiometry extend to Joseph Thomson's discovery, in 1905, that a sodium-potassium alloy was emitting negative particles; N. R. Campbell and Alexander Wood

*Faure (1977) presents a complete discussion of the many advantages of the isotope dilution method of dating.

extended his finding in 1906 and concluded that beta particles (electrons arising from the breakup of neutrons) were emitted in proportion to the amount of potassium present. In 1921, Francis Aston examined potassium, among several elements he was concerned with, and found two potassium isotopes, of masses 41 and 39, neither of which was thought responsible for beta emissions. Thus F. H. Newmann and H. J. Walke and independently Otto Klemperer were led in 1935 to postulate a potassium isotope of mass 40 that decayed to calcium 40 by the loss of a beta particle from the nucleus.* That same year, Alfred Nier, using a greatly refined machine, not only demonstrated the existence of potassium 40 but determined its abundance as 0.011 ± 0.001 percent of natural potassium. Further work by William R. Smythe and Arthur Hemmendinger in 1937 showed that potassium 40 produced all of the emissions from potassium. By 1940, then, three isotopes of potassium were known (of atomic weights 39, 40, and 41), but little was understood about potassium 40 decay constants (rate of decay), and virtually no research toward that end was undertaken until after World War II.[15]

Brilliantly reasoning through the facts attending the abundance of argon 40 in the atmosphere, Carl F. V. Weizsäcker in 1937 postulated that this isotope was derived from the decay of potassium 40 in the earth's crust and suggested that argon 40 in minerals be used as a tool of geochronology. Weizsäcker's belief that argon was produced by electron capture was validated in 1943 by F. C. Thompson and S. Rowlands, who urged others to seek excess argon 40 in geologically old potassium-rich minerals. Five years later Lyman T. Aldrich and the by-then-renowned Alfred Nier, using a machine of Nier's design, determined for the first time that the ratio of argon 40 to argon 36 was greater in rock (two feldspars and two evaporites) than in the atmosphere, and also suggested that argon had been lost from the minerals. They also demonstrated the dual decay of potassium—some of it yielding calcium, some argon. Thus by 1948 the use of the potassium-argon method appeared promising for dating, but neither the methods for extracting and handling argon nor knowledge of the half-life of potassium 40 was refined enough to permit exact age determinations.[16] Nonetheless, Aldrich and Nier's exciting demonstration prompted research—on both the handling and the half-life questions—that was to make potassium-argon dating a widely used method by the mid-1950's.[17]

*Meyer and Schweidler (1927) treat the early history of beta-activity studies of potassium.

The first potassium-argon date that appeared trustworthy was produced in Germany by Wolfgang Gentner and his coworkers. Friedrich G. Houtermans refers to their success in selecting samples for analysis as a "good guess."[18] A more likely explanation is that Gentner was painstaking in his choice of samples, for he was aware that Hans Suess had obtained spurious results in 1948. Having relied upon the advice of mineralogical chemists in his sample selection, Suess had been provided with recrystallized specimens that were quite useless for dating purposes. Two years later, Friedolf Smits and Gentner dated an Oligocene sylvite (a potassium chloride mineral), from Buggingen in the northern Rhine Valley, at 20 million years.[19] Concerned that this figure was not in accord with dates obtained by other methods, Gentner looked to another problem area, the loss of argon gas by diffusion from the solid material. They found that small mineral grains contained less argon by weight than large grains, presumably as a result of diffusion. By 1953 they succeeded in correlating argon diffusion with the results of the uranium-helium dating method to correct their figure to 26 ± 4 million years.[20] This historically significant success was due mainly to a greatly improved method of extracting argon from the sample and careful selection of samples. These two factors remained important when later refinements in methodology led to success in the dating of whole mafic rocks, which are mixtures of minerals rather than monomineralic rocks like sylvite.

There were of course many causes of imprecision in the early measurements. The problems for future research that seemed most important to potassium-argon geochronologists during the early 1950's were outlined at the Conference on Nuclear Processes in Geologic Settings held at Williams Bay, Wisconsin, in September 1953. Much of the discussion concerned experimental techniques, including memory effects in the spectrometer (contamination from previous samples); the effect of stopcock grease on argon loss; the role of fluxes in the liberation of argon from the sample; the probabilities of atmospheric argon contamination and how to overcome it; and a host of other laboratory problems that plagued the early experimenters.

The next year, John P. Marble prepared a *Summary Report of the Committee on the Measurement of Geologic Time for 1952–53*, and noted that the field was growing—at least as measured by the large increase in numbers of papers concerned with radiometry. The valuable work on the decay of potassium 40 looked particularly promising—yet much work remained to be done; Marble warned that the total decay constant and the half-life appear to be fairly well known, but we are not really sure of the branching ratio between the two decays. Methods for accurate determination of the potassium content of minerals of various types should receive further study. The precise measurements of small amounts of calcium should be investigated. We also need further work on the extraction of argon from minerals and its complete separation from other gases extracted at the same time. Further precise studies of the isotope ratios of potassium, calcium, and argon are also needed.[21]

Next in importance to establishing the decay constant was the quest for precision in the branching ratio—the relative proportions of calcium 40 and argon 40 yielded by the dual decay of potassium 40. In 1950, Mark G. Inghram and his coworkers at the University of Chicago obtained a sample of sylvite from Strassfurt, Germany, measured its content of calcium 40 and argon 40, and determined a value of 0.06 for the branching ratio. Abul Mousuf in 1952 and Richard Russell's group in 1953, basing their calculations on microcline feldspars, concurred in that estimate. In 1954, using one of the same samples as the Russell group, but different techniques, Gerald Wasserburg and Richard Hayden demonstrated a much higher proportion of argon 40. Harry Shillibeer showed in 1954 that the use of sodium hydroxide as a flux in melting the sample was instrumental to greater argon 40 yields. Wasserburg and Hayden, who had also fluxed with sodium hydroxide, later analyzed potassium feldspars from four sites that had been dated accurately by the uranium-lead method, and concluded in 1955 that the argon-40 yields indicated a branching ratio of 0.085. (Modern measurements indicate that 89.5 percent of the potassium 40 atoms decays to calcium 40, whereas 10.5 percent decays to argon 40, which yields a branching ratio of 0.117.)

In concurrent research, George Wetherill, Lyman Aldrich, and Gordon Davis in 1955 showed that mica and feldspar from the same rock had different ratios of argon 40 to potassium 40, the mica indicating a branching ratio of 0.105. Robert Folinsbee, Joseph Lipson, and John Reynolds in 1956 showed the same effect in rocks from the Canadian Yellowknife region. In commenting on these results, Donald Carr and John Kulp in 1957 attributed the differences to argon leakage from the feldspar; in contrast, Lipson noted the following year that "the discrepancy cannot be explained simply by diffusion."[22] Thus the problem of retentivity or diffusion was still a formidable one in the late 1950's. As late as 1957, John Marble, in the *Report of the Committee on the Measurement of Geologic Time*, opined that "the [uranium-] lead method is the standard against which all others are judged, the [uranium/thorium-] helium method is largely discredited, and the rubidium-strontium and potassium-argon methods are still under development."[23]

Up to that time, the potassium-argon daters had been looking at old rocks: most of the early efforts concerned the measurement of radiogenic argon in Precambrian and Paleozoic horizons.[24] Why was that so? What delayed the rise of young-rock dating, which became so indispensable? Much of the early research was in part guided by the fact that the older rocks were found to contain far more radiogenic argon than did the younger rocks. All early work (with the solitary exception of Mark Inghram's group in 1950), until Wasserburg and Hayden in 1954, involved experimental measurements of argon that required relatively large argon samples (more than 0.001 cubic centimeter). A small sample of Precambrian rock easily yielded such a quantity, but for Tertiary rocks pieces larger than 150 grams were required. Large argon samples provided the advantage of yielding strong signals, or peaks, in the mass spectrometer, which reduced the error from background noise in the measurement, particularly in the case of the weak signal for argon 36, the nuclide necessary for the correction for atmospheric argon. Because of the many other problems that plagued the early researchers, it was not possible to take on the further difficulties inherent in attempting to work with younger rocks.

Happily, those ancient rocks, which were the least difficult to date, represented the part of the geological column whose age was in greatest doubt and generally excited the most interest among geologists. Most of earth history lies in the Precambrian eon, where fossils are extremely few and where difficulties in interregional and intercontinental correlation are enormous. The early potassium-argon daters saw a new tool that could be applied to a large number of rocks that did not lend themselves well to other methods of isotopic dating. To supply a date for an unreckoned interval was regarded as an important contribution, and in that context importance seemed directly proportionate to the age of the sample.[25]

Even after the method was refined, radiometrists tended to continue their concentration on the older rocks. They may have done so, at least in part, because they had only limited understanding of the host of geological problems they would have had to face in the dating of young rocks. After all, geologists had learned or postulated much about younger rock formations, whereas in contrast the old Precambrian terranes, such as the region of central Canada called the Canadian shield, were so complex that only a radiometrist could say something definitive about age and correlation; thus the early isotopic daters, most of them geologically untutored, could operate in old terranes with near impunity. Those vast regions of the continents comprising the oldest rocks represented the most exciting frontier: because so little was known of them, they offered hope of discoveries beyond the limits of other methodologies. Moreover, there had arisen in the history of geology a certain pride in demonstrating an increasingly older earth; this had been so since 1650, the time of Archbishop Ussher's largely biblically derived notion of a world created in 4004 B.C.

Despite the advantages of working with older rocks, they also present certain intrinsic problems, most of which derive from the loss of varying amounts of radiogenic argon with time. Some argon escapes even at low temperatures by diffusion through the crystals, especially by natural pathways such as dislocations and cracks in the crystal lattice. Moreover, different minerals have quite different diffusion rates. Older rocks have generally been subject to more complicated histories, including deep burial, folding, heating, and attendant metamorphism, which contribute greatly to argon loss; thus great care in petrologic examination and in working out the history of the sample is essential in dealing with very old rocks.

Predictably, the refinement and widening application of potassium-argon dating during the early and middle 1950's, aimed always at older rocks, came to encompass the oldest rocks known: meteorites. In contrast, geologists at Berkeley sought to date the youngest rocks, and in their efforts would write a chapter vital to this history.

Reynolds and the Origins of Young-Rock Dating

For the University of California, Berkeley, as for so many other institutions, the years following World War II were a time of ferment: new funding, new appointments, expanded research, and a welcome increase in interdepartmental associations. The faculty of geology, then under the chairmanship of Howel Williams, was enhanced by the arrival of Francis J. Turner in 1946 and John Verhoogen in 1947. Both were destined to election to the National Academy of Sciences and other signal honors; they were outgoing men, and quickly established professional relationships and

John Verhoogen of Berkeley was among the prestigious group who discussed the evolution of the earth, and the value of emerging programs in radioactive age measurement, at the Rancho Santa Fe Conference, convened by the National Academy of Sciences in October 1949. The conference considered the growing numbers of well-dated rock specimens from important Precambrian-age structures in North America, as well as orderly large-scale patterns of rock dates; radioactive dating had clearly arrived at the forefront of earth science.

Row 1, left to right: Edward Teller (seated); Beno Gutenberg; Harold C. Urey; Norman L. Bowen; Louis B. Slichter; William Latimer. *Row 2*: Perry Byerly; unidentified; James Gilluly; Fred Whipple; Harrison Brown; Edward Goldberg (hidden); William W. Rubey (hand in pocket). *Row 3*: H. P. Robertson; Carl Eckart; Linus Pauling; David Griggs; Oliver Wulff; Verhoogen. *Row 4*: Roger Revelle; Maurice Ewing; Adolph Knopf; Patrick M. Hurley.

friendships with members of the Mathematics and Physics Departments. Among those with whom Turner was close socially and personally was Francis A. ("Pan") Jenkins of the physics faculty. Jenkins was charged in 1950 with filling several new faculty vacancies upon completion of the new annex to Le Conte Hall, the main physics building. Before departing for the East to interview candidates—an unusual venture, since Berkeley's reputation in physics normally obviated the need to recruit—Jenkins asked Turner what sort of physicists might be useful to geologists. Turner, who was "really interested in tracing the origin of rocks and magmas by isotopic composition rather than dating" per se, replied that an isotopic physicist skilled in mass spectroscopy would be a useful adjunct to the Geology Department.[26] At that time Turner had been reading about and was concerned with the potential applications of isotope work to geology.[27] He made his interests known to the geology faculty, who met for lunch frequently to discuss departmental and intellectual concerns and research projects. Turner's interest in isotopic work was later reinforced by that of John Verhoogen, who soon began work in the isotopic composition of silicon in quartz.*

When Jenkins arrived at the University of Chicago in 1950 during his recruiting trip, he met John H. Reynolds, a former electronics physics major at Harvard (B.A., 1943) who was just completing his Ph.D. dissertation. Reynolds seemed an excellent fit, and Jenkins soon hired him. His work centered on mass-spectroscopic studies of branching ratios in copper, bromine, and iodine; he had also studied double beta decay.[28] Though he formally pursued these studies under A. J. Dempster, he was actually working more closely with Mark Inghram, who had just started teaching as a young assistant professor. During World War II, Inghram had worked on the isotopic assays of uranium samples, which involved precise mass-spectroscopic analysis. Reynolds recalled that Inghram showed him "how to be fearless in baking out vacuum systems" and that he "had an instinct for good simple design that was cleaner, more sensitive and more precise" than he had known before. Inghram was also "persevering, aggressive, and technically oriented" and "at the top of the list" of those to whom Reynolds owed an intellectual debt: "certainly the dominant person in my experience at Chicago."[29] Reynolds was first introduced to rare-gas mass spectroscopy by Inghram, and it was Enrico Fermi's colloquium talk at Chicago on double beta decay that most stimulated his interest during his graduate days. After working on an experiment devised by Inghram to measure double beta decay using an old mineral, Reynolds turned to an experiment on xenon, which provided the

*Francis Turner was told by Harrison Brown that Verhoogen's early work in silicon isotopes also prompted the California Institute of Technology to undertake isotope studies (Turner, t. 1, s. 1, ii-1-78). For the results of that work, see Reynolds and Verhoogen (1953).

familiarity with rare-gas techniques that was crucial to his later work on potassium-argon and xenon isotopes.[30]

Reynolds, then 27 years old, arrived at Berkeley in July 1950 as a new assistant professor in a department comprising one of the most illustrious groups of high-energy physicists ever assembled. Mass spectroscopy was then as new to Berkeley as was John Reynolds. He recalled: "I came to Berkeley and built a conventional machine similar to the one at Chicago as soon as I arrived."[31] His plan to begin research at Berkeley was formulated even before he departed Chicago; drawings were made for a magnet design, and several important parts and instruments were ordered from Chicago, including iron forgings, a vibrating-reed electrometer, copper metal for the flight tube, and much else. "The machine I built here first from the very beginning was an interim machine I could get something started with, but I knew that soon I would try to do something that was more original. . . . The first machine was a workhorse that we used for all sorts of work."[32] The Nier-type machine was assembled and fully functional by May 1951—fast work in view of Reynolds's teaching duties and other concerns.[33] From then until 1953 the machine was turned to a series of investigations touching on the "Surface Ionization of Lanthanum," published in 1952, and to isotopic studies of silicon, germanium, and hafnium.* As late as August 24, 1953, Reynolds still was seemingly committed to a research program based on the use of the conventional 9-inch-radius, 60°-deflection Nier-type spectrometer, as evidenced in his letter to Fred Mohler of the National Bureau of Standards, to whom he made known his intention to pursue studies "probably in the near future in uranium and lead."[34] Although he had been working since October 1952 under a contract from the U.S. Office of Naval Research, he gave, as yet, no hint to those beyond the campus of a new machine he had in mind, designed for small amounts of the rare gases.† In a conversation in 1978, Reynolds said:

*For the isotopic studies see Reynolds, 1953; Reynolds and Verhoogen, 1953; and Reynolds and Ypsilantis, 1953. In 1979, Verhoogen recalled: "The work on silicon started only after John Reynolds's arrival in Berkeley. My role in it was only to provide Reynolds with samples selected to give the greatest variations in isotopic ratios. I remember a lunch at the faculty club with Turner and Jenkins to discuss Reynolds's appointment, at which I expressed concern about the usefulness of the proposed silicon work in the absence of a theoretician in statistical mechanics who could interpret isotopic ratios in minerals to deduce conditions of formation. . . . Pan Jenkins did not seem to understand. As it turned out, Reynolds's silicon data have never been put to any use, as far as I know" (J. Verhoogen, written comm., vii-17-79).

†Funding for research undertaken during the three-year period beginning x-1-52 was $18,990 under contract "Nonr 222(20), NR 024-175," U.S. Office of Naval Research, Nuclear Physics Branch.

The incentive for trying to build a static machine came from my observation of [Silvio J.] Balestrini, my first graduate student, who did a problem in which he irradiated iodine targets in a cyclotron and then took the xenon [created by transmutation] out and made isotopic measurements on it, and he did it in that big old conventional machine that we put together just after I arrived [at Berkeley]. He had enough gas [sample] so that he made out all right, but I think somewhere along there I must have really come to think that this is an awfully clumsy, wasteful way to be doing this and the approach ought to be something smaller, cleaner, and bakable, that you could seal off and do these measurements in. I think it was somewhere watching this work that Balestrini did on the rare gases that I came to this idea. Then it was also clear that there were a whole lot of problems that could be tackled by it. Argon dating was a very obvious one, together with these problems in xenon that I'd gotten into.[35]

In Reynolds's successful proposal of July 1955 for continued support from the Office of Naval Research, he showed why it had made sense to build such a machine.[36] Its chief virtue would be to make possible the analysis of previously unanalyzable, extremely

John H. Reynolds, at right, with H. Ewald, Mark Inghram, John Hipple, Alfred Nier, and J. Kistemaker (left to right) on May 24, 1955, in San Francisco, following Reynolds's first public talk on his new static-mode mass spectrometer, which he had completed by late 1954. The occasion was the Third Annual meeting of the American Society for Testing and Materials, Committee E-14 on Mass Spectrometry.

small samples of the rare gases. Twenty years later, he compared the new model with the old.

> The obvious limitations of design, where you have a great big metal tube that was supported in the middle of a high metal framework, was that it was clumsy to get ovens around it. I think somewhere along the line I thought that if you really want to get something baked out you don't want to have any cold places. You can bake the hell out of one part, but if you have a little cold place somewhere else, the stuff [contaminants] migrates over there; then when you cool everything down and start running again, that's the source of poisoning. So I had a clear idea that the thing ought to be something you could put an oven down [on] and blast the whole thing so that the first cold part would be outside of the valve you were going to shut off, and so everything would be clean. That suggested making something rather small that mounted in such a way that it fitted easily inside an oven. I read some papers by Homer Hagstrom of Bell Laboratories, who did have a small glass machine, but he didn't seal it off. It looked like the kind of machine that we ought to be using for the work we wanted to do.[37]

The possibility of a static-mode machine in which a small sample was sealed off within the mass analyzer, and its atoms counted for a long time, as contrasted with the conventional, dynamic-mode spectrometer, in which the sample is run through quickly and used up, had been discussed in the University of Chicago group, but many vagaries attended the subject and nothing came of it.[38] Reynolds does not regard his graduate training as the source of the essential particulars that were uniquely combined to build his static-mode, glass machine.

> I don't know if it's an intellectual thing; I was an ambitious young guy who realized that if you are going to make your mark in this thing you can't just do whatever everybody else has been doing—you've got to do your own thing. I was searching around to find how I could establish my scientific identity and get started on a line of work which was really my own. That's really what made it go. . . . I felt confident enough in my technical ability to start something new . . . and I felt it almost necessary to start something new.[39]

Gerald J. Wasserburg and Richard J. Hayden, both contemporaries of Reynolds in the University of Chicago Physics Department and later elected to the National Academy of Sciences for their contributions in mass spectroscopy, agreed by the early 1950's that the serious technical question was, in Wasserburg's words, "Do you try to beat background by increasing the intensity of the

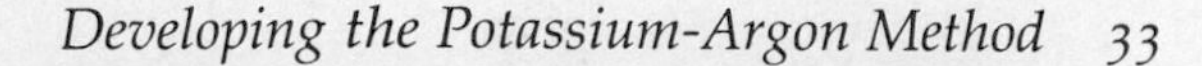

John Reynolds at Berkeley in 1981 with a recent version of his mass spectrometer.

ion beam or can you beat it by counting forever at low intensities? . . . The competition between these two issues is what Reynolds had to address."[40] The approach that had been prevalent, and was used by Inghram, Hayden, Wasserburg, and others, was to increase the effective signal by introducing the sample as fast as possible in order to achieve a strong signal for a short time so that the background noise was overwhelmed or swamped out. Such a technique, if applied to a large enough sample, could be used to date fairly young rocks.

But it was apparently not clear to the spectrometrists whether one could ever achieve success by running in the static mode. A static-mode mass spectrometer may be likened to a sealed box in which atoms can be recounted a great number of times, thus yielding increasingly better statistics. The great difficulty with the procedure is that interfering atoms that have accumulated from earlier analytical runs come out of the walls of the machine, where they were embedded. This is called the memory effect. The prevailing treatment consisted of pumping a strong vacuum in the machine to draw off the interfering atoms so that they would not build up pressure and preclude measurement. Static analysis makes sense especially for rare gases, which are run in very small samples, and require a low background. A small sample can in turn be extracted by means of a simpler extraction system (train) with fewer parts, a fact that would prove to be of singular importance to the success of the new machine.*

The Static-Mode Prototype

Early in 1953, Wulf Kunkel, then a graduate student at Berkeley (and later a professor of plasma physics there), went to the Westinghouse Laboratories in East Pittsburgh, Pennsylvania, and visited with Daniel Alpert, who is commonly regarded as the principal formulator of ultrahigh-vacuum technology and the inventor of the most efficient bakable high-vacuum valve.[41] Upon his return to Berkeley, Kunkel delivered a talk to the department, and when Reynolds heard it he knew he should "marry high-vacuum techniques to static analysis."[42]

Achieving a high vacuum is crucial to successful static analysis, since a high vacuum reduces the contaminants that interfere with the path of the ions being accelerated down the flight tube toward

*In 1981, Alfred O. Nier in "Some Reminiscences of Isotopes, Geochronology, and Mass Spectrometry" wrote that in 1948 he and L. T. Aldrich (Aldrich and Nier, 1948b) had run their "mass spectrometer in the static mode for the first time . . . in argon and helium investigations" (p. 12). Nier's loose use of the term "static mode" would appear to contradict my version of the history of the development of the static-mode mass spectrometer. However, in conversation in June 1981, Nier remarked that he should have used the term "quasi static-mode" in referring to his instrument of 1948 (t. 1, s. 1). Nier and Aldrich had in fact used a recirculating system in which they replaced the forepump by a charcoal trap refrigerated by liquid nitrogen. The gas sample was actually drawn from the mass spectrometer through the diffusion pump to the charcoal trap, back into the sample bulb and then reentered into the mass spectrometer. In contrast to such a recirculating system, a true static-mode machine has the diffusion pump cut off by a valve from the mass spectrometer; the mass spectrometer is thus isolated during analysis. Nier's "quasi static-mode" system was designed uniquely for the study of smaller amounts of helium than were required for normal operation in the dynamic mode. The recirculating system did not permit as high an operating pressure in the spectrometer and thus reduced its sensitivity. Nier also remarked in the same conversation that his machine was metal with wax joints and could not be baked to reduce contaminants; it "was not a clean system" (t. 1, s. 2). The charcoal trap in that system permitted only the return of helium to the mass spectrometer for analysis during recirculation; other elements, including argon, could therefore not be analyzed. In a telephone interview in June 1981, Lyman T. Aldrich recalled that in comparison with the Reynolds mass spectrometer of 1954, the recirculating machine that he and Nier used in 1948 "was a very primitive thing." Aldrich also remarked that his argon determinations in 1948 were done on a conventional dynamic-mode, Nier machine, *not* the "quasi static-mode" instrument (t. 1, s. 1).

the collecting area. In a high vacuum the ion streams of a given isotope are focused with less dispersion and produce sharper peak shapes; this is especially true of the argon 36 peak, which is small and difficult to measure. When a small sample is analyzed in a dynamic machine, it is impossible to get sharp signals, because the density of contaminating particles interferes with ions of the isotopes being measured.

By May 1953, the plan for the machine was complete, its construction well under way, and intentions clear concerning its earliest applications, although the question of the branching ratio in potassium decay was still emphatically open. On May 19, 1953, Reynolds wrote to Richard D. Russell of the University of Toronto spectrometry group, and said in part:

> As Verhoogen may have told you, we are not planning to use our conventional Nier-type instrument in the K-A[r] investigation but are assembling a new apparatus which, we hope, will enable us to operate with somewhat smaller samples than you are working with. . . . The machine is to be a 4.5 in. radius, 60° type with an all-glass envelope. The ion source will be conventional, but the collector will comprise a 9-stage Ag-Mg-dynode electron multiplier for increasing sensitivity. We shall incorporate the new Alpert vacuum techniques, developed at Westinghouse, in hopes of suppressing background by an order of magnitude. Thus we shall use all-metal vacuum valves (in both the spectrometer and extraction apparatus); we will provide for baking all our systems at 450°C, and will experiment with final pumping with the Bayard-Alpert ionization gauge—having first isolated the tube and sample system from the diffusion pumps. We are aware that these methods are by no means essential in K-A[r] work, but have hopes that we can achieve enough extra sensitivity to make feasible other rare gas investigations of interest.
>
> We are indeed anxious to get more samples from you. Tentatively, I could plan to use from 2–5 grams of your assumed 10% K, 10^9-year material per run. Thus, I suppose 50 grams would be a reasonable amount to request. If our hopes with regard to background suppression are not realized, we might have to come back for more. We hope to run a great many samples when the apparatus is complete and thus my attitude is: The more specimens the better. If you can spare samples of your entire age sequence, they would be welcome, as would samples of a common age but varying potassium content.
>
> We read all your reports with great interest. Have you heard of Hayden's work (at Argonne) on K-A[r] in Holburton County granite? He checks you in that, although he favors the higher branching ratio and thus interprets the results as "50% leakage." I try to keep an open mind on this baffling question of the branching ratio.[43]

Reynolds decided to build in glass because it was then generally believed that an ultrahigh vacuum could not be achieved in a metal machine involving flanges, and that metal could not be as easily baked and made clean. His decision was undoubtedly influenced as well by the much lower cost of glass and by the presence of a master glassblower—Morley Corbett, the fabricator of the glass apparatus used in the Berkeley Physics Department. Corbett performed the superlative glass work on the first Reynolds machine and many others that followed. "The fact that the Alpert valves had been used earlier in glass systems also reinforced my decision to work in glass,"[44] Reynolds said. Moreover, for the elements he wanted to minimize in his intended studies in xenon and argon, the background contamination was lower in a glass system at that time than in a metal system.

Reynolds also had solid technical support that allowed him repeatedly to design and fabricate new or modified versions of the many components. The metal parts for the Reynolds spectrometer were provided largely by William Brower, who ran the machine shop for the Physics Department and designed and built the modified versions of the Alpert valves while Clifford Grandt constructed the intricate source and detector parts.[45] The Alpert valves proved difficult to build.[46]

Why someone had not earlier built a static machine is a reasonable question, for the machine Reynolds built did not employ any major components that could not have been made accessible to any enterprising experimental physicist trained in mass spectrometry and familiar with its instrumentation. The explanation most commonly suggested is that there was no clear need for a static machine in order to overcome background noise during a sample run; but as researchers had progressed to analyzing ever smaller samples (down to one ten-billionth of a cubic centimeter), the problem mounted. As Wasserburg put it, "John [Reynolds] obviously had the damn good sense to pick this other way [static analysis] and make it work—and it really did. . . . John's a very good experimental physicist—very few like him in the field. . . . Most mass spectroscopists could not produce new instruments."* Whoever would produce a new ultrahigh-vacuum machine had to come to the task from experimental physics and have both a feeling for instrumentation and the intuition to select the proper experiments

*Wasserburg's remarks (t. 1, s. 1) were buttressed in conversations with G. H. Curtis, J. F. Evernden, G. W. Wetherill, R. W. Kistler, J. Obradovich, and others.

during the course of the machine's development. Joseph Lipson later said that Reynolds was "extremely methodical, with a sense of the essential experiment to be done. . . . He arranged everything necessary both materially and conceptually. . . . Everyone thought that Reynolds's experiments were too difficult to do"; and he was regarded as "bold in order to achieve new capabilities."[47] In fact, the backgrounds of those who have been responsible for technical innovations in mass spectrometry, such as Reynolds and Nier, were largely or exclusively in experimental physics; generally they lacked training in geology per se, although some worthy exceptions would appear.

Joseph Lipson had arrived in Berkeley at the same time as Reynolds, as a graduate student with a B.S. in physics from Yale, but the two did not meet until after Lipson's preliminary doctoral examination in 1952. Lipson had originally planned to work with Chaim Richman, his professor and a member of his pre-oral committee, who thought well of him. Richman asked Lipson in 1951 to work with him on his doctoral project at Berkeley's Lawrence Radiation Laboratory, but Lipson was denied security clearance because of a family political background that appeared questionable in the charged atmosphere of the McCarthy era. Reynolds, moved by Lipson's plight, interceded on his behalf, though unsuccessfully. Later, hearing that Reynolds was looking for a graduate student, Lipson responded quickly.

Late in 1952, when Reynolds had needed help with the construction of his static machine, he urgently prepared Lipson for work in mass spectrometry. Lipson designed the external source magnet that focuses the ionizing electron beam; the main magnet that focuses the ions in the flight tube (the 60° sector magnet) was planned by Reynolds. Lipson also assisted Morley Corbett in the glass work and the coating of the inside of the tube with stannous chloride—the first attempt to make the flight tube electrically conducting, with a silver compound, had ended in failure.[48]

After the failure with silver coating, Reynolds searched for an alternative and learned that the General Electric Corporation had used stannous chloride successfully in the coating of a betatron torus (a doughnut-shaped flight tube that accelerates electrons at a constant radius). Reynolds thus knew that such a coating was stable in a high vacuum. Although the information was proprietary, General Electric sent him their recipe. "Until we tried it we didn't know if it would work. . . . It was the only known technique that was stable under oxidizing conditions."[49] The objections to the technique were obvious and strong: Wasserburg explained in 1978 that "it was a joke that stannous chloride coating in a tube would work. Chlorine was mass 35 and 37, argon was 36 and 38—they were too close [in mass]; therefore we worked to get chlorine out of the machine, not in."[50]

Technically, Lipson was highly competent. He worked closely with Reynolds in the acquisition of parts and the design of several modified versions of existing spectrometer components, including a "multicircuit control for ultrahigh vacuum gauges."[51] His contributions to the first successful dating were many; perhaps most important, he built the first very clean gas-extraction system at Berkeley, using handmade Alpert valves that facilitated complete bakeout of the line to drive off contaminants. The vacuum valves

Joseph Lipson in Le Conte Hall in 1958, with some of the mass-spectrometry instruments he helped to assemble and operate during the early, critical years of the potassium-argon dating program at Berkeley.

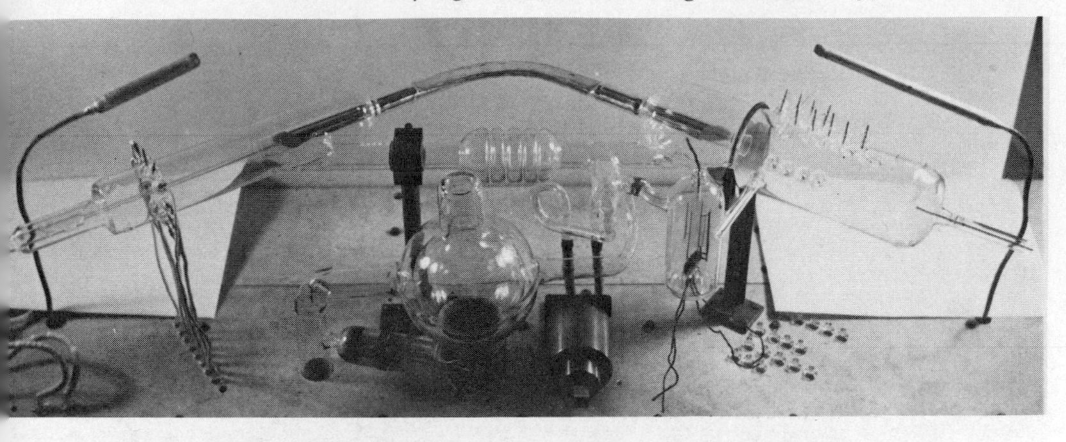

The first model of the Reynolds glass mass spectrometer, shown here partially assembled, used a silver coating in the curved central mass analyzer, or flight tube. Later Reynolds and Lipson replaced the silver compound with a stannous chloride coating.

represented a great improvement over the mercury cutoffs and greased stopcocks used elsewhere; the line was carefully planned to achieve simplicity and mechanical strength. The Alpert valves were convenient to operate, clean, and accident-free, and thus the line could easily be brought to a pre-reaction vacuum of 10^{-8} mm (about 0.00000000001 times atmospheric pressure) by pumping overnight.

The first Reynolds glass mass spectrometer was calibrated and the first samples run in late 1954. In the early work of 1954, samples of 7 to 20 grams were typically run.[52] Thus Berkeley had the "cleanest line around" in the mid-1950's.[53] Shortly after dating had begun, it became increasingly clear that a clean gas-extraction line was just as important as the mass spectrometer in successfully dating young rocks.[54]

The first results from the completed static-mode machine were derived from three granitic rocks supplied by John Verhoogen.* Reynolds requested that Verhoogen supply samples of something young with a reasonable amount of potassium and not reveal to

*Reynolds's Laboratory Notebook no. 3, p. 40, identified them as KA-8 (a Johnson granite from Soda Springs, Yosemite National Park, dated at 92 million years), KA-9 (a phenocryst of potassium feldspar in the Cathedral Peak Granite, Yosemite, dated at 86 million years), and KA-10 (Cathedral Peak Granite, west end of Tuolumne Meadows, Yosemite, dated at 108-116 million years). The results of this dating did not appear in print until 1956 (Folinsbee, Lipson, and Reynolds). The dates were revised in the published presentation from the figures in Reynolds's notebook and listed in two columns based on alternative decay constants.

The original, fully assembled, Reynolds mass spectrometer in Room 181, Le Conte Hall, Berkeley campus, 1954.

him the age of the rocks, so that a true blind test could be performed.[55] The nature of the blind test was announced by Verhoogen to the meeting of the University of California's Institute of Geophysics Technical Conference on December 3, 1954.[56] At the same meeting, Reynolds made the historic first formal announcement of his and Lipson's success in building and operating a static machine. In the course of his presentation, entitled "Argon Potassium Age Determinations," he also reported the first potassium-argon dates from Berkeley; his calculations had been completed only the night before.*

Lipson began the next project on the machine as part of his doc-

*The proceedings of the conference were not published. Seemingly portentous, immediately above Reynolds's name and paper title in that announcement appears the name of Jack Evernden reporting on "Variation in Mantle Velocities": Evernden was soon to depart seismology and enter isotopic dating with enthusiasm.

toral research. The subject, the dating of glauconite, was chosen as a result of a conference with Francis Turner, of the Geology Department at Berkeley, who "conceived the form of the study" and provided assistance.[57] "Lipson was very gregarious—often when I gave him a problem to work on he'd go and talk to other people [to secure information]," said Reynolds.[58] Such was the case when Lipson and Gerald Wasserburg, at the Geological Society of America meeting in Los Angeles in 1954, spoke of their mutual interest in attempting to date glauconite.[59] It is noteworthy that the earliest potassium-argon dates from glauconite were reported by Wasserburg and Hayden in September 1955 and published July 31, 1956;[60] Lipson's were published in August. Wasserburg and Hayden retained their priority in publication by one day. The ten glauconite samples for Lipson's dissertation were supplied by Turner, who had written to New Zealand for them; Lipson's dates, among the youngest obtained by the potassium-argon method at that time, ranged from 16.3 to 49.2 million years. Wasserburg and Hayden's three glauconites were considerably older at 67.6 to 574.0 million years.

Folinsbee and the First Uses of the Reynolds Machine

From just before the beginning of the 1950–51 academic year, when Reynolds and Lipson arrived in Berkeley, until December 1954, when Reynolds and the geologist Robert Folinsbee first met, the mass spectrometry effort at Berkeley was conducted with no direct participant who had a geological background. From the outset Reynolds felt less than comfortable in dating rocks, because both he and Lipson lacked full understanding of their origins and context.*

Garniss Curtis might have filled the bill. Reynolds was first introduced to Curtis in 1952 on the back steps of the geology building, Bacon Hall, near the Le Conte physics annex, by John Halsey, a graduate student in geology under Charles Gilbert who had been a naval officer colleague of Reynolds's. Reynolds and Curtis sometimes met during lunch meetings of the geology faculty, generally held in Verhoogen's office around the old table inherited from Andrew Lawson, Berkeley's late patriarch of geology. Indeed by 1953

*Reynolds, t. 5, s. 2, vi 14 78; J. J. Lipson, oral comm., ii 4 78. Reynolds twice commented on his disquietude in a letter of vii-15-55 to William Oke of the California Institute of Technology in which he discussed a number of points, including the work of Folinsbee (Document No. 4B).

Bacon Hall, on the Berkeley campus, was designed by John A. Remer and built in 1881 at a cost of $77,000, of which $25,000 had been donated by Henry Douglas Bacon. First named the Bacon Art and Library Building, it was remodeled and renamed Bacon Hall and occupied by the Geology Department in 1911 when the Library moved to its new quarters. The clock tower was removed in 1925 because it was considered an earthquake hazard. The building was razed in 1961 to clear the site for Birge Hall, the physics annex.

Reynolds's need for a geologist in his spectrometry work prompted him to invite Curtis to join in. But the 1912 eruption of Mount Katmai, Alaska, had come to occupy Curtis's interest. He declined Reynolds's offer and left for Alaska, arriving in the Valley of Ten Thousand Smokes on June 21, 1953,[61] with Howel Williams, a Berkeley volcanologist.[62] In September, following the accidental death during fieldwork of Werner Juhle, who had undertaken a study of Mount Katmai for the U.S. Geological Survey,[63] Curtis replaced him and continued the Katmai study on a part-time basis until the fall of 1955. Only then, as he grew dissatisfied with the Katmai project, did he turn his attention to potassium-argon dating.

During Curtis's Katmai involvement, Robert Folinsbee had heard Reynolds speak of his first successful potassium-argon dates, in his historic presentation to the Institute of Geophysics Technical Conference at Berkeley, in December 1954. A Canadian, Folinsbee had

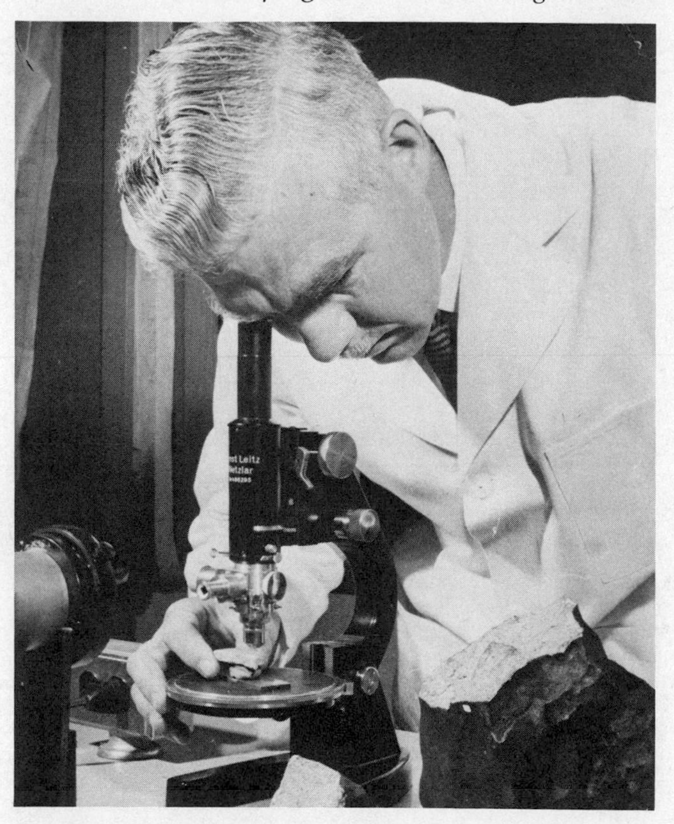

Robert E. Folinsbee, F.R.S.C., was the first geologist to participate directly in the potassium-argon dating program at Berkeley (in 1954 and 1955, during his sabbatical leave from the University of Alberta).

arrived in Berkeley in September of that year on sabbatical leave from his professorship in the Department of Geology at the University of Alberta in Edmonton. His undergraduate training in geology at Alberta had been followed by work under Frank F. Grout and Reuben Ellestad in the rock analysis laboratory at the University of Minnesota, work that led in 1942 to a Ph.D. in petrology.* During 1945 and 1946, after three years of service in the RCAF, he had done postdoctoral work at Harvard with Esper Larsen, who was at the time using zircons as a helium source in

*Samuel Goldich, an important character in Chapter 3, also worked there. Folinsbee noted that during his Minnesota graduate studies (1938–42), Alfred O. Nier, in the Physics Department, was running uraninite samples for Arthur Holmes on his mass spectrometer, unknown to those in the Geology Department.

dating. Before leaving Harvard for Edmonton, Folinsbee had attempted to date an Archean monazite[64] from the Northwest Territories and had even contacted the U.S. Geological Survey in the hope that they could date it by the lead-alpha method.*

Folinsbee was thus well suited by background to recognize the exciting implications of Reynolds's presentation. He had originally come to Berkeley to work with Francis Turner in metamorphic petrology, but shortly after his arrival he began to work more closely with Verhoogen, who was very interested in the dating problem. When Reynolds came to the Geology Department seeking someone with a background in quantitative rock analysis, Verhoogen brought the two together. Thus Folinsbee departed petrography and field studies to embark on what was to become a lifelong career in isotopic dating, a career marked by numerous honors, including election to the Royal Society of Canada.[65]

Returning briefly to his office in Edmonton after joining Reynolds's group, Folinsbee gathered up some previously collected granite samples from the Canadian shield and some potassium salts from the Elk Point Basin; back at Berkeley he began the mineral separations necessary for dating, and the flame-photometer measurements necessary for potassium analyses.† By April 29, 1955, at the meeting of the Cordilleran Section of the Geological Society of America, Folinsbee, Lipson, and Reynolds reported that ratios of argon 40 are higher for micas than for feldspars from geologically related rocks, and that of the feldspar group of minerals, the potassium-rich ones (orthoclase, microcline, and perthite) show higher and more consistent ratios than the plagioclase feldspars.[66] Their results were in complete agreement with those of George Wetherill, Lyman Aldrich, and Gordon Davis of the Carnegie Institution of Washington, who actually were somewhat behind the Berkeley group in their studies but gained priority in publication.[67] These reports prompted further studies of argon content and diffusion in mica and feldspar;[68] but those later studies tended

*Folinsbee, 1955. Zircon is a mineral ($ZrSiO_4$) common as an accessory in silica-rich igneous rocks. The Archean is the early part of Precambrian time (early earth history); monazite (Ce, La, Nd, Th) (PO_4, SiO_4) is widely disseminated as an accessory mineral in granitic rocks.

†The flame photometer is an analytical instrument based on the principle that all elements emit a characteristic radiation when excited by extremely high temperatures; however, the alkali metals will emit radiation when burned in a gas-air flame. The wave length emitted by any one element is characteristic of that element, and the intensity of the radiation is proportional to the amount of the element present. The flame photometer compares visible radiation from known solutions with the intensity of radiation from an unknown sample.

only to confirm that the basic facts concerning diffusion had been largely explicated by the time of the April 1955 meeting. These earliest inquiries were naturally addressed to the potassium-rich minerals; the diffusion behavior of plagioclase feldspars and the common mafic (magnesium- and iron-rich) minerals other than biotite remained in question. Later other minerals were found suitable for dating, and by the 1960's the effort would lead to the crucial ability to date whole basic rock.

It was not until 1956 that the Berkeley group published a full account of the first work done on the only static-mode mass spectrometer then in existence.[69] The first dates from Berkeley were not young dates but rather reflected the background, influence, and interests of the first geological resident in the dating laboratory—for Folinsbee, who both educationally and professionally was steeped in older geological terranes and problems, had provided old Canadian rock samples for dating "from the drawer" in his office. These samples yielded eight of the 12 dates reported in that first paper; seven were Precambrian granites and one was a sylvite of Middle Devonian age.

But aside from Folinsbee's influence in sample selection, the earliest efforts at dating rocks using Reynolds's new, sensitive machine were not directed immediately to young rocks, although they would have provided a suitable test of the spectrometer's limits. Apparently the uncertainties that invariably attend a new machine and method compelled the Berkeley group to attempt less demanding tasks first, yielding verifiable results and establishing a measure of confidence, before attacking problems uniquely within the potential of the Reynolds spectrometer.

In addition to three Cretaceous granites provided by Verhoogen, the group dated a chondritic meteorite from the Forest City fall of Iowa, another specimen of which had been dated in 1955 by Wasserburg and Hayden at Chicago; the two specimens matched to within 1 percent in their potassium content and argon-potassium branching ratio.

One of the samples that Folinsbee brought to Berkeley was from the Crowsnest pyroclastic volcanic rocks of Alberta, which are rich in the mineral sanidine. Tests on this sample provided the first indication that high-temperature volcanic feldspars retained their argon, because the date determined for it, 94.5 million years, was precisely the established date. Once it was found that sanidine in pyroclastic rocks lent itself well to potassium-argon dating, the method was expanded to a great number of rock units.* (The epochal dating of the ancient East African man *Zinjanthropus* by Garniss Curtis and Jack Evernden, discussed in the next chapter, used sanidine from pyroclastic rocks.) It was because of the usefulness of sanidine that Folinsbee went on to date many ash falls in the western Canada basin.[70]

Folinsbee worked closely with Reynolds and Lipson until his sabbatical ended in June 1955. During his half-year participation he played a highly significant role in bringing to them a geological perspective in the selection of suitable problems. He also performed analyses on the only available flame photometer, located in the Civil Engineering Department, and analyzed the data from the runs of the mass spectrometer. Reynolds generally performed the runs, and Lipson did most of the argon extractions and separations; a rough division of labor prevailed among the trio.[71]

During the time of Folinsbee's participation, Reynolds was also recovering argon and xenon from meteorites for dating. Through a series of important discoveries concerning the age of those elements in meteorites, he won international recognition. By the time of his first sabbatical leave in 1956, Reynolds had started cosmic-ray work on meteorites because, he said twenty years later, "I was too far afield in geology and didn't feel my control of the geology area in experiments on dating rocks was adequate." He noted that "evaluation of the work lay too much in geology" and that "*meteorites were more comfortable, since the geologists knew little more than I.*"† Thus Reynolds's many publications after 1956 were concerned almost exclusively with rare gases and meteorites, and include only three papers about the dating of terrestrial or lunar rocks by the potassium-argon method.[72]

During the spring semester of 1955, Reynolds and Folinsbee often attended the Geology Department's daily luncheons, where they joined Verhoogen, Turner, Charles Gilbert, Perry Byerly, Garniss Curtis, Jack Evernden, Charles Meyer, and Adolph Pabst. The Geology Department was thus continuously informed of the im-

*Folinsbee, t. 1, s. 1. This clue to much future dating, especially of very young rocks, was announced on iv-29-55 at the Cordilleran Section meeting of the Geological Society of America; only the title was published in the proceedings volume (Folinsbee, Lipson, and Reynolds, 1956).

†Reynolds, t. 5, s. 1, vi-12-78 (emphasis added). Reynolds's remark here is clearly congruent with the conclusion drawn earlier about why old rocks were the first object of study. Meteorites are old rocks, and dating them rids the geologically uninformed radiometrist of the need to treat a host of geological particulars required when dealing with younger rocks.

portant progress being made by the dating group, but none at that time were moved to enter dating themselves.[73]

Export to Edmonton

When Folinsbee returned to Edmonton in June 1955, at the close of his sabbatical year at Berkeley, he took with him a new-found research program and an eagerness to apply it to the many problems of his native Canada. Shortly after his return, he built a gas-extraction line and a mass spectrometer. Folinsbee later recalled that "Verhoogen was the real figure in that year that I had in Berkeley and really got me going in science—he was the major influence in getting Garniss Curtis and Jack Evernden into potassium-argon dating. . . . When I got back I managed to raise money from the Geological Society of America and the National Research Council of Canada by 1956 and started immediately to work on the dating of volcanic sanidine that was reported in 1957 at the American Geophysical Union meeting in Toronto." The spectrometer tube at Edmonton was acquired from Buddy Thorness of Nier's laboratory at Minnesota and used break-seals in place of Alpert valves. Folinsbee arranged the employment of both Joseph Lipson, in the Physics Department, and Halfdan "Bud" Baadsgaard, in the Geology Department, the latter arriving from the University of Minnesota's isotope lab in the fall of 1957. The three worked together on an extremely clean and accurate machine and derived dates from ash falls that are acceptable today, more than 20 years later.[74] "I lured Baadsgaard here away from Minnesota; it was the best thing we ever did here. . . . All the good analytic work really is Baadsgaard's work; I've never been a really good analyst. . . . Baadsgaard really took over from Lipson," Folinsbee said in 1978.[75] The youngest date produced by the Edmonton group was a Pleistocene sanidine from the Rhine Valley, dated at one-half million years in 1960. This was never published,[76] however; the major dating concern of Folinsbee's laboratory was older rocks.[77] Folinsbee has specifically noted that "with our equipment we were able to date half-million-year-old samples in 1957. . . . You can check on this with Baadsgaard, and we've got the runs in our record here."[78]

This potential for dating young rocks in 1957 at Edmonton, if coupled with the presence of a rock magnetist interested in polarity reversals, might have brought about early the marriage that would later prove so fruitful. And in fact, a rock magnetist was there: "In Alberta at that time there was a physicist from Holland [Jan J. Hospers] who had done the original paleomagnetic work in Iceland. . . . Hospers was here in physics and interested in paleomagnetics. . . . He was important in getting the physics group to bring in Joe Lipson, which was essential for my setting up a potassium-argon lab. . . . We were interested even then in doing Cretaceous magnetostratigraphy with potassium-argon [age] control."[79]

Jan Hospers was indeed a notable contributor to early magnetostratigraphic studies. Thus at Edmonton a likely paleomagnetist was in close communication with a radiometry group capable of dating young rocks. Folinsbee later noted that Hospers believed they could date magnetically oriented tuff beds by the potassium-argon method and that he had spoken of his interest in both reversal and directionality studies. No collaboration took place, however, and Hospers made no further effort to date magnetically determined sequences before he left in 1960 to assume a professorship at the University of Amsterdam.[80] It was not until the mid-1960's that the Netherlands acquired capabilities in mass spectrometry; thus Hospers lacked easy access to radiometric dating during the early 1960's.*

The following exchange, between Folinsbee and me during an interview in 1978, sheds more light on why the marriage did not take place at Edmonton.

"In 1960 we both [Folinsbee's group and Curtis and Evernden at Berkeley] presented a paper . . . in New York. . . . They did the Tertiary and we did the Cretaceous and we were working very closely together. They knew exactly what we were doing in potassium-argon dating [ash] falls. I knew what they were doing; they stuck to the Tertiary and we stuck to the Cretaceous."

"Why?" I asked Folinsbee.

"Because we had a Cretaceous section here in Alberta and they had a Tertiary section. . . . We did date some Tertiary stuff."

"So you're telling me that it was really proximity, or the geography of the respective institutions, that dictated the puzzle-solving field?"

Folinsbee replied, "That's right; we had no particular interest in very late Tertiary because we simply didn't have it."[81]

*The lack of isotopic dating capability in the Netherlands during that time, attested by Tjeerd H. Van Andel (oral comm., vi-9-78), was also a likely deterrent to further efforts by Martin Rutten after his first, premature attempt to formulate a polarity-reversal time scale in 1959 (treated in detail in Chapter 4).

Thus it happened that the potassium-argon lab at Berkeley was for more than five years the only one concerned with very young rocks. The group at Edmonton was well suited in terms of both men and equipment to pursue young-rock problems, but did not. Moreover, they were tantalizingly capable of a collaborative effort toward resolution of magnetostratigraphic problems during Jan Hospers's several years of residence, but nothing came of it.

Chapter 2

Dating Increasingly Younger Rocks

BEFORE departing Berkeley at the close of his sabbatical year in the spring of 1955, Robert Folinsbee invited Garniss Curtis of the Geology Department to join John Reynolds and Joseph Lipson in their mass spectrometry efforts. Folinsbee explained that he much wanted to be able to tell Reynolds that he had found his replacement before he left for Edmonton. Folinsbee told Curtis of the virtues of potassium-argon dating, calling it "a fantastic boon to geology" that permitted attack on a great number of problems beyond the uranium-lead method. Curtis was still deeply involved at that point in the Mount Katmai project with the U.S. Geological Survey, and had formulated a hypothesis concerning the volcano's origin that he hoped to present at the next meeting of the Geological Society of America that fall. He was greatly attracted to the offer—the dating method was becoming widely known and had a broad spectrum of potential geological applications—but he declined and offered to try to enlist Jack Evernden, his close friend and colleague, in his stead.

Evernden and Curtis Take Up the Task

Garniss Curtis and Jack Evernden had both been awarded their doctoral degrees at Berkeley in 1951, Curtis's in geology and Evernden's in geophysics. The basis for their future collaboration was the several years of friendship that had grown from commuting together from off-campus residences. Curtis explained to Evernden, whom he knew "was unhappy and preparing to go back to work for Standard Oil," that he would be the sole person from geology working in Reynolds's dating lab if he took the offer, and he "sold him on the idea."[1] Evernden felt he was "spinning his wheels" at Bacon Hall and was too tightly controlled in his work by Perry Byerly; thus he was open to Reynolds's offer. He immediately tried to persuade Curtis to defer his project and work with him as a team, but Curtis declined.

During the summer of 1955 Evernden again asked Curtis to join him, Reynolds, and Lipson, and remarked, "They need someone who knows how to separate the minerals from a rock. . . . We want to get mineral concentrates." Curtis, although still working on the Katmai project, made some mineral separates for the group. In conversation in 1978, Curtis said: "Having done that I wanted to see what the results were . . . so I followed through and slowly got sucked into that, because Jack [Evernden] is not one to let up; once I'd given him a mineral concentrate, he wanted another and another and another. . . . I did a little glassblowing on the lines one day. . . . I hated that glassblowing but felt it was one of those things you had to do. . . . You had to learn the whole technique."[2]

Curtis's involvement in potassium-argon dating grew, and soon began to displace his other activities. The commitment to learning what was necessary to effectively apply the powerful new tool to geological problems meant discontinuing the Katmai project. Curtis continued:

> That was throwing away a big opportunity, because I was making some big, big changes in the whole picture . . . about the eruption of Katmai . . . that was then prevalent. I presented my original hypothesis . . . at the fall meeting of the Geological Society of America in 1955 in New Orleans. . . . [After that] I had no money to continue with the Katmai project. . . . I'd worked assiduously on the project for a year very hard, and I was extremely honest with the amount of time that I had put in and reported to the Survey; the following fall I got a report from the Survey that they thought that I was not putting in enough time and thus didn't approve a raise in my rank. . . . That really boiled me because I had been putting in every minute that I possibly could, weekends included. . . . It was very easy then to move into the K-Ar dating when Jack kept coming and coming to me. I'd already presented my results in two different meetings, one at the Geological Society of America and one at the A.G.U. [American Geophysical Union] back in Washington [D.C.], so I had it before the public actually. . . . Naturally I became more and more intrigued with the possibilities of potassium-argon dating for other problems. . . . Intellectually I'd solved everything I'd set about to do with Katmai.[3]

Thus by the fall of 1955, both Curtis and Evernden were fully participating in the work of the Reynolds laboratory.

The professional and educational backgrounds of both Curtis and Evernden help to explain their intellectual attraction to potassium-argon dating, their ability to apply it to a broad range of geological problems, and their eventual concern with very young rocks. Intellectually and personally, they were well suited for fruitful collaboration.

Jack F. Evernden was born in Florida in 1922 but raised in California, attending Berkeley city schools and the University of California at Berkeley. His undergraduate education in mining engineering was interrupted by four years of service during World War II, but he returned to complete his bachelor's degree in mining geology in 1948. The decision to pursue studies in geophysics in graduate school was a natural outgrowth of his ability in mathematics and the growing importance of geophysical methods in mining exploration. He was surprised, he said in 1978, to find that the major program in geophysics under Perry Byerly turned out to be largely seismology: "There was essentially nothing in that major that was applicable to geophysical exploration for mineral deposits." A job offer as a faculty member at Berkeley after completion of the Ph.D. was tendered contingent upon his going to work for Standard Oil of California as a geologist, in order to acquire "a more professional geological background." Byerly arranged for Evernden's two-year job with Standard Oil, and then in 1953, Evernden took up his faculty post at Berkeley.[4]

Evernden's thesis in seismology, published in 1951, was concerned with the behavior of surface seismic waves, and it was his and his colleagues' expectation that he continue such studies; but, he said, "I had a lack of intellectual motivation to pursue problems in seismology. . . . I just didn't find anything particularly interesting to do at the time. . . . I was dreadfully bored. . . . Seismology, at that time, had not exploded into the thing that it is today. . . . [Reynolds's] time machine came along, and it absolutely enchanted me with the potentialities for getting numbers on time scales and events that we didn't have any on before."[5]

His decision to change research disciplines was a profound one because the new field required that he laboriously learn new techniques and methods, and because Perry Byerly, by whom he was hired and under whom he worked, had quite explicit expectations

The major driving force behind young-rock potassium-argon dating at Berkeley, Jack F. Evernden carried that program to the Australian National University in Canberra, and later played a crucial role in developing seismological methods for discrimination of underground nuclear explosions.

of him in seismology. "I just decided to do it [change research areas], I didn't ask Byerly. . . . Byerly didn't talk to me for ten years because of it."* There was apparently "no discouragement to go into K-Ar work" from the other Berkeley faculty.[6] In fact, he learned later "that there was a meeting in the department of the senior types, Frank [Turner] and so forth, and they were absolutely considering giving instructions to Garniss [Curtis] and myself to work in this area . . . but they never had to do it because . . . I was tremendously motivated right from the beginning."[7]

Although Evernden's undergraduate degree was in mining geology in the College of Engineering, he had taken the full complement of geology courses required of geology majors and ventured into problems of paleontology and biostratigraphy. "Incidentally I was a geophysicist, but I think emotionally and intellectually I have always been more of a geologist than a geophysicist."[8] When I interviewed him in 1978, he spoke at length about his interests in a

*Perry Byerly, after that lapse of ten years, called Evernden up to his office and told him that as a young man he had gone to the departmental chairman, Andrew Lawson, to request funds for a gravimeter in order to search for the root of the Sierra Nevada. Lawson had told Byerly, "Cobbler, stick to your last" (Evernden, t. 1, s. 2, iii-10-78).

great range of geological problems far afield from geophysics and especially about his concern with the lack of agreement among biostratigraphers in attempting to draw time boundaries based on different fossil groups; he was aware, early and keenly, of the extent to which biostratigraphy might profit from the virtues of the "time machine."

Garniss H. Curtis was born in San Rafael, California, in 1919 and educated near there in Marin County before entering the University of California at Berkeley, also to major in mining engineering. He was interested in science but not in a specific science. He chose mining engineering because it was likely to involve work out of doors and seemed compatible with his fondness for hiking and camping and the things he "really wanted in life." He found that he "hated every moment of the mining engineering for two or three years," until he "started taking some geology courses." He sensed that mining was a "dead" area and felt that engineers were not concerned with the why of anything, merely the how. Contact with geology through courses taught by Carleton Hulin, Charles Gilbert, and Adolph Pabst led him to take many more elective courses, including paleontology, and to change his major from mining engineering to economic geology in his junior year, which required him to take an additional year of undergraduate study; thus he completed the course requirements for two majors by graduation, in 1942. Hulin's influence was instrumental in Curtis's decision, made while he was still an undergraduate, to return to Berkeley for a doctorate when the war was over. Upon graduation he was deferred from military service by his employment, first at a small copper mine in Arizona and then at Shell Oil in Ventura, California, where he did field mapping. In January 1946 he entered graduate school at Berkeley and began studies under Hulin. Soon, however, Hulin left for a position in industry, and Curtis, while still at work on his doctorate, was hired to replace him, which required that he simultaneously teach full-time.

Curtis's basic interests and experience had led him to study both metalliferous deposits and the volcanological processes of their genesis. With his broad experience in field geology, he might best have been described as close to a general geologist, especially in view of his rich undergraduate background. When he took up radiometry, Curtis was a fully endowed product of the Berkeley earth sciences tradition: he was all one might hope for in a geologist. Yet he would be required to make decisions about how best to apply an

Garniss H. Curtis in his Berkeley office in 1979. During his tenure, from 1957 to 1961, as a Research Professor in the Miller Institute for Basic Research at Berkeley with Jack Evernden, they supplied "approximately one-half of all data [then available] for establishing the geological time scale from 70 to 500 million years," extended the applicability of the potassium-argon dating method such as to overlap radio-carbon dating at 30,000 years, and by dating *Zinjanthropus* demonstrated a greatly increased antiquity for man as well as errors of up to 30 million years in the age of rock beds dated by fossil plants.

esoteric dating tool, not widely known and still less understood in his profession.

During the 1955–57 academic years, Joseph Lipson was primarily responsible for instructing Curtis and Evernden in the use of the Reynolds mass spectrometer and the gas-extraction line.[9] During those years and subsequently, John Reynolds also provided advice and technical support.* Once they had learned to do extractions and run the spectrometer, Reynolds noted, "Curtis and Evernden were self-propelled. . . . I merely supplied technical backup. . . . Evernden had amazing push and drive that made things happen."[10]

During the academic year 1955–56, Lipson was engaged in the lab finishing his thesis on the pioneer dating of glauconites (pub-

*Funding for research in mass spectrometry, including Lipson's salary, during this period was from the Office of Naval Research. Reynolds's research proposal is Document No. 5C. Funds were also provided by the U.S. Atomic Energy Commission, Division of Research, Physics Branch. The renewal proposal and cover letters, submitted by Reynolds, are Document No. 5D.

lished in 1958), which included aid in the geological questions "generously" provided by Curtis and Evernden. Lipson later recalled that Curtis and Evernden "knew the extraction systems and how to get glauconites from rocks . . . and essentially supplied the [geological] materials. . . ." Reynolds, who "never did run the gas-extraction at any time," was instead concerned mainly with experiments using the mass spectrometer.[11]

Industry played a part in Berkeley's radiometry program. Lipson believes that his employment to continue radiometry during the 1956–57 academic year, with the title of assistant research physicist, resulted from the visit of executives of the Shell Oil Company to the university at about the time that he was completing his thesis in the spring of 1956. Later in the day of that visit Francis Turner, then the department chairman, called Lipson to his office to ask how much money it would take to keep the potassium-argon laboratory running during the following year. Shell had offered to fund Lipson to stay on following his graduation.* Their financial commitment speaks for the potential that industrial-exploration geology saw in potassium-argon radiometry. Lipson was unaware that Turner had earlier trained a U.C.L.A. student, Donald Higgs, for work in high-pressure rock deformation, at the request of Shell Oil. Shell thus owed Turner a favor, and when Curtis and Evernden approached Turner for funds to assemble another potassium-argon laboratory, Turner requested $30,000 from Shell. The University received a check for that amount within a week, to be spent over a two-year period.[12]

Curtis and Evernden worked with Lipson during their first two years in the potassium-argon laboratory, from 1955 to 1957. They started with igneous micas, dated as young as a million years.[13] They then attacked a long-standing geological problem in California—the ages of the granite bodies forming the Sierran intrusive sequence, a subject that attracted much attention, especially among western geologists.† Influenced by geological problems of their region, by the traditional concerns of the earth sciences

*The funding by Shell Oil is confirmed on p. 3 of Curtis and Evernden's research proposal to the Miller Institute on x-5-56 (Document No. 6), which states: "A mass spectrometer for determination of potassium and four potassium-argon separation lines . . . are maintained and operated during the current year by a grant of $15,000.00 from the Shell Oil Company. This grant was provided for testing the [potassium-argon dating] method."

†See Evernden, Curtis, and Lipson, 1957; Curtis et al., 1958. Their results were impressive and conformable with structural interpretations of the sequence of intrusions; however, they were later proved incorrect. The extent to which reheating rocks *in situ* by geologic processes could influence a radiometric age was not appreciated at that time.

at Berkeley, and by their contact with European anthropologists, Evernden and Curtis turned to increasingly younger rocks in ensuing studies.

The Berkeley Research Tradition

In contrast to the concern with old rocks demonstrated by early potassium-argon geochronologists in general, the intellectual history and the major research interests of the Berkeley earth science departments led researchers there to different lines of inquiry.

"The great strength and emphasis of the Berkeley geology department, extending back to the 1920's, lay in field studies," according to Francis Turner. "[Nicholas L.] Taliaferro was a fanatic for field work. . . . Everybody who was anybody chose a field problem, and Charles Gilbert's main driving force was field geology."* The emphasis on field geology extended to the undergraduate curriculum and reflected a lack of specialization; extensive hand-lens work and field mapping were required. In 1939, Professor Adolph Knopf of Yale was prompted to remark to Francis Turner, then a fellow at Yale, that field geology training at Berkeley was "unparalleled in the United States." It had arrived at that state through the long-term efforts of men such as Andrew Lawson, George D. Louderback, "Tucky" Taliaferro, and others. Turner thought that Garniss Curtis, as a result, was practiced in field mapping, "well trained in the use of the microscope, and could see what his rocks meant. . . . He wanted to put the story together."[14]

In a coherent geologic study, all else hangs on the linchpin of time; petrology, structure, geomorphology, and the host of particulars accumulated in almost any field study are set aside until they can be chronologically ordered. In the main, the research interests of the department were temporally focused by the nature of the regional geology of the West Coast and were apparent in the vast majority of departmental publications centered on it. Extreme, but indicative, was Taliaferro's oft-quoted remark that for him, geology "stops when you get to the Nevada border." Turner noted: "If he'd been in Ontario, it would have been the Precambrian. It was the surroundings" that determined research directions. Evernden had been influenced mainly by Byerly, who wanted him to become a general geologist before taking up his professorship; he was also

*Taliaferro joined the Berkeley faculty in 1918, retired in 1958, and died in 1961. Charles M. Gilbert taught there from 1938 until his retirement in 1977.

strongly influenced by Taliaferro, who urged him to pursue field studies. In Turner's words, Curtis and Evernden "developed the tool for the specific purpose of trying to use it to date young rocks, which grew from their field-oriented perspective. . . . Curtis was originally interested in K-Ar [radiometry] in order to get dates for his field work." It would, of course, be especially useful as a means of "getting some dates onto the younger history, where the [time-diagnostic] fossils were few and [where] anyhow the fossil evidence [often] conflicted."[15] Many of the young-rock problems grew out of Sierran studies concerned with determining rates of uplift, tracing the history of glaciation, quantifying erosion, and calculating intervals between volcanic episodes. "They were interested in process rates in young geology; that's why [K-Ar at Berkeley] took that direction. If they had been working in the Precambrian it would not have gone like that."

Rocks as young as 30 to 40 million years could be dated easily at the time Curtis and Evernden entered radiometry. Curtis initially wanted to learn the potassium-argon method in 1955 so that he could date still younger rocks in order to examine the sequence of events in the Pleistocene epoch; he recognized explicitly that the dating system ought to be pushed "to as young as possible." Thus began the Sutter Buttes project of 1956—a search for very young dates that could be checked biostratigraphically. The Buttes are a small group of hills in the middle of the flat Sacramento Valley. They formed when magma (molten rock) intruded the Sutter Formation, a series of fossil-bearing Pliocene sedimentary beds that had earlier been dated by conventional means by Charles Gilbert of the Berkeley faculty. Also, Curtis later noted, "I'd been working on the Sutter Buttes with Howel Williams for a long time. . . . [Williams] had published on the Buttes back in 1929, but he'd made some mistakes which I recognized, so that when we had an opportunity to date the Sutter Buttes, I immediately got started, because there were both sanidine and biotite (potassium-rich minerals) in some of the rocks there."[16] Curtis and Evernden also dated the Bishop Tuff, another important Pleistocene rock unit with stratigraphic control.*

"I wanted to see how young we could date things," said Curtis. "Both Jack [Evernden] and Joe Lipson were enthusiastic about find-

*See Evernden, Curtis, and Kistler, 1957. The age of the Bishop Tuff is controlled by its relationship with the Sherwin Till of the Sierran glacial sequence. The redating of the Bishop Tuff was crucial in formulating the polarity-reversal scale discussed in Chapter 6.

ing young things to date. . . . I saw that if we could date very young things we could get to the younger part of the geologic time scale. In my mind I knew that the Pleistocene, which by [Frederick E.] Zeuner had been given an age of 500,000 years, had to be older. . . . Just too much happened in the Pleistocene to have happened in 500,000 years."[17] Evernden, Curtis, and Lipson's first effort, published in 1956, was aimed at defining the Pliocene-Pleistocene boundary and demonstrating that the Pleistocene's duration was greater than that generally ascribed to it. "I was not thinking about the origins of man [work that Curtis later did] at that time. I was thinking about this in the geological and paleontological sense. . . . Jack got us both involved in the paleoanthropological through his contacts with [Alberto C.] Blanc and [Louis S. B.] Leakey."[18]

Ronald W. Kistler, who began a research assistantship with the group in February 1956, commented in 1978: "Most of the initial impetus and real hope for that young dating was to put some physical time scale related to climates, mostly [for] glacial epochs. . . . The interest at that time was to date Pleistocene events. . . . A standing major problem in Pleistocene geology is the correlation between continental glaciations [of North America] and the more easily determined Alpine glaciations from the classic sections of Europe." Because of that interest, members of the group began reading in Pleistocene geology to define the limits of such major problems and began to speak to Pleistocene specialists. The dating effort on the Bishop Tuff was undertaken because of its relationship with Sierran glacial deposits. Some of the first dates implied that the standard estimates of the lengths of the glacial epochs might be incorrect. "At that time, this was a very new technique, and we didn't know if [the discrepancy lay with] the technique or was only in the minds of the Pleistocene geologists. . . . We got pretty deeply into the Pleistocene literature." That research provided background for the group's first publication of 1957, centered on the Pleistocene.[19]

The influence of paleontologists on geologists in the matter of dating is of considerable importance. Curiously, the Paleontology Department at Berkeley was unique in that it had been split off from the Geology Department since 1929: in the 1950's it was separated physically as well as administratively. That separation somewhat limited the communication between the two groups and tended to channel them according to specialties. Biostratigraphically oriented paleontologists, who could supply crucial dates to the geologists, were necessarily more concerned with physical stratigraphy and attendant geologic problems and naturally tended to form alliances with the geologists. It thus came about early that the major paleontologic influence on the Geology Department was from biostratigraphers working in younger rocks.

The younger rocks, late Mesozoic and Cenozoic, lent themselves by their availability and fossil record to the growth of the biostratigraphic tradition at Berkeley. In 1978 Wyatt Durham remarked, "I think that the chance location of the University [of California] in the midst of a whole bunch of marine Tertiary and Cretaceous [rocks] determined the scope of the early interests in stratigraphy and paleontology."[20] The early pattern of almost exclusive concern with young rocks by the biostratigraphically centered paleontologists continued until the 1950's, when Ralph Langenheim joined the Paleontology Department and began his Paleozoic studies.* Those faculty members treating the lower vertebrate groups pursued studies in older rocks, but were biologically rather than biostratigraphically oriented. In contrast, the mammalian researchers, because of the nature of the animal groups involved and their fossil record, grew better able to demonstrate their use in chronostratigraphy, and by 1941, formal North American land-mammal ages were definable.[21]

Curtis's great interest in dating stemmed in large measure from his own undergraduate interest in paleontology, which led him to take a number of elective courses in it; also, perhaps more than any other member of the geology faculty, he sustained friendships with a number of paleontologists, especially J. Wyatt Durham and Donald E. Savage, both of whom were biostratigraphically oriented. Curtis was thus keenly aware of the historically difficult and often futile attempts by biostratigraphers to classify fossil-bearing layered rocks according to their time of deposition. He was also aware that a disciplined radiometric assault on the Pleistocene should be guided in part by biostratigraphers who could direct him to volcanic rocks interbedded with sediments bearing time-diagnostic fossils. Thus when Curtis and Evernden began their program of worldwide age correlation by use of potassium-argon in 1956, they turned to Durham and Savage for aid. Savage's broad knowledge of well-dated terrestrial mammal sequences in rocks that included

*An exception is C. W. Merriam, who began work on the Silurian-Devonian rocks of the Roberts Mountains of Nevada in the early 1930's.

Donald E. Savage's broad knowledge of well-dated terrestrial fossil mammal deposits interbedded with potassium-rich volcanic rocks made him a natural collaborator at Berkeley with Garniss Curtis and Jack Evernden.

high-potassium volcanic rocks made him a natural collaborator; Durham, however, specialized in marine sediments, in which such a rock combination is rare. Therefore Savage provided advice and later, in 1964, coauthored an important paper with Evernden, Curtis, and Gideon T. James that demonstrated essentially perfect congruence between K-Ar ages and the mammalian age designations of biostratigraphers. Charles A. Repenning, an authority on mammals with the U.S. Geological Survey, wrote:

Savage was (and is) very influential and a major stimulant because he knew (and knows) the mammalian geochronology very well and could specify older and younger faunas as few others could. This was needed to confirm the K-Ar dates. Although Evernden, Savage, Curtis, and James (1964) was in intent and effect a paper dating mammalian geochronology, it was also the first paper to prove, through agreement of results with mammalian geochronology, that K-Ar was a fantastically useful tool—and the mammalian geochronology was the proof.* The K-Ar guys admitted it

*"Many paleontologists, including the older ones, at the U.S. Geological Survey [and elsewhere], felt quite threatened by K-Ar dating, acting as though they feared it would eliminate their jobs." Unattributed by request, this surmise was volunteered by more than one interviewee of note.

. . . by word of mouth to me. The mammal geochronology really put K-Ar in business in 1964.[22]

There is little doubt about the great dependence of potassium-argon daters on biostratigraphy, not only for aid in sample selection but, more important still, for approbation from a discipline that had held sway in matters temporal since Cuvier and Brongniart's and William Smith's works early in the nineteenth century, which first showed that earth's history is decipherable from the record of life entombed in its crust.

Thus, the Berkeley earth sciences departments appear to have had a sustained, long-term concern with young rocks. When the unique capabilities of the Reynolds mass spectrometer, in consort with an almost singular high-vacuum argon-extraction line, were placed in the hands of such geologists, one might have anticipated what came to pass.* The method, the scientists who refined and applied it, and the firm support their institution provided made potassium-argon dating one of the most exciting research frontiers of the 1950's and 1960's. Before the method could reach its potential, however, old techniques had to be refined and crucial new ones devised.

Cleaning Up the Act

By 1956 the important problems of potassium 40's branching ratio and its decay constant had been largely solved by pioneering work in other research centers, and only small changes have been made since.[23] In contrast, questions of retentivity or diffusion of argon in many minerals remained problematic, and diffusion studies continued for many years. However, the retentivity of some of the common rock-forming minerals, especially certain widely prevalent micas and feldspars, had been clarified by 1957.[24] In that year, the major unresolved questions facing researchers working toward younger dates were how to select appropriate materials, optimize analytical procedures, and, most important, reduce contamination,

*Another hint of the possible influence of instrumentation is suggested by the commentary of Bruce Doe, chief of the Isotope Geology Branch of the U.S. Geological Survey, who visited several research centers in the Soviet Union in 1976, including the University of Moscow and the nearby Vernadsky Institute and the University of Leningrad. "The Russians have done more things first," he noted, "but their equipment is not good enough to stay with United States science. . . . [They] concentrate on the Precambrian because their equipment is primitive" (personal comm., v-18-78). Is it equally likely that their equipment remained primitive because they concentrated on the old, argon-rich rocks of the Precambrian, which did not spur development of more sensitive instruments?

especially that caused by the adsorption of atmospheric argon onto the surfaces of crystal grains.

The contaminating atmospheric argon was able to mask the much smaller radiogenic argon fraction in samples less than 1 million years old. For example, Evernden noted in 1957 that in two biotite samples about 1.5 million years old, the argon 40 content of the argon sample derived by older procedures was 96 percent atmospheric in one and 76 percent in the other.[25] When analyzed, the high atmospheric argon fraction of the first sample would cause a 24 percent error in the derived age; this was the result of only a 1 percent error in the determination, caused by the presence of atmospheric argon. Thus it was crucial that techniques for eliminating most of the atmospheric argon be developed that did not concomitantly alter the radiogenic argon content. In Evernden's words, "[To] learn to get a very young age was to learn to run the [extraction] line and to treat the sample in a way that absolutely minimized contamination by atmospheric argon. In some of the first runs, we didn't know about this problem; we didn't know the source of atmospheric argon and . . . how it was getting into the sample."[26] According to Curtis, "The mass spectrometer was beautiful[ly clean]; the problem was in the extraction line. We recognized that until we could get low air argon contamination, we couldn't date very young materials. Jack [Evernden] and Joe Lipson applied themselves, and I'll give credit to Jack particularly for eliminating, one after the other, the sources of air argon."[27]

The extraction system contained several features that were thought to be such sources. Most obvious, and early recognized, was the use of a sodium hydroxide flux in the fusion of the sample. Curtis and Evernden inherited a standard procedure in which the flux and sample were melted in a nickel or stainless steel bomb placed in a resistance heater of limited temperature.[28] The flux introduced atmospheric argon and also produced water in the reaction; the water got into the extraction line and had to be removed by freezing, along with carbon dioxide, which some workers feared might contribute to the formation of chlorine compounds that would further interfere with analytic procedures. The flux was also extremely corrosive and would destroy the metal bomb every few weeks, causing lengthy delays in the operation of the extraction line.

The technology of directly fusing a mineral in a high vacuum was truly in its infancy in 1956, and any departure from accepted procedures was experimental. The first effort to eliminate the flux was through the use of a new high-temperature (1,400°C) resistance heater made in Sweden by the Kanthol company; the heater eliminated the need for the flux and improved the contamination problem but still involved the reuse of the metal bomb, which was difficult to clean. The Kanthol heater was used for about six months, during which time the first Pleistocene samples were run at Berkeley.[29]

In 1957, Curtis received word from John L. Kulp, before Kulp's publication of his findings, about an induction heater that replaced the troublesome metal bomb with a molybdenum crucible hung on a molybdenum wire in a glass bottle.[30] The melting of the sample in glass by induction (electromagnetic) heating further reduced contamination: as Curtis noted, "It was an important step in achieving reliability in analytical runs."[31] The Physics Department helped build an induction heater that was considered superior to those commercially available. With it the K-Ar group "could then do a sample in a few hours instead of two days."[32] Surprisingly, induction heaters that were suited to isotopic dating had been described in 1950 by W. G. Gouldner and A. L. Beach, but these were not used in K-Ar dating until 1957, by Donald Carr and John Kulp. "There were two extraction lines in operation such that fluxing and the use of the Kanthol resistance heater overlapped," recalled Kistler. "There was even overlap in the use of all three—the fluxing method, Kanthol heater, and induction heating."[33]

In an effort to drive adsorbed atmospheric argon from the surface of the mineral grains, they were baked, but fear that too intense or too prolonged heating would result in significant loss of the radiogenic argon led Evernden to pursue lengthy diffusion experiments that occupied the efforts of the laboratory for weeks on end.[34]

Sanidine, a common mineral of proven value in dating by 1957, usually occurs in volcanic rocks such as rhyolite, which has devitrified glass shards adhering tightly to its surface. The glass is a repository for atmospheric argon. In 1957, Curtis began to use the technique of dissolving the glass in hydrofluoric acid, which proved immediately successful.* This practice was extended to all young feldspars in order to remove the outer portions of the crys-

*The samples were etched in dilute (7 to 10 percent) hydrofluoric acid at 50°C for 20 to 30 minutes (Evernden and Curtis, 1965).

tals, in which the adsorbed atmospheric argon was concentrated. The samples were then washed in twice-distilled water and frozen to preclude further air contamination. According to Curtis, "the freezing didn't improve things over the acid leaching, which was the key. We found that the longer the specimens were exposed to air, the more contaminated they became." Curtis therefore placed each sample in a pre-fired extraction line immediately after leaching by acid. Leaching lowered the atmospheric component of the argon 40 from 80 or 90 percent to less than 40 percent.[35]

The Berkeley group was aware that a large amount of the atmospheric argon present at the end of the clean-up came from a device then in use called a calcium getter, in which calcium metal chemically removes extraneous gases, mainly nitrogen and oxygen. As late as September 1955, at the Second Conference on Nuclear Processes in Geologic Settings, Gerald Wasserburg asked John Reynolds, "Is there any notice that some of this [extraneous atmospheric argon] is coming from the calcium metal?" Reynolds's response indicated that the available data did not locate the air argon source; of equal note in his response was the fact that the group had not yet determined "how important, if at all, it is to outgas [purge] the extraction equipment."[36] Calcium metal is manufactured in an argon atmosphere to prevent its combustion; it thus becomes highly contaminated. The calcium getter could not be purged at a higher temperature than that required for a sample run, without destroying the calcium; thus the atmospheric argon could not be eliminated from the calcium during the pre-extraction bakeout of the entire extraction line.

In 1958, Jack Evernden devised the technique of heating the calcium until a film of calcium was deposited on the upper glass wall of the furnace. The evolved gases were then pumped out, including the atmospheric argon driven from the calcium metal turnings. The calcium furnace was then turned off during the sample run and a torch applied to the upper glass surface of the furnace to activate the clean calcium film. This method of purifying the getter effected an appreciable reduction in atmospheric argon contamination. It worked so well that when titanium was first used as a getter at Berkeley in 1961, several years after it had been in use at other institutions, including Cal Tech and M.I.T., the reason was experimental convenience rather than improvement with respect to atmospheric argon contamination; titanium can be outgassed at high temperatures and used repeatedly.* Calcium had to be replaced after each sample run. Curtis recalled, "Jack said it's a lot easier to use [the titanium getter], but there is no difference in the quality of the runs."[37]

Just before the introduction of the titanium getter, further improvement had been realized when the mineral samples were baked for 18 to 24 hours, to drive off atmospheric argon, at temperatures calibrated to match their diffusion behaviors; sanidine was baked at 400°C, coarse biotite at 200°C, and fine-grained biotite at 100°C.

The Berkeley laboratory also began the practice of placing heaters around the charcoal rods, called fingers, used to separate the noble gases from each other by fractional adsorption. Heating the fingers during the extraction kept them from adsorbing gases until the experimental run.[38] The precise extent to which this improvement contributed to reduction of air-argon contamination was not clear, but it was one of the many innovations implemented by the Berkeley group during the late 1950's in their exhaustive efforts to measure accurately ever-smaller amounts of radiogenic argon and thereby to overlap the time range of the carbon 14 method, which extends from the recent to about 70,000 years ago.† By 1960 most of their innovations in the laboratory had been made, and in fact few new techniques have since been introduced there or elsewhere. Evernden told me, "We were able to get the atmospheric argon . . . down to a content of vanishing significance; it just had no relevance to anything. . . . We could date a rock that had a few percent potassium . . . at 10,000 years and not be significantly perturbed by atmospheric argon."[39]

Evernden attended the Fifth International Congress for Prehistory and Protohistory at Hamburg in August 1958 "with the sole intent to deliver the message that we were dating some young rocks; we were interested in meeting the carbon 14 time scale and interested in any correspondence about such dates where volcanic rocks are intercalated with rocks that have carbon 14 dates on them, any-

*G. H. Curtis, J. F. Evernden, R. W. Kistler, and J. H. Reynolds, commented orally concerning the titanium getter on vi-29-78; Kirsten's publication of 1966 also treats it. G. B. Dalrymple wrote to me on iv-18-81 that the "titanium was first used at Berkeley by John Reynolds. Curtis and I replaced the calcium furnaces with titanium while Jack [Evernden] was in Australia. . . . They worked well."

†Evernden and Curtis indeed achieved their aim of overlapping the carbon 14 method. Their former student, Brent Dalrymple, dated 11 sanidine samples from rhyolite domes of the Mono and Inyo Craters of California; his 22 calculated ages, which appear to be the youngest reliable extant ones by the method, ranged from 2,900 to 12,400 years (Dalrymple, 1967).

where in the world." As a result of that plea to an international audience, Evernden and Curtis received not a single response.

Dating Whole Basic Rocks

Igneous rocks, which form by the cooling and crystallization of molten rock (either lava or magma), are divided by their composition into two main categories. Basic rocks, among which basalt and gabbro are common, are rich in iron and magnesium and poor in potassium; conversely, acidic rocks, including common granite and rhyolite, are poor in iron and magnesium and rich in potassium. Basic rocks are widely distributed both temporally and geographically, and are commonly found within well-dated stratigraphic sections. Such widespread availability of basalts dated by their positions between fossil-bearing beds made them early candidates for attempts by the geochronologists.

Curtis and Evernden discussed the dating of "whole-rock" basalt in the mid-1950's, before Joseph Lipson left for Edmonton in 1957.[40] Compared to the rocks they had been dating—the granites and the larger crystal fractions from tuffs—basalts contain smaller mineral grains.* Evernden and Curtis spent considerable time examining many samples from well-dated stratigraphic sequences, but finally eschewed basalts for a number of reasons. Basalts are filled with minute cracks and crevices that contain extremely small secondary or gas-phase crystals. They feared that baking the whole rock to clean it would affect those finely grained gas-stage minerals, some of which were rich in potassium, and that the radiogenic argon could be lost. Evernden recalled: "We started early on trying whole-rock basalts. We tried to determine their retentivity and found that most basalts were coming out too young. We concluded that there were very few basalts on which we could get good dates.† If we could not get a crystal concentrate out of the basalt, then we would just not date it. If you [examined a] thin section [of] the rock and determined that there is absolutely no devitrification [alteration] anywhere . . . then it was legitimate to try and date it."[41]

During the mid- and late 1950's, Curtis and Evernden did not re-

*Tuff is a compacted deposit of volcanic ash and dust that may contain appreciable amounts of sediments such as clay or sand.

†The microprobe, which would have been very useful in such determinations, did not arrive in Berkeley until about 1962 after others elsewhere (discussed later) had successfully dated whole basic rocks.

quire many whole-rock dates for their purposes. Because their central concern then was a paleontologically controlled time scale, they worked mainly with stratified tuffs in connection with various kinds of fossil materials. (In contrast, the paleomagnetists, as we shall shortly learn, were crucially dependent on basalts.) Curtis recalled, "We hadn't exhausted the material that we had, and didn't feel that we had to go to inferior things like whole rocks."[42]

Basalts or their separated major minerals had been investigated in the 1940's by workers using the uranium-helium method and had been found to undergo large losses of helium;[43] thus no further work on basalts was undertaken using those isotopes. By the late 1950's, little systematic work had been done on the low-potassium minerals that occur in basic igneous rocks to explicate those properties, especially retentivity of argon, that are essential to disciplined dating practice.

In 1958, Paul Damon and John Kulp obtained experimental results that led to their hypothesis that hornblendes contain excess argon; many workers were therefore dissuaded from dating hornblendes or hornblende-bearing whole rocks.* However, in 1959, Khabibulla I. Amirkhanov and his coworkers measured the diffusion coefficient of argon in a pyroxene mineral that is compositionally and structurally similar to hornblende, and found it to be extremely small at geologic temperatures. This discovery moved Stanley Hart and a group at the Massachusetts Institute of Technology to date a suite of ten hornblendes and three pyroxenes, using a Reynolds-type mass spectrometer. The ages of the dated samples, when compared with the ages of the coexisting biotite, feldspar, and zircon that had been previously determined at M.I.T. and the Carnegie Institution, showed "good geological concordance."[44] This demonstration of the suitability of hornblende for K-Ar dating moved Hart in 1961 to examine the retentivities of hornblendes, other amphiboles, pyroxenes, and micas. Immediately after the presentation of his results he remarked, in response to Paul E. Damon during a discussion, that thus far he had found "no argon ages on hornblendes that suggest excess argon."†

*Hornblende [Ca_2 (Mg, Fe)$_5$ $(OH)_2$ (Al, Si)$_8$ O_{22}] is a member of the amphibole group of rock-forming minerals that are closely related chemically and crystallographically to the minerals of the pyroxene group. Amphiboles and pyroxenes together make up about 17 percent of the igneous rocks.

†Quote from Hart, 1961. Although hornblendes are still considered reliable minerals, Hart and Dodd (1962) later demonstrated that some pyroxenes from deeply buried metamorphic rocks do take up excess argon 40 during crystallization and thus yield anomalously old ages.

Hart's paper of 1961 motivated Glen P. Erickson and John Kulp of Columbia University to try to date the nearby Palisades later that year because it contains several distinctly different petrologic phases and types and a hornfels rock. Furthermore, the Palisades tabular igneous intrusion (sill) is stratigraphically dated as Upper Triassic, but "most important is the fact that certain phases of the sill contain small concentrations of biotite, which may give the actual age of the formation."[45] With the biotite age and its well-defined stratigraphic position as checks, Erickson and Kulp concluded that the chilled smaller mineral grains of the border or marginal area of the Palisades sill, which cooled more quickly from the molten state, appeared to hold all their radiogenic argon, and the medium and coarse phases lost about 15 percent. They noted, in keeping with all earlier admonitions regarding petrologic criteria for dating, that the coarser rock shows more secondary alteration than the fine-grained rock, contains much more potassium feldspar as a distinct mineral, and shows other mineralogical differences as well. Their conclusion, that "in some cases the argon retentivity of certain types of basaltic rocks may be as high as that of biotite," encouraged further studies of basalt.

Simultaneously, at the Australian National University, Ian McDougall (who would soon make major contributions to young-rock radiometry) had been prevailed upon by John Jaeger, his departmental head, to take up a year-long postdoctoral fellowship at Berkeley under Evernden and Curtis in order to learn potassium-argon dating. During the 1960–61 academic year, McDougall, like Erickson and Kulp, was moved by Hart's paper of 1960 to "investigate the relative argon-retention properties of pyroxene, plagioclase, and sanidine in two specimens from a Tasmanian dolerite body" in the hope of dating the intrusion of the dolerite. He was especially encouraged by the possibility of dating a whole-rock sample from the edge, or contact, of the dolerite where the rock had chilled rapidly during its emplacement.*

The Tasmanian dolerites, unlike the Palisades sill, were not well controlled stratigraphically, but they could be placed within a Jurassic to Tertiary time interval with good probability.[46] McDougall's calculated ages based on pyroxene (two samples) and pla-

*It is curious that early warnings that altered rock specimens were unsuitable for dating did not deter some from dating dolerite, which in British usage is defined as an intrusive igneous rock of the composition of diabase (U.S. usage) that has been much altered by the decomposition of feldspars and mafic minerals (*Glossary of Geology*, 1972, p. 206, Am. Geol. Inst.).

gioclase (four samples), which ranged from 168 to 195 million years, suggested high argon retention, since the two minerals are very dissimilar in crystalline structure. The single whole-rock specimen, a chilled dolerite, was calculated at 148 million years. That was regarded as "a surprisingly good age," because the rock was mainly devitrified glass and obviously had retained more radiogenic argon than expected. McDougall's results, along with the almost simultaneous studies of Erickson and Kulp, went far to encourage the dating of whole basic rocks.

Several other investigators published on whole basic rock-dating during the early 1960's.[47] In 1963, John Miller and Alan E. Mussett described results obtained from seven samples of dolerite from the Carboniferous age Whin Sill of northern England, and confirmed the marked correlation between the "apparent ages of the samples and the relative amount of alteration of the plagioclase feldspar." Earlier studies—by J. R. Stevens and Harry Shillibeer in 1956 and Samuel Goldich, Bud Baadsgaard, and Alfred Nier in 1957—had suggested that argon is retained at the boundaries of the grains. Miller and Mussett also tested this suggestion, by grinding a dolerite sample and comparing the age derived from the whole rock with that from powder. A demonstrated 7.5 percent argon loss was in keeping with the idea.

These studies thus demonstrated that some fine-grained volcanic rocks retained their argon and could be dated. They also increasingly affirmed the precaution urged by the earliest workers in the field; namely, that a careful selection of samples to guard against alteration, with its concomitant loss of radiogenic argon, is a crucial preface to any dating attempt. It took discipline for radiometrists to reject unsuitable rocks that nevertheless held the portent of answering profound geological questions.

By the early 1960's the unique strengths of the young-rock potassium-argon method—in the age range and the rock types accessible to it—made its few practitioners especially sought after. And, among those seeking potassium-argon dates, the paleomagnetists who were pursuing polarity-reversal studies and in critical need of dates, unfortunately were not reckoned as high-priority clients by the radiometrists in those early years. Much of this book, in fact, concerns the often-frustrated efforts of rock magnetists to acquire dates on young rocks.

The Berkeley dating method spread first to Folinsbee's laboratory at Edmonton; however, it would be two other research centers—one distant, at the Australian National University in Canberra, and one close by at the U.S. Geological Survey's western headquarters in Menlo Park, California—that would play crucial parts in the development of the polarity-reversal time scale, a story that unfolds in the chapters that follow.

Chapter 3

Confirming and Extending the Tradition

THE CRUCIAL links of historical chains are often forged of pure opportunity. Such was the case when the young-rock, potassium-argon research program, still in its infancy, found application in a host of problems of earth science: some old, some new.

The Miller Institute Project

Early in 1956, Adolph Sprague Miller, who had established the Miller Institute for Basic Research in Science for the appointment of research professorships on the Berkeley campus, made a substantial increase in the endowment upon the death of his wife Mary.[1] A faculty-wide call for research proposals ensued, which prompted Garniss Curtis to suggest to Jack Evernden that they jointly request support to develop a worldwide time scale, using potassium-argon dating. Such a scale, which would make possible intercontinental correlation of rock bodies, had been an aim of geologists since their science had been given its name in the mid-seventeenth century.

Although Curtis and Evernden's goal was an old one, they elaborated on several points that promised new results, beyond those offered by other methodologies. Potassium-argon dating, unlike the uranium-lead method, could be applied to a great range of rock types and ages, including some as young as one-half million years. (At that time, the vast majority of reliable absolute dates came from rocks at least tens of millions of years old.*) Because the increased

*In regard to my earlier conclusion that physics-oriented mass spectrometrists were mainly interested in very old rather than young dates, Curtis noted that during the interview by the Miller Institute Evaluation Committee, the atomic scientist Glenn Seaborg asked him how old he thought the universe was (Curtis, t. 2, s. 1, iv-19-78).

sensitivity of the Reynolds machine permitted the dating of igneous rocks and formerly undatable sedimentary rocks, a major section of Curtis and Evernden's proposal concerned the acquisition of rock samples—"the critical mineral samples necessary for a reliable absolute time scale." These samples had to contain one of the three minerals known to be reliable (unaltered biotite, glauconite, or potassium feldspar) and had to be datable independently by "ordinary stratigraphic means," essentially biostratigraphic. They thus aimed their search at fossiliferous sedimentary beds containing datable glauconite, and biotite- or potash-feldspar-bearing volcanic rocks interbedded with sediments holding time-diagnostic fossils. A global travel plan was outlined for collecting by "those in charge of the project" to ensure reliable samples. By the time of their proposal, in October 1956, four argon-extraction lines were in operation at Berkeley, which greatly increased their capacity for dating, since the time required to perform an extraction is several times greater than that required to analyze the samples on the mass spectrometer. To guarantee the usefulness of the samples, the proposal stressed that "the persons who interpret the analyses must collect the samples, for they must have unquestioned knowledge of their geologic meaning and validity; otherwise the precision of the results is suspect."[2]

Unlike almost all other radiometrists, who had learned mass spectrometry and related techniques as physicists and were generally unapprised of the geological requirements attending potassium-argon experiments on rocks, Curtis and Evernden were geologists of broad perspective. They recognized the profound potential of an emergent method residing in another discipline and possessed those personal attributes—and a supportive research environment—that enabled them to acquire and apply that method in an imaginative and dynamic way. At the time of their entry into radiometry, there were no textbooks or formal courses in the subject. Departments of geology, with the exception of those in proximity to a mass-spectrometry group, were generally not aware of ongoing developments in potassium-argon dating. The unusual circumstances on the Berkeley campus that had brought John Reynolds from Chicago to build a static-mode machine and put him in close contact with a group of geologists who were favorably disposed toward new lines of research, however far removed from their own, had also provided Evernden and Curtis with what may now be reckoned a unique opportunity in that decade. Their proposal, supported unanimously by the senior departmental staff members, was submitted early in October; notice of the award, on January 25, 1957, confirmed funding of the project for two years, including salaries for both men, travel and subsistence abroad for collecting, and two years of aid by a junior research geologist.[3]

During the first two years of the project, Curtis and Evernden acquired many datable samples, most derived from Western Europe and rocks of correlative age from Africa, Australia, Japan, New Zealand, and South America. Almost from the start of their dating efforts, Curtis and Evernden, who had been partly trained in and greatly influenced by the biostratigraphic tradition at Berkeley, generally did not attempt to date a sample unless it could be clearly demonstrated in the field that the sample was derived from volcanic rocks bounded by sediments bearing time-diagnostic fossils.[4] A number of publications resulted from the work of that first two-year grant period, including dating of California granitic rocks, Pleistocene volcanics, and sedimentary rocks.[5] In addition, papers delivered by Curtis at the Geological Society of America meeting in Oregon in March 1958 discussed the ages of California intrusive rocks, the ages of Tertiary rocks from California, Nevada, and the Gulf Coast, and the rate at which the sea eroded the land—all based on the new technique. Foreign presentations were made by Evernden at Hamburg in August 1958 on the dating of Pleistocene rocks, and he announced further Pleistocene and some Tertiary dates in November, in London. Perhaps the most important paper resulting from the project's early work was the "Diffusion of Argon in Biotite, Glauconite, and Sanidine and Its Bearing on the Dating of Young Geologic Samples," which Evernden presented in Berkeley in November 1958 to the National Academy of Sciences, with an expanded version published in 1960.[6] Implicit in that paper was the important fact that geologists not only had acquired a tool from another discipline but were hard at work refining it and making excursions in new areas of technique and methodology; a desire to date increasingly younger rocks provided much of the impetus.

The initial grant proposal had stipulated a division of duties such that "Professor Evernden . . . be freed of departmental duties during the Fall semester and Professor Curtis during the Spring semester for the duration of the Project." Thus freed of his fall teaching load and other campus responsibilities, Evernden had embarked on his first sample-collecting trip in the summer of 1957, and had headed off to Europe to gather Tertiary age rocks. His first stop was to meet Kenneth Oakley, the eminent anthropologist, in the British Museum of Natural History. During World War II, Oakley had studied glauconite deposits in England, because they were a commercial source of iron; it was his knowledge of glauconite that attracted Evernden to him in the hope of getting directions to datable European rock localities. Twenty years later Evernden recalled that "at that time . . . [Oakley] was solely interested in matters anthropological . . . and said he'd tell me about the glauconite but [asked] what were the potentialities of using the potassium-argon technique for dating early man." There were then no radiometric techniques for dating fossil man earlier than the limits of the carbon 14 method—about 50,000 years. Evernden responded to Oakley that he didn't "know how young we can date but I'm sure going to find out."[7]

Evernden was strongly influenced by Oakley. The anthropologist not only suggested workable collecting sites and introduced him to people who eventually helped immensely in the sample collecting, but also impressed Evernden with his ideas about the usefulness of radiometry to human paleontology. Evernden remarked that up to the point of meeting Oakley,

> we hadn't attempted to optimize our capability. Ultimately when we really got serious about trying to date young rocks, we had to find out where the atmospheric argon [contaminant] was coming from, how it was entering the sample, how we could eliminate it, [and] how we could minimize the contamination. The learning of this was in order to achieve the capability to date as young a rock as possible. The motivation to do it was the desire to date young rocks, but before I met Oakley we just really hadn't had that desire yet; we really hadn't started down that road.[8]

Evernden has reiterated this important point, noting that he had neither an interest in fossil man nor a desire to work toward the very youngest dates before he met Oakley in 1957.[9] Oakley suggested that Evernden contact Alberto C. Blanc, a professor of anthropology at the University of Rome, because of his knowledge of the Italian early-man sites and of volcanic rocks bearing potassium-rich minerals, and also Louis S. B. Leakey, for his knowledge of the East African Rift Valley, which contains important fossil-man sites stratigraphically bracketed by volcanic beds.

Following his early summer meeting with Evernden, Oakley wrote to both Leakey and Blanc to introduce Evernden and to explain his intended visits and his mission. After a three-month sample-collecting trip through northern Europe in search of datable rocks from the classic Tertiary-age type sections, Evernden met Blanc in Rome, and the two journeyed through Italy collecting samples.* By the time Evernden returned to England from Italy, Oakley had received a latter from Leakey expressing his wish to aid in the work of the Berkeley group. Although the potassium content and petrography of the volcanic rocks in the Italian deposits were known, for the African rocks there was uncertainty concerning potassium content and the availability of discrete mineral grains that would be large enough to date; insufficient information had been published.

Without expectations, Evernden visited Leakey in pursuit of datable rocks from Olduvai Gorge in East Africa. At the time, no bones had been found at the stratigraphic levels, or horizons, from which Evernden would collect samples, but there was a "tool culture" known as Olduwan about which Leakey had developed an elaborate interpretation; thus anthropologists were anxious to have an age for the tool-bearing beds. Evernden visited several sites in the Gorge and happily found "easily datable crystal-bearing . . . nuée ardent explosive tuffs," which formed from a rapidly flowing, turbulent, fiery gaseous cloud carrying ash and other volcanic fragments. He collected samples bearing crystals of the potassium-rich feldspar mineral anorthoclase, "which were intended to date nearly the base of the sequence at Olduvai Gorge [from] a horizon interspersed with the strata bearing Olduwan tools. . . ." This was in 1957; only two years later, at that same level, in that same site, Leakey found the bones he named *Zinjanthropus*, then the oldest known ape-like man. Evernden later said: "As a matter of fact, that horizon was the one that buried Zinj on his living site . . . and so all our subsequent dates simply confirmed the date we got on that horizon at that time. . . . We didn't know it, but essentially the first thing we dated at Olduvai Gorge was *Zinjanthropus*."[10]

In a conversation in 1978, Curtis recalled the same event: "That fall [1957] was when we first dated Bed I from Olduvai, which had been collected with Louis Leakey in Africa, for which we got a date of 1.75 million years. . . . That was before the famous skull *Zinjanthropus* was found by Mary Leakey. Louis was not certain at the

*Type sections are sequences of rock beds that serve as standards of reference for formal geologic time intervals. Three of the samples from Italy were dated at Berkeley and reported by Evernden in 1958 at the Fifth International Congress of Prehistory and Protohistory in Hamburg (Blanc, 1958). Those Pleistocene dates were important in the formulation of the first radiometrically disciplined geomagnetic polarity-reversal time scale, treated in detail in Chapter 4.

time where the skull was found in relation to the dated bed; the date was obtained from a volcanic ash further down the gorge from where *Zinjanthropus* was found. Leakey wanted us to come back."[11]

Thus, in order to add confirmation to that first date, Curtis visited Olduvai two years later, carefully assessed the stratigraphic relations, and collected samples from other horizons. The impact of that date on human paleontologists was great indeed. According to Sherwood Washburn of Berkeley, "It was four times older than previously thought at that time . . . [and it] changed the conception of the rate of evolution regarding fossil man."[12]

The samples that Evernden was collecting abroad in 1957 were sent to Berkeley to be dated by Curtis, Lipson, and Ronald Kistler. Kistler was a geology student whom Howel Williams had encouraged to enter the Berkeley graduate school after two years of military service in February 1956. Kistler's main interests were not originally in geochronology; he entered a research assistantship in the K-Ar laboratory only through economic expediency and had not yet formulated plans for graduate study. Thus he was open to learning the potassium-argon dating method. He noted that "almost the total impetus was toward young rocks from when I first entered the Berkeley K-Ar laboratory."

The laboratory, said Kistler, was located

> in a little room in the physics building [Le Conte Hall] where the argon extractions were done; the mass spectrometer was in another little room (about 10 by 15 feet). Curtis did his work over in Bacon Hall [the geology building, now demolished] down in the basement. Curtis dealt most directly with the mineral separation and the potassium analyses. There was a pretty good separation of the two parts of the effort. The two [Curtis and Evernden] traded off, but mostly there was a division of labor. Because of the separation of work space, group interaction took place mostly outside the laboratories. Evernden was the major driver in the group, very energetic.[13]

Garniss Curtis went abroad in the spring of 1958 to collect minerals for additional dates to use in elaborating a potassium-argon time scale. He visited Ceylon, Egypt, and Europe but did most of his collecting in the Tertiary rocks of New Zealand, Australia, and Japan; a political crisis in Ceylon precluded his planned field work there. Curtis also visited the Australian National University, where he laid the groundwork for the transplant, from Berkeley to Canberra, of a young-rock, potassium-argon laboratory. It was during this interval that some conflict arose between Curtis and Evernden that diminished their collaboration. They apparently held different views regarding research aims and the ordering of priorities in selecting specimens.[14]

In December 1958, Curtis and Evernden submitted an application to renew their Miller Institute professorships. Entitled "Continued Development of the Potassium-Argon Dating Method," the document stressed the growing importance of the new dating method and the numerous areas of geology to which it had been and could be applied.* It reported that both Ronald Kistler and John D. Obradovich were at work on their doctorates, and William C. Eisenhardt and M. E. Woakes had largely completed their master's theses in potassium-argon dating. Accompanying the application was a letter from Francis Turner, who had recently assumed the chairmanship of the Geology Department. In the letter Turner makes it clear that the new research program had won praise at Berkeley, and his mention of a sizable grant from the National Science Foundation suggests that its value was recognized elsewhere:

> It is a pleasure to support in the strongest terms the applications of Garniss H. Curtis and Jack F. Evernden for reappointment. . . . The application has the enthusiastic support of other senior members of the Department of Geology. . . . Space and facilities for the program will continue to be available in 1959–61 as at present. The National Science Foundation has just awarded a grant of $58,000 to Professors Curtis and Evernden for purchase of new equipment.
>
> I enclose copies of a statement drawn up by Acting Chairman C. M. Gilbert, 5 October 1956, on the occasion of the initial application by Professors Curtis and Evernden. This is a fair statement of the project. During the three semesters during which Professors Curtis and Evernden have worked under the auspices of the Institute, they have more than fulfilled the hopes expressed in Professor Gilbert's letter. . . . By diligent and imaginative investigation tempered with a critical and cautious approach to what is a most complex problem, they have produced results of fundamental significance and of very great interest to geologists and to students of the early history of man. Already their work has achieved international recognition. Moreover, it has proved a great source of inspiration to students and faculty in our Department, who are kept in constant touch with progress in this field by informal seminars conducted by Professors Curtis and Evernden.
>
> The project is of fundamental importance in geology. It requires expen-

*The application especially noted Curtis and Evernden's hope of extending their work to clay minerals in sedimentary rocks; the results of their student John Obradovich, who was to demonstrate in 1964 the unreliability of authigenic sedimentary minerals in dating, were not yet known. For the application, see Document No. 8.

sive equipment which fortunately has been made available by courtesy of the Department of Physics and by large grants from outside sources (such as the National Science Foundation). It requires the cooperation and advice of a physicist expert in techniques of mass-spectrometry; and in this respect we are fortunate in our association with Professor John Reynolds of the Department of Physics. The project also requires highly specialized research personnel and abundant time entirely free for research purposes. In Professors Curtis and Evernden we have ideal personnel. Their appointment to the Institute for an additional two-year period (one semester each per year) would give them the necessary time and travel facilities to prosecute their research program adequately and fruitfully.*

Curtis and Evernden had requested funds, in their own words, specifically to "pursue the solution of these three problems: (1) dating of shales, (2) dating Pleistocene rock (less than one million years), and (3) clarification of the geologic time scale for the past 500,000,000 years of earth history." Their request was approved on May 15, 1959.[15]

The Australian Connection

In April 1958, Curtis visited the earth science faculty at the Australian National University at Canberra for two weeks and went on field trips with a number of the university's scientists, including the paleomagnetist Edward M. (Ted) Irving. He also delivered a lecture on the work of the Berkeley group in potassium-argon geochronology. The Australians showed great interest in the form of a concrete proposal from Professor John C. Jaeger, who was head of the Department of Geophysics.[16]

Jaeger's proposal reflected his university's ties with Berkeley as well as his own style of leadership. Those ties dated from 1951, when Francis Turner of Berkeley first visited the A.N.U. on a Guggenheim Fellowship and began what would become a sustained exchange of people between Canberra and California. Turner was impressed with the work of Mervyn Patterson, who later came to work with David Griggs at the University of California at Los Angeles on rock deformation on Turner's recommendation.[17] In July 1956, Turner returned to visit several Australian universities, including Canberra.

*Turner's letter is part of Document No. 8; Gilbert's is Document No. 9. In addition to the informal seminars on the Berkeley campus, Curtis and Evernden also conducted seminars at the U.S. Geological Survey in Menlo Park; thus scientists there were closely apprised of developments at Berkeley.

John C. Jaeger (in 1960), late head of the Department of Geophysics at the Australian National University at Canberra. Sir Mark L. Oliphant had had a free hand when he first developed the Research School of Physical Sciences, as its first director at the A.N.U. in the late 1940's; he appointed Jaeger its first professor and head. Ian McDougall recalled that Oliphant may have been persuaded to set up the Geophysics Department only because of the availability of Jaeger, whom he regarded as outstanding.

John Jaeger was a man of great intellectual breadth; although originally trained and practiced in applied mathematics, he had done notable research in a number of other areas, including heat flow and paleomagnetism. When called to his post as professor and head of geophysics at the Australian National University, he was charged with reorganizing the department.* During a telephone interview in 1978, Ted Irving, a former staff member at the

*Jaeger was appointed the first head of the Department of Geophysics in 1952 when the department was established by Sir Mark L. Oliphant, the first director of the Research School of Physical Sciences in the late 1940's immediately following the Act of Parliament that established the Australian National University. Ian McDougall noted that "the A.N.U. was conceived in part as an institution for basic research that would attract back expatriate Australians of note; it has superior research facilities and benefits to staff. . . . Oliphant had a free hand in developing the Research School of Physical Sciences and may have been persuaded to set up Geophysics because of Jaeger's availability" (t. 1, s. 2, viii-12-78). The Radiochemistry Department was set up at the same time, but "was not very successful; in the late fifties the school decided to disband the Radiochemistry Department. Part of it was incorporated into Jaeger's Department of Geophysics; this included [J. R.] Richards and the MS2 mass spectrometer, H. (Bill) Berry, who had been doing some Pb-X work on zircon, and a number of technical staff" (I. McDougall, written comm., viii-16-78).

A.N.U., trained in paleomagnetism under Keith Runcorn at Cambridge University, recalled Jaeger's style:

> Jaeger had enormous virtues over a wide range, [and] extraordinary administrative sense. . . . He chose very good people to join the [geophysics] group at Canberra. . . . The work that was being done was engineering or geology or rock chemistry, rather than pure geophysics. When I'd just arrived there, I asked Jaeger why he appointed the people who were there. One of the staff members was a geologist, one was an engineer, and there was myself, a sort of geologist *cum* geophysicist, and there was himself, an applied mathematician who was supposed to be running a geophysics department. He said to me: "Well, it's crazy to get into the classical areas of geophysics," that is, seismology and gravity and geomagnetism, the classical tripos of geophysics, "because if you get into that field you need enormous amounts of money and at least ten years before you become the least bit competitive. So what you should do is get into areas where people aren't." Of course this is what attracted him to paleomagnetism. . . . He had a very broad view in looking at the earth.[18]

Many former associates have recalled Jaeger's colorful, highly competitive character and acuity in assessing potentially fruitful investments in research.[19] His force of will in implementing programs he deemed worthy was demonstrated once when he requested a large amount of money from the vice-chancellor of the university, who told him that such an amount was unreasonable and impossible to fund. Jaeger casually replied that the money had already been spent.[20]

During Curtis's Australian visit in 1958, Jaeger told him that he was much interested in setting up a potassium-argon laboratory at Canberra in order to supplement the work of John R. Richards in lead-isotope dating and to supply dates for Irving's paleomagnetic work;* he went on to ask if Curtis could return to Canberra to help set up a K-Ar laboratory and play a role in making it functional. Jaeger suggested that the A.N.U. send someone to Berkeley to be trained while the laboratory was being assembled. During Curtis's two weeks at Canberra, staff member Mervyn Paterson introduced him to Ian McDougall, a graduate student who was studying the petrology of the Tasmanian dolerites under Germaine Joplin.

Nothing further came of Jaeger's proposal to Curtis until the spring of 1960, when Jaeger visited Berkeley and they spent a day together examining the geology of the nearby site of the Stanford

*"Richards was not doing U-Pb dating, but examining "common" lead (galenas) from ore deposits. 'Model' ages can be derived only" (I. McDougall, written comm., viii-16-78).

Edward (Ted) Irving, about 1960. Irving's work in the Torridonian sandstones of Scotland during the mid-1950's, and later at the Australian National University in Canberra, provided solid evidence for field reversals, although he was concerned primarily with directional data in plotting ancient magnetic-pole positions.

Linear Accelerator. Jaeger returned to the subject of acquiring a K-Ar laboratory and also asked if Curtis thought Ian McDougall suitable to undertake a year-long postdoctoral fellowship at Berkeley in order to learn the dating method. Curtis had earlier spoken to Evernden about Jaeger's 1958 proposal, and the two had decided that if the Australian connection were to be established, Evernden would go to set up the laboratory.[21] The deal was struck, and Jack Evernden left for Australia in August 1960.

The contract between Berkeley and Canberra was an imprimatur, by Australia's most prestigious earth science group, on Berkeley's potassium-argon radiometry program; it also brought forth a bud that was to become a highly competitive branch, under Ian McDougall's leadership.

Of this period at Canberra and the events leading up to it, Ted Irving recalls that "the school of Physics decided to break up . . . the [faltering] Department of Radiochemistry and put [parts of it] into Geophysics under Jaeger"; this "meant that he got a pot of money, and he also got . . . [John] Richards" and technical staff.

Jaeger's problem was to import the technology as rapidly as possible. According to Irving,

> Jaeger went immediately for the potentially geologically important thing, that is, measuring the basic rocks and [setting up] the Reynolds mass spectrometer; so he got money from the University to get Jack Evernden over for six months [August 1960–January 1961]. . . . Jack's thing was to bring the mass spectrometer with him and personally install it. . . . During his time there Jack had absolute power in the lab; everything was laid out for him, as far as it could be technically, and the preparation of rock samples [was provided]. Evernden had the mass spectrometer going within a few weeks. Jaeger gave him absolute priority in the lab. Evernden was interested in the time scale work.[22]

Evernden was able to get the mass spectrometer operating so quickly because, as he later recalled, "the components of the mass spectrometer were put together in the physics shops at U.C. Berkeley; Morley Corbett did the glassware. . . . The diffusion pumps, Alpert valves, sample bottles, the whole main spectrometer tube and metal work inside . . . that whole business came right out of the Le Conte Hall physics shops."[23] Evernden simply put together all those parts for the entire gas-extraction system and mass spectrometer upon his arrival at Canberra. In this way the Australian National University came to have instrumentation identical to that at Berkeley.*

By the time of Evernden's arrival, Irving had been collecting stratigraphically well-dated Australian rocks for five years in order to plot ancient positions of the magnetic poles; thus the kinds of rocks Curtis and Evernden were seeking for their refined time scale were conveniently available.

John Richards had been doing lead-isotope studies on an English-made Metropolitan-Vickers (MS2) mass spectrometer since 1952; upon the arrival of William Compston at the A.N.U. in 1960, the machine was to be converted to use solid rather than gaseous sources for strontium and rubidium studies.† When Evernden came to Canberra, Richards "became another pair of hands helping him set up the potassium-argon system." Richards further remembers that "the system came out in crates from Berkeley—

*Evernden's activities at Canberra in assembling the K-Ar dating laboratory and his interaction with Curtis are in part detailed in a series of eight letters from Evernden to Curtis from ix-4-60 through xi-20-60 (Documents No. 12A–I).

†The MS2 was purchased new from England in 1952 when Richards had been appointed to the Department of Radiochemistry (I. McDougall, written comm., viii-16-78).

Evernden arrived in August of 1960 and spent six months with us. It was Jaeger's decision that I should move on to argon dating away from the lead isotopes, which was generally my main occupation. . . . I worked on the argon system for a year or so until McDougall came back from his postdoctoral year at Berkeley."[24]

With Allan W. Webb of the Bureau of Mineral Resources, Richards and Evernden worked together running samples during Evernden's half-year stay at Canberra. Evernden kept up his usual feverish work pace, and in that short time he and Richards produced more than 50 potassium-argon dates on Australian granitic Paleozoic and Mesozoic rocks.[25] In 1978 he recalled that at his departure "there was only one guy [in Canberra] who was skilled enough to operate the potassium-argon system and that was John Richards. Ian [McDougall] was at Berkeley while I was in Canberra, and he was still in Berkeley when I got back."[26] After Evernden returned to Berkeley, his collaboration with Curtis continued, resulting in a number of publications, but it was a period of diminishing personal closeness between the pair that ended their efforts as a team.[27]

The written record of their achievements by 1961, prepared at the end of their second two-year appointment in the Miller Institute, was cast largely in terms of demonstrating the profound impact of K-Ar dating on disciplines within and beyond earth science.[28] Noting that their primary effort during their tenure had been to collect rock specimens, so that analytic work would continue on many samples, the pair summarized their many accomplishments from 1957 to 1961, some of which were truly remarkable. They had derived virtually all the age data available on rocks of less than 70 million years, and approximately one-half of all data for establishing the geological time scale from 70 to 500 million years. They had also extended the potassium-argon technique to overlap carbon 14 dating at 30,000 years, demonstrated a greatly increased antiquity for man, and made adjustments of as much as 30 million years in chronostratigraphy based on fossil plants.

Curtis and Evernden had in fact demonstrated the imaginative refinement and use of a borrowed tool in answering a number of profound questions in various disciplines, and in so doing had developed and brought approval to a powerful new program that they had then carried abroad. The dating method was imported to Australia mainly through the efforts of John Jaeger, for whom its broad potential could not then have included its possible use in the formulation of a polarity-reversal time scale. Jaeger apparently never mentioned such a use, and it was not one of the central concerns of the Canberra paleomagnetist Ted Irving. Irving undoubtedly influenced Jaeger in the matter of acquiring K-Ar dating capability; however, Irving's main concern since early in his Cambridge training had been dating older rocks in attempts to trace the wanderings of magnetic poles for the different continents in elucidation of continental drift. As late as the months just before publication of the first polarity-reversal time scale by Cox, Doell, and Dalrymple in 1963, Irving on more than one occasion counseled them to pursue studies in polar wander rather than polarity reversal as a more probably rewarding area of paleomagnetic research.[29]

The Federal Scene

At the Australian National University, Jaeger, an assertive academic administrator, was able to almost single-handedly implement the importation of the new research program in potassium-argon dating. At the United States Geological Survey, however, the program could not be imported with Jaeger-like efficiency; a large, diversely hierarchical federal agency, the Survey was, by most standards, conservative in scientific and administrative practice.

By 1958, Curtis and Evernden had visited the U.S. Geological Survey in Menlo Park, California, a number of times to speak about their work in potassium-argon dating. Their talks favorably impressed geologists S. Cyrus Creasey and Paul C. Bateman, among others, with the method's possibilities, particularly for their own research. Bateman had long been concerned with securing dates on rocks in the Sierra Nevada, and Creasey wanted to date intrusive igneous rock structures in Arizona (porphyry stocks), on which he had published extensively; he believed that the rocks had not been reheated since the time of their original cooling from the molten state and would therefore lend themselves to dating. Creasey and Bateman had been friends since they were roommates at the University of California at Los Angeles in 1941 and 1942, and thus became natural collaborators in their not quite officially sanctioned efforts to gain the first K-Ar capability at Menlo Park.

Cyrus Creasey recalled: "There was pressure in Menlo Park in the late 1950's to get dating—it was a conversation topic at most lunch gatherings."[30] There were many more requests from project geologists than it was possible to accommodate in the Survey's limited dating facilities. "Geologists were clamoring for K-Ar dates.

Branch chiefs were making requests for dates, but no one submitted a proposal for a lab."[31] Henry Faul was the only potassium-argon radiometrist in the Survey at that time; his laboratory was in Washington, D.C., and he was apparently concerned mainly with generating dates in elucidation of the geologic time scale and thus not disposed to providing dates for those engaged in field studies. Most geological field studies, then as now, could be greatly enhanced by radiometric dating of rock units. The limited capability and great demand engendered competition in securing dates, and serious questions arose over the ordering of priorities.

Creasey and Bateman, however, discussed the acquisition of a laboratory for several years before moving to build it. In December 1958, Bateman, on Curtis's recommendation, hired Ronald Kistler in part-time "WAE" status in the Mineral Deposits Branch to pursue field studies in the Sierra Nevada; it was planned that Kistler would become a full-time Survey employee upon his graduation in 1960 and help assemble a potassium-argon laboratory at Menlo Park, if funds could be found.* During 1959 Bateman and Creasey began equipping a small mineral-separation laboratory after encountering great difficulty in acquiring space, as well as a fume hood and other equipment.

Kistler, after receiving his doctorate, came to Menlo Park in June 1960. The Survey had just granted $1,500 each to Creasey and Bateman, who proceeded to more fully equip the mineral-separation laboratory with Creasey's funds and build a gas-extraction line with Bateman's. Glassware and equipment were purchased following the advice of Garniss Curtis, who agreed that because the Survey laboratory lacked a mass spectrometer their extracted gas samples would be run on the Berkeley instrument. When Kistler arrived, Creasey had almost completed the mineral-separation lab. Kistler began to assemble the gas-extraction line in the separation lab after acquiring essential glass components and Alpert valves from the technicians at Berkeley who had provided them for the Berkeley and Canberra instruments. On October 6, 1960, according to Kistler's laboratory notes, they ran the first gas extractions on a sample of biotite from the San Manuel Porphyry of Arizona that had been collected from Creasey's field study area. Kistler taught Creasey to perform mineral separations and to blow glass. The line

*WAE stands for "While Actually Employed," an employment status that permits the Survey to avail itself of the part-time services of those employed elsewhere and also of promising students.

functioned for almost two years, during which time samples were analyzed at Berkeley. John Obradovich, a graduate student and research assistant under Curtis and Evernden at Berkeley, performed many of the analyses on these samples and played a major role in the training of Ian McDougall during his postdoctoral fellowship. Obradovich also instructed Brent Dalrymple, who was to become a central figure through his dating contributions to the development of the polarity-reversal scale.

Although the operation at Menlo Park has been characterized as "bootleg," Charles A. Anderson, chief of the Mineral Deposits Branch of the Geological Survey in Washington, D.C., had in fact approved Bateman's hiring of Kistler and the construction of the argon-extraction line.* In 1960 the Survey suffered a cut in funding, which required unpopular administrative decisions regarding assignment of personnel. In particular, a large joint project, undertaken with the state of Kentucky, caused dozens of geologists from several federal centers to be relocated on a long-term basis. Morale fell, and it was natural at such a time for geologists engaged in more familiar studies to question expenditures to acquire isotopic dating capabilities, which were still considered esoteric. The presence of the extraction line also evidently violated departmental guidelines.†

The argon line was perhaps also regarded as "bootleg" because in 1961 the K-Ar dating method was still considered unproven. Many within the Survey and elsewhere, including the then chief of the Geochemistry and Petrology Branch, William Pecora, believed the method still experimental, requiring further cross-checks and thus not ready for mass use on a regular basis. Bateman was saved from some of these doubts because he was then a field geologist; he recalled, "I had no real appreciation of the technical limitations of K-Ar methodology; I knew that there were limitations but didn't know enough to be dissuaded from trying, especially because I had an important need for radiometry to date my Sierran granites."[32]

In 1960 the U.S. Geological Survey had a poor reputation in iso-

*Several interviewees used the term "bootleg" in referring to the Menlo Park lab.

†David M. Hopkins recalled: "I think it was 'bootleg' because, even then, geochronological activities were thought to belong in a topical branch; at that time it would have been Geochemistry and Petrology, headed by [William] Pecora, who was equal in status to Anderson. Anderson was sanctioning and participating in bootlegging, an activity which was seen (by both Emperors) as belonging in someone else's Empire. The radiocarbon lab and the lead-alpha lab were both in the G&P Branch. The mapping branches [including Mineral Deposits] weren't supposed to have labs of any sort beyond what you could manage in your own office" (D. M. Hopkins, written comm., xii-5-78).

topic geochronometry. Anderson, by then promoted to chief geologist, asked Samuel S. Goldich, who had spent more than a decade with Alfred Nier on the faculty at the University of Minnesota, to become chief of the newly formed Isotope Geology Branch in Washington, D.C.* Goldich's training and research concerns were mainly in old rocks, and he was practiced in both the potassium-argon and rubidium-strontium methods.† He was charged with integrating and expanding the facilities in order to build a dating organization in keeping with the generally outstanding performance of the Survey.[33] According to Goldich, "In 1959 and 1960, the Survey's reputation was poor due to the lead-alpha dates being off. . . . I thought the lead-alpha dates stunk." The Geophysics group was often then unable to attract people they wanted, because the civil service grades they could offer were not competitive.[34]

Shortly after assuming leadership of the Isotope Geology Branch, Goldich visited Bateman, Creasey, and Kistler in Menlo Park. The three feared that the visit was a prelude to a proposal that the argon-extraction line at Menlo Park be transferred to the jurisdiction of the Isotope Geology Branch; that request came within weeks of Goldich's return to Washington. Goldich decided to purchase a metal Nier mass spectrometer from Minnesota because of his own familiarity with it, rather than a glass Reynolds machine from Berkeley. "I was afraid of not being able to get good replacement parts for the Reynolds machine because there were few glassblowers around who could duplicate parts with Morley Corbett's precision."[35] In addition, Alfred Nier had offered to sell him a machine at cost.‡

Goldich hired John Obradovich, a graduate student at Berkeley, to work with Ron Kistler.§ Obradovich had received an under-

*The decision to form a separate Isotope Geology Branch in the Survey derived in part from concern about the need to control activities involving radioactive materials (oral comm., C. Wahrhaftig, 1-6-80).

†See the References under Goldich for his publications on those subjects.

‡The spectrometer, which arrived in Menlo Park in 1961, was equipped with valves that permitted it to be run in the static mode. Document No. 12J shows the date of acquisition and purchase price of $12,500. Before the formation of the Isotope Branch, Creasey and Bateman were required to use sodium hydroxide fluxing on their extraction line for lack of funds to acquire an induction heater, which was by then in use at several radiometry centers and was considered a very necessary component for young-rock dating. Creasey and Bateman's dating needs did not include young rocks; a Berkeley-type extraction line (without induction heating) and a Berkeley-trained radiometrist gave the Menlo Park facility the potential for dating young rocks, but that was neither their need nor their goal at that early date.

§Kistler spent the first six months of that year in Washington, D.C., building a rubidium-strontium laboratory for the Survey; he resumed his potassium-argon studies upon return to Menlo Park in June.

John Obradovich, Samuel S. Goldich, Ronald Kistler, and John Richards, from left to right, at a party at Richard Doell's home in Menlo Park in August 1963, following a meeting of the International Union of Geodesy and Geophysics.

graduate degree in physics at Berkeley and subsequently had gone on in geophysics to write a dissertation on K-Ar dating of glauconite and other authigenic minerals. Although Obradovich did not finish his dissertation until 1964, he began working with Kistler in Menlo Park in December 1961, which necessitated his commuting to Berkeley from then until 1964. Obradovich dated the first young rock, a 3-million-year-old glauconite from the Lomita Marl, on the Nier machine. He reminisced: "I would rather have gotten a Reynolds machine, but it was something that I had no voice in at the time." Surprisingly, "the Nier machine was just as sensitive as the Reynolds machine."[36]

Within the Survey, from long before Goldich's tenure, there was the important question of how much freedom to allow certain specialists such as radiometrists or paleontologists. On the one hand, their abilities might permit them to treat theoretical questions without the strictures of application, but on the other hand, they were most immediately useful as service personnel providing data to others attacking broader geological problems. Goldich noted, "If scientists are going to work on problems, they have to be allowed to work on their own problems and thus do their best work." The implication is that they need to have full understanding of the purport and contextual significance of their problems and research results.* Asking scientists to provide a service to other scientists without participating as research partners was not in keeping with Goldich's idea of the proper philosophical and functional framework for the Isotope Geology Branch. "We had to be free to tell people that we can do it or cannot do it."

Subsequent developments show that his view was widely prevalent among radiometrists; in retrospect, certain frictions that arose within the Survey appear attributable to the conflict of basically incongruent organizational philosophies and needs rather than personal disagreements. The radiometrists within the Survey feared being reduced to providing service to others in whose research they had little interest. They often, with great effort, provided dates and then were not invited to share published authorship. This view of the radiometrists' role, and these frictions, are an important element in the story, told in Part II, concerning the difficulties of the Survey's paleomagnetists in securing dates for their needs.

Before 1962, the young-rock daters at Berkeley were not concerned with the resolution of questions in paleomagnetism of any sort, and certainly not with polarity-reversal studies. The Australians, in contrast, did intend to date magnetized rocks by the potassium-argon method, but not yet for the purpose of addressing the question of polarity reversals. At the Geological Survey in Menlo Park by 1962 both a laboratory and two Berkeley-trained potassium-argon radiometrists were in residence, but because requests for rock dates on important conventional studies far exceeded their capabilities, the radiometrists were not inclined to date young rocks for esoteric and apparently unpromising studies in polarity reversal.

The growth and spread of dating young rocks by the potassium-argon method was clearly separated from polarity-reversal studies. The first substantial marriage of the two research programs resulted from courtship by the paleomagnetists of a radiometric maiden; both were raised in the same Berkeley residence, but for several years one seemed almost oblivious to the other.

*Although Goldich was not explicit, his position was clear and was shared by other geochronologists, who did express this view.

PART II

Uncovering the Key:

Geomagnetic Polarity-Reversal Scales

Chapter 4

The Birth and Growth of Paleomagnetics

POTASSIUM-ARGON dating of young rocks was the key to the development of the polarity-reversal time scale, just as the scale was the key to the confirmation of seafloor spreading. Because potassium-argon radiometrists did not perceive polarity-reversal studies as a potentially rewarding area of application until about 1963, the two research programs were quite discrete before that date. They thus lend themselves, with the exception only of Martin Rutten's work of 1959, to separate historical treatment up to the beginnings of their sustained collaboration at that time.

Geomagnetic polarity-reversal studies emerged slowly from an earth-magnetics and paleomagnetics research program that was fraught with theoretical uncertainty, considered esoteric, little understood, and viewed with skepticism bordering on disdain by much of the geological community. By the late 1950's, groups in several countries were pursuing questions of ancient magnetic-pole positions (the "directionalists"), secular variations of the magnetic field, polarity reversals (the "reversalists"), self-reversals, and stability of remanent magnetism.* The paleomagnetic data of the

*During the 1950's, the most important were the English (Blackett, 1952; Runcorn, 1955a, b, 1956a, b; Nairn, 1956, 1957a, b, 1960a, b; Bullard, 1949, 1955; Creer, 1958, 1959; Clegg et al., 1954a, b, 1956, 1957, 1958; Irving 1956a, b, 1957a, b, c, 1958a, b); French (Neel, 1949, 1951, 1955; Roche, 1950a, b, 1951, 1953, 1954, 1956, 1957, 1958, 1960; Roche et al., 1959; Roquet, 1954; Thellier, 1951; Thellier and Thellier, 1959); Icelanders (Einarsson and Sigurgeirsson, 1955; Einarsson, 1957a, b, c; Brynjolfsson, 1957); Japanese (Nagata et al., 1952a, b, c, 1953a, b, 1954, 1957a, b; Uyeda, 1958); Dutch (Hospers, 1951, 1953, 1954a, b, 1955; Rutten et al., 1957, 1958, 1959, 1960); Russians (Popov, 1947; Khramov, 1958; Valiev, 1960a, b); and Americans (Graham, 1953, 1954a, b, 1955, 1956; Balsley and Buddington, 1954, 1957a, b, 1958, 1960; Verhoogen, 1956, 1959; Cox, 1957; Doell, 1955a, b, 1956, 1957). Most of those named continued work into the 1960's and beyond.

1950's had been derived largely through efforts to resolve the question of continental drift and had attracted most attention within that context; the data had not, however, provided definitive evidence for drift nor a unanimity of opinion regarding interpretation, even among the paleomagnetists themselves. For more than a decade, drift had been a problematic bastion against the not-quite-successful attack of those directionalists who comprised the bulk of the rock-magnetism community.*

The geologists of the 1950's and early 1960's who did follow the rapid growth of paleomagnetic research focused little of their attention on the highly problematic subject of reversely magnetized rocks. Although such rocks were first recognized near the start of this century,[1] their study did not attract many investigators. Almost suddenly and quite unexpectedly, even to those pursuing them, studies of reversely magnetized rocks during the early 1960's by a group in California and another in Australia emerged to overshadow the directionality program. The reversalists produced data

*Directionalist studies were singularly influential in reviving the drift question and surely deserve a separate book-length history. They began largely as an outgrowth of the building of the first astatic magnetometer by Patrick Blackett (1952); he built it not to study rocks, but for use in an unsuccessful experimental attempt to relate a rotating mass to the generation of magnetism. The English school of rock magnetism and paleomagnetism arose during the 1950's around Blackett's magnetometer at Imperial College of the University of London, and their research concerns became increasingly focused on rock magnetic direction, which showed systematic differences with the age of the rocks studied. Their Triassic rock magnetic data indicated a clockwise rotation of 34° and southward position for England during that period (Clegg et al., 1954). Other data from the Deccan Plateau suggested that the Indian subcontinent had drifted from 20° south to 20° north in the last 80 million years and had also rotated 20° counterclockwise (Clegg et al., 1956; Irving, 1956b). The London group's evidence was interpreted differently by the magnetists at Cambridge: Keith Runcorn at first held that it instead implied southward wandering of the north magnetic pole. He clearly was against continental drift, and cited (1955b, pp. 283–84) extra-magnetic evidence against it, presented arguments for polar wandering, and concluded that the paleomagnetic data at hand were not in keeping with "appreciable amounts of continental drift" (1955b, p. 289). Runcorn, like many others, viewed the physical difficulties implicit in moving continents as greater than those attending polar wander, but his study (1959) of North American rocks (with Neil Opdyke as his field assistant) yielded data incongruent, in terms of polar wander, with those from Europe; therefore, he came to subscribe to continental drift (Opdyke, t. 1, s. 1, v-11-79; E. M. Irving, oral comm., xi-30-78). The directionalists at London and at Cambridge were at odds also over how far to interpret ancient longitudes from the magnetic data; the London group, wary of possible continental rotation, favored restricting interpretations to latitude only, while the Cambridge group less conservatively estimated longitudes to schedule the opening of the Atlantic. By 1960, the majority of the English paleomagnetists were "drifters," but had not yet convinced most magnetists in the United States of the correctness of their views, as evidenced in the influential "Review of Paleomagnetism" by Allan Cox and Richard Doell. Cox and Doell cautiously stated that "paleomagnetic results for the Mesozoic and early Tertiary might be explained more plausibly by a relatively rapidly changing magnetic field, with or without wandering of the rotational pole, than by large-scale continental drift" (1960, p. 645). It was an exciting time in which much evidence mounted and paleomagnetics drew worldwide attention. By 1961, Doell and Cox were moved to conclude in a second influential major review that "it is difficult to explain all of the presently available paleomagnetic data without invoking continental drift" (1961a, p. 302).

S. Keith Runcorn played the most important role among the English paleomagnetists. Runcorn's studies at Cambridge and Newcastle were decisive in reopening the question of continental drift during the 1950's. The quantitative paleomagnetic evidence that he, Edward Irving, and other directionalists derived converted many to a mobilist view of the earth and greatly influenced Richard Doell, Allan Cox, and others in the world paleomagnetics community.

that serendipitously contributed to the solution of seafloor spreading, a question that subsumed that of continental drift.

Allan Cox and Richard Doell's review of the paleomagnetic data accumulated by 1960 attests to the rapidly growing number of mainly directional studies listed among their 174 cited references. The explosive growth of research in all of rock magnetism can be read in Edward Irving's bibliography of only three years later, which had swollen to approximately 550 entries. But the success of the reversalists in providing an interpretive key for the confirmation of seafloor spreading had a singular impact on the growth of paleomagnetic research, evidenced in Michael McElhinny's note of 1973 that by the end of 1970, the number of independent investigations published had grown to 1,500.[2]

By 1954, growing interest in rock magnetism and paleomagnetism led the National Science Foundation to sponsor a conference on the "Anomalous Magnetization of Rocks." According to the NSF's *Fifth Annual Report* (1955), the conference, which met August 7–9 at the Institute of Geophysics, University of California, Los Angeles, "focused on certain rock formations which have a magnetic polarization that does not conform to present-day magnetic field flux. Participants included a number of outstanding American scientists as well as experts from England, Canada, Japan, and France. Discussions emphasized the differences of opinion between the physicists and the geologists in the field of rock magnetization."

The conference agenda instructed that: "The basic question motivating the organization of this conference is 'To what extent is it possible to trace the history of the changes of the earth's magnetic field by studies of the magnetic properties of rocks?'

Before this question can be answered, it will be necessary to consider the questions 'What are the origins and natures of the residual external magnetic moments of rocks' and 'Do we have acceptable proof that the earth's magnetic field reverses?' It is on these last two questions that the conference may yield answers or at least indicate new avenues to the answers."

Row 1, left to right (kneeling): Charles Kittel, John W. Graham, unidentified, Ernest H. Vestine, S. Keith Runcorn, Louis B. Slichter, Francis Bitter, Ronald G. Mason, unidentified. *Row 2* (standing): David T. Griggs, Takesi Nagata, Walter M. Elsasser, Linus Pauling, J. A. Clegg, John Verhoogen, Emile Thellier, Carl Eckart, Gustaf O. S. Arrhenius, James R. Balsley, unidentified, C. Duncan Campbell(?), John C. Belshe, Lawrence W. Morley, Arthur F. Buddington, Philip M. Du Bois.

Early Studies in Geomagnetism

The earliest commentary on earth magnetism is found in Chinese records of the second century B.C., which record the invention of the magnetic compass.[3] The Chinese knew about magnetic declination (the angular difference between true and magnetic north) by 720 A.D., some 800 years before this knowledge reached the West.[4] By 1269, Petrus Peregrinus demonstrated the dipolar character of magnets and the verticality and greatest strength of their force at the poles.[5] Magnetic dip or inclination was observed by Georg Hartmann in 1544, but until it was rediscovered and publicized by Robert Norman in 1576, few knew of it. The influential and widely known *De Magnete* of William Gilbert, published in 1600, appears not to depart greatly from Peregrinus and other earlier works, but his careful experimentation and clear summary of what was then known of the earth as a magnet lent impetus to studies by others.[6]

The discovery that magnetic declination changed with time at any fixed location was made by Henry Gellibrand in 1634;[7] this change of the field with time is called secular variation. Such variations have been mapped and shown to be most rapid in certain areas that move about irregularly and grow and decay over periods of roughly a century. There are a number of such foci at which secular changes are most rapid. In 1839, Karl F. Gauss performed a spherical harmonic analysis of the earth's magnetic field. The main dipole, which is the dominant field, is subtracted from the field observed over the whole earth's surface to yield the nondipole field; the two fields are then examined separately in studies of secular variation. A great number of such investigations have been undertaken since the time of Gauss in attempts to explicate the basic causal mechanism of the earth's magnetic field. These were somewhat accelerated after the rise of the reversal question, which lies at the center of this unfolding history.

Why the earth behaves as a huge magnet maintaining a field of which the main component is a dipole, presently inclined at an angle of 11½ degrees to the earth's rotational axis, is poorly understood. It appears that the rotational and magnetic poles have been nearly coincident (within 15°) throughout much of geologic time. The earth's high internal temperature does not permit permanent magnetization of the entire earth. The magnetic field strength measured at the earth's surface is stronger than could arise from permanent magnetization of any known crustal rocks. A permanently magnetized earth would also fail to explain secular variations and the near-coincidence of the rotational and magnetic poles; it is also incompatible with reversals of the entire magnetic field.

In 1919, Joseph Larmor first suggested that the sun's magnetic field is induced, or generated, through a self-exciting dynamo effect. In 1946, Walter Elsasser postulated that the region of liquid iron and nickel in the earth termed the outer core conducts electricity and acts as a self-exciting dynamo that induces the geomagnetic field. Edward Bullard, in 1949, extended Elsasser's theoretical construct into a model of a dynamo aimed at demonstrating how a great magnetic field could have grown from an initially small, random one.

In 1955 a study by Elsasser treated the paths of particles in a rotating flow system within the earth's core to show how a toroidal (ring-shaped) field can be produced from an original poloidal (in meridional planes) field, and Bullard and H. Gellman attempted in 1954 to test by computer the probability that a particular flow pattern would produce a dynamo. Neither study offered an explanation of how the earth's magnetic field could reverse itself. Thus by the mid-1950's, many students of paleomagnetism faced with a reversely magnetized rock were wont to attribute its reversed character to some self-reversing property of the rock itself; to most the notion that the earth's magnetic field could reverse itself was tantamount to the suspension of a universal constant. Tsuneji Rikitake in 1958, and Douglas W. Allan in the same year, coupled disc dynamos to demonstrate spontaneous reversals of the external field; Allan's example, showing the computed current in two coupled disc generators, did resemble the record of polarity of the earth's field as it later became known. However, their influence was limited, since any analogy between the very complex and highly theoretical magnetohydrodynamic scheme of the earth's interior and their simple models was bound to be tenuous.

Robert Boyle may have been the first to recognize, in 1691, that solid objects that have cooled from an extremely heated state may acquire remanent or residual magnetism.[8] Baron Alexander von Humboldt correctly speculated in 1797 that lightning had induced points of intense magnetization that he encountered in a magnetic survey of a mountaintop in the Palatinate. Such lightning-magnetized rocks attracted the attention of a number of nineteenth-century scientists, and by mid-century instruments had been improved enough to show that lightning-induced magnetism is a near-surface phenomenon distinct from the general geomagnetic field. The demonstration by Achille Delesse in 1849 that recent lava flows were uniformly magnetized parallel to the earth's field set off the lightning-struck *punti distinti* with their random magnetic orientations as clearly different, and may have prompted Macedonio Melloni's study in 1853 of the lavas of the Phlegraean Fields and Mount Vesuvius.

Melloni found that the remanent magnetism of the lava flows was indeed roughly parallel to the field of the earth. He fired volcanic-rock specimens in his laboratory to red heat in various spatial orientations, then found that on cooling they all acquired a magnetization parallel to the earth's field; he thus surmised that lavas acquired their natural remanent magnetism as they solidified. Additional field work near Rome by Giuseppe Folgheraiter in 1899 verified Melloni's conclusion. Folgheraiter extended his studies to include bricks and pottery—Etruscan, Greek, and Roman vases, urns, and amphoras—which he came to believe could be guides to ancient inclinations of the early magnetic field if their positions at the time of firing were known. He showed that remanent magnetism is highly stable when he found that the bricks in a Roman wall were magnetically scattered and that 2,000-year-old, randomly buried Aretine vases, when stood on their bases (the position in which they were presumably fired), showed a consistent angle of magnetic inclination.[9]

The years around the turn of the century saw several important discoveries in the physics of materials that may have given impetus to the unfolding developments in rock magnetics based on field studies. Pierre Curie's law showing that paramagnetic susceptibility is inversely proportional to absolute temperature was published in 1895. And the theory of Paul Langevin in 1910 gave the name paramagnetism to the phenomenon in which the application of a magnetic field to atoms having a magnetic moment tends to align these dipole moments along the direction of the field to effect an increase in magnetism. In the following year Pierre-Ernst Weiss advanced his theory of ferromagnetism, which derives the type of magnetism found in household magnets from the interaction between electron spins in nearby atoms in solid material. These were only some of the efforts of the time that helped explicate the behavior of magnetized rocks.

In 1904, Pierre David deduced, from the pattern of consistent inclination with random declinations, that the slabs of fine-grained volcanic rock (trachyte) used in the construction of the 2,000-year-old Temple of Mercury atop the almost mile-high summit of the Puy-de-Dôme in France had been quarried in a horizontal position. He also determined that the same trachyte in place in the quarry had a magnetic direction and inclination almost identical to the earth's present field. He thus buttressed the conclusions of Folgheraiter.

In 1909, G. E. Allen noted that the thermoremanent magnetization of volcanic rocks is much more intense than remanence acquired in the earth's field without heating (that is, isothermal remanent magnetism).

The earliest Japanese effort along these lines was apparently made in 1912 by Seiji Nakamura and S. Kikuchi, who cooled rocks in the laboratory from high temperatures and showed that they acquired strong remanent magnetism in the exact direction of the prevailing geomagnetic field. They did not, however, understand the details of what was later defined as thermoremanent magnetization.

The other crucial line of inquiry in connecting the clues gleaned from fired artifacts and volcanic rocks began in the early 1900's with the earlier mentioned study of Brunhes and David of the magnetism of several lava flows and the baked clay beneath them in the Auvergne region of France—an extension of Folgheraiter's works.[10] They provided a good summary of several minor papers during the first decade of the century and documented the first clear evidence that the baked-clay beds agree in magnetic orientation with their overlying flows. Most important, they found that the directions in the Pontfarein flow and its underlying baked-clay bed were almost exactly opposite to that of the present magnetic field. This discovery marked the beginning of the subject of reversely magnetized rocks and the controversy over their meaning that went on for decades afterward.

R. Chevallier's contribution of 1925 lay mainly in his careful correlation of the remanent magnetizations of lava flows on Mount Etna with records of the earth's magnetic field at the time of their eruption. Such records, available from observatories close to the eruptions from the twelfth to the seventeenth centuries, provided discipline to another dimension of the growing inquiry.

Almost simultaneously, Paul L. Mercanton suggested that if re-

Motonori Matuyama, after whom the Matuyama epoch of reversed earth magnetism was named, was the first paleomagnetist to investigate the timing of magnetic reversals by examining the ages of rocks. After an eminent career as a geophysicist, he became a university president, and also began acting in the Noh theater, the classic Japanese dance-drama.

versely magnetized rocks were caused by a reversed earth field, such reversals should be recorded worldwide. He then collected from the northern and southern hemispheres rock samples that were normally and reversely magnetized.[11] He was the first to anticipate the usefulness of such studies in testing the questions of polar wandering and continental drift, then being hotly debated because of the near-coincidence of the rotational and geomagnetic axes.

Several Soviet studies on magnetic properties of rocks in about 1926 arose from the discovery of a large iron-quartzite ore body near Kursk in the U.S.S.R. based on magnetic prospecting.* And in 1927, in a series of "Experiments in Magnetization of Rocks During Heating," Frantz Levinson-Lessing "proved the stability of the residual magnetism of some igneous rocks."[12]

Although the field-reversal hypothesis had been touched on earlier, it was not set in a temporally disciplined framework of any sort

*Noted in the 1961 text, *Rock Magnetism*, by Takesi Nagata.

until the classic field studies of Motonori Matuyama of Kyoto Imperial University, who in April 1926 began to collect and examine more than 100 basalt specimens from 38 locations in Manchuria and Japan. He found that they were all magnetized either close to the present field or opposite to it. His acute observation that the polarity of a rock clearly correlated with its stratigraphic position or age led him to remark that in the early part of the Pleistocene epoch, the earth's field had been reversed, and that it changed later to produce the normal (present-day) magnetization found in the younger group of rocks.[13] His accurate surmise has since been thoroughly confirmed.

By 1933 a number of salient points had been clarified through field and laboratory studies relating to the magnetic stability of and coercive forces acting upon rocks, bricks, and fired ceramics; reversed magnetism had been identified and correlated with relative age, and both Matuyama in 1929 and Mercanton in 1933 had suggested that the geomagnetic field was once opposite in direction to the present field.

The Character of Rock Magnetism

In 1933, Johann G. Koenigsberger delivered an important paper to the Sixteenth International Geological Congress, reporting the results of a series of experiments on the effects on magnetism in rocks of various chemical compositions, temperatures, conditions of formation, and ages.[14] He found that the iron minerals magnetite and maghemite were the important carriers of magnetism, measured carefully the natural remanent magnetism acquired by rocks on cooling in the earth's field, and first advanced a theory of thermoremanence that became generally accepted. Aleksei N. Khramov considers that Koenigsberger "was the first to state the problems scientifically, to examine previous work critically, and also to carry out numerous experiments to determine the magnetic properties of many igneous and metamorphic rocks."[15]

Koenigsberger's efforts may be said to mark the beginning of a period of accelerated research on the remanent magnetism of the common rock-forming ferromagnetic minerals, the research aimed in particular at improving and evaluating the reliability of paleomagnetic data. Koenigsberger obviously prompted other important early studies, including those of Emile and Odette Thellier in 1938, Takesi Nagata in 1943, John W. Graham in 1949, Naoto Kawai in 1954, and Juliette Roquet in 1954. The Thelliers used the bricks of historical monuments to determine the secular variations of the geomagnetic field during the past 2,500 years and performed experiments that proved beyond question that the residual magnetism of baked clays, both natural and in fired artifacts, was stable and could be used in reconstructing the history of the geomagnetic field.[16] Nagata demonstrated that ferromagnetic mineral grains acquire a stable type of remanent magnetism in both igneous rocks and baked earths as they cool, even in the weak magnetic field of the earth, and amplified Koenigsberger's theory of thermoremanence. Working with volcanic rocks, Nagata did comprehensive and highly systematic studies and thus was able to explain many properties of thermoremanent magnetism; these included the important effects of thermal and magnetic demagnetization, which were crucial to later work in "cleaning" rocks of interfering magnetic components that often mask the important components of magnetic "memory."[17]

Louis E. F. Néel's theoretical studies of 1951 and 1955 on magnetism, especially with regard to the means by which a rock might undergo self-reversal, were crucial in influencing the community and are treated later in this chapter.

The method of demagnetizing rocks by continuous or stepwise heating was first applied by Emile Thellier in 1938, but the use of steady magnetic fields for such demagnetization was not introduced until Ellis Johnson, Thomas Murphy, and Oscar Torreson's work of 1948. Emile Thellier and Francine Rimbert first applied alternating magnetic fields to demagnetize rocks in 1954.[18]

Gustaf Ising's pioneering investigation of the varved clays of Sweden in 1943 was an attempt to determine if sedimentary rocks could become magnetized in the direction of the ambient field at the time of deposition. A sequence of varves (annual layers) spanning 350 years from Viby, Sweden, showed a 10° difference in declination for the oldest as compared to the youngest, which agreed with the present field; the oldest had an inclination of 50° and the youngest 30°.

Paleomagnetic studies in sedimentary rocks of North America began with that of Alvin McNish and Ellis Johnson in 1938 on the seafloor sediments off the coast of Labrador and the varved clays of New England, some of which are 20,000 years old. The magnetic directions measured differed from the present field, and McNish and Johnson were led to postulate that such data would permit the reconstruction of the geomagnetic field during the time of deposition of such sediments, but the state of the paleomagnetic art may be read in their closing remark: "Until many more data have been obtained, judgment must be withheld as to whether or not these measurements represent the direction of the earth's magnetic field at the time the sediments were formed; it is the writers' opinion that they do." Continued study by Johnson, Murphy, and Torreson, of the Department of Terrestrial Magnetism of the Carnegie Institution, led in 1948 to a refined picture of the secular variations in New England from 15,000 to 9,000 B.C., which they based on varved clays; the ancient magnetic field seemed to behave much as the modern one, which they based on empirical data (from Sweden and elsewhere) concerning the past 400 years.[19] This study did much to attract the attention of many outside the field; notable among them was the Nobel laureate Patrick M. S. Blackett, who remarked, "My own interest in rock magnetism was aroused initially by the . . . publication in 1948 of a paper [by Johnson, Murphy, and Torreson] entitled 'Pre-history of the Earth's Magnetic Field.'"[20] The varved clays examined in that study and others were not con-

Patrick M. S. Blackett entering the physics department in the Schuster building at the University of Manchester in 1949. Reading theories of the magnetism of stars prompted his attempt to calculate the deflection of a cosmic ray in traversing the galaxy, for which he won the Nobel Prize in 1948. Rock magnetic studies later began at Imperial College in London with the use of Blackett's astatic magnetometer, which he had built for an attempt to generate a magnetic field by rotating a mass of gold; the experiment was unsuccessful.

solidated enough to be called rock and thus did not clarify many of the questions relating to sedimentary rocks, such as the effects on magnetism of compaction, compression, and attendant heating and chemical changes.[21]

A brief summary of how magnetism is acquired in rocks will be helpful before we turn to a discussion of petrological characteristics and their possible bearing on magnetic behavior. Sedimentary rocks acquire magnetism in two principal ways. First, minute grains of magnetic minerals such as magnetite (Fe_3O_4), when deposited in water, tend to align themselves to some extent with the ambient magnetic field before lithification (compaction and cementation) traps them, more or less, in their positions of burial. Sedimentary rocks that contain magnetic material, when lithified under uniform conditions and undeformed, thus may show weak but significant magnetism. Second, chemical changes known to produce magnetism can occur: ferrous carbonates, derived through weathering of silicate minerals, precipitate as a ferric hydroxide colloid (jelly) between grains in sand deposits and eventually dry to become hematite. Hematite sandstones are magnetic, but almost nothing was known about the details of the mechanism of their magnetization in the 1950's, and little more is known today.

The majority of the rocks studied by early workers, as well as the rocks that led to the formulation of the polarity-reversal scale, were not sedimentary but igneous, once molten; they derive their magnetism from various iron-titanium oxide minerals that have a wide range of magnetic properties. One iron sulfide (pyrrhotite) is also important in a few special rock types. As molten rock cools in the earth's magnetic field, it first crystallizes, then acquires thermoremanent magnetization as its temperature drops below the Curie points of its ferromagnetic minerals. The Curie point, named after Pierre Curie, is the temperature above which thermal agitation prevents spontaneous magnetic alignment or ordering of atoms, which is characteristic of the ferromagnetic state; it is 675°C for hematite, 585°C for pure magnetite, and ranges from 585° to −100°C in titanomagnetites, decreasing with increasing titanium content. Thermoremanence is mostly acquired while the temperature is within a few degrees to a few tens of degrees below the Curie point.

It had been repeatedly demonstrated by the 1930's that in rocks with randomly oriented magnetic grains, thermal remanence is parallel to the field that prevails during cooling. The intensity of such remanent magnetization is normally proportional to that of the field if it is less than about 1 oersted.* The earth's magnetic field is thought to have varied from approximately 0.05 to 1.0 oersted, but probably was never appreciably stronger. Therefore, igneous rocks should give accurate and reliable information about the intensity and direction of the field in which they cooled, provided they did not subsequently undergo physical or chemical changes that altered the magnetism acquired at their formation. Most rocks are stable during minor changes in temperature or magnetic fields, but a small number of igneous rocks are now known to be of no paleomagnetic value because they are intrinsically partially unstable.

The magnetism acquired by passage through the Curie point is not like that acquired by a magnetic material in a weak field at constant temperature; such isothermal magnetization is very much less intense than thermoremanence induced in the same field. Isothermal magnetization acquired at ordinary temperatures in rocks exposed to a weak magnetic field like that of the earth for a short time is thus generally negligible. Exposure for periods of thousands of years, however, induces a "viscous" magnetization that grows roughly as the logarithm of the exposure time. Many rocks thus acquire a viscous magnetization parallel to the recent field that may tend to obscure any earlier thermoremanence acquired when the field had a different direction. The viscous component caused serious interpretative problems in the early days of paleomagnetism, before it was learned how to magnetically clean rocks.

Another major difficulty in interpreting the magnetic data arises from the chemical, physical, and magnetic changes that occur in rocks after their formation. Rocks can be altered by heat and pressure during long periods of deep burial; furthermore, groundwater can induce oxidizing or reducing conditions. Of course, the severe effects of metamorphism are usually easily recognizable and preclude the present use of metamorphic rocks, which have suffered great heat and pressure, in magnetic studies.

A number of experimentalists had explicated several properties

*The unit of magnetic-field intensity equal to a force of 1 dyne exerted on a unit magnet pole. A dyne is a force that, applied to mass of one gram for one second, would give it a velocity of one centimeter a second.

of rock magnetism by the mid-1950's, but the lack of understanding about the natural alteration of magnetic properties with time was especially ominous to the skeptics. It was found that a 500-million-year-old rock, undisturbed, possessed a thermal remanence only one-fifth that displayed by the same rock after it was heated and cooled through the Curie point in the laboratory.[22] This result suggested that the earth's field 500 million years ago was only one-fifth the present one, or that the magnetism of the rock had decayed by 80 percent, or that some combination of the two had taken place. It was known that some minerals, stable at high temperatures and pressures, could lose their remanence as they slowly changed phase in less severe conditions. The stability of rock magnetism, however well proven for the majority of rocks in various studies, was attacked in general by many who pointed to some rocks that had changed their magnetism when left in a new position for only a few days.

The few exceptions to the magnetic reliability of rocks in general tended to be exaggerated. As late as 1954, Blackett warned: "The complexity and variety of the phenomena of the magnetization process in rocks, and our relative ignorance of much of the basic physics and chemistry of the processes, have made it essential to adopt a cautious attitude in the interpretation of the observed facts."[23] His admonition probably reflected the increasing number of studies that by the mid-1950's had highlighted a crucial question: Why are some rocks reversely magnetized?

Brunhes and David first recognized magnetically reversed rocks at the start of this century in a sequence of lava flows in France; they described a basalt layer and the baked clay beneath it, both of which displayed reversed magnetism. Mercanton in 1926 and Matuyama in 1929 found more magnetically reversed rocks and suggested that the earth's magnetic field had once been reversed.[24] A large number of subsequent studies (to be discussed shortly), done by various nationals on different continents, showed that of all rocks carefully examined, approximately half were normally magnetized (also termed "positive") and half reversed ("negative"). The emergent central question concerning the cause of reverse magnetization in rocks took this form: Did the earth's dipolar magnetic field periodically undergo reversal by some unknown mechanism? Or did something within the rocks themselves cause the magnetization to lie in a direction opposite to that of the field which prevailed when the rocks underwent a process that we suppose caused them to become magnetized? I will refer to the first type as field reversals and the second as self-reversals.

The Case for Field Reversals

Irving in 1964, McElhinny in 1973, and others have succinctly summarized the effects to be anticipated as evidence that field reversals had actually taken place:

1. Regardless of petrologic type, contemporaneous rocks that are not self-reversing should have the same polarity.
2. In certain more complete stratigraphic sections, transitions from one polarity to another, captured during polarity reversals, may be found.
3. The chemical, physical, and mineralogical properties of normally and reversely magnetized rocks should show no systematic differences.
4. Igneous rocks should have the same polarity as the baked adjacent rocks.
5. Data should be acceptable only from rocks carefully tested against self-reversal in the laboratory.[25]

We may segregate the various lines of evidence that bear on the reversal question in terms of these five criteria.

It is appropriate to view the line of evidence first presented by Brunhes himself, the paleomagnetism of baked contact rocks. Roderic L. Wilson in 1962,[26] Edward Irving in 1964, and Michael McElhinny in 1973 summarized a great deal of previously published data.* A tabulation of the data shows 49 cases in which the polarity of the invading igneous rock agrees with that of the baked contact rock and only three cases of disagreement. (Also, Einarsson and Sigurgeirsson in 1955 reported an unspecified number of contacts showing agreement.) By 1960, the data strongly suggested the validity of field reversals. Minor further evidence could be drawn from the magnetizations of the baked and unbaked parts of the same rock unit. Polarities in baked and unbaked contact rocks do not have to agree if it is the baking that induces the reversal. Agreement was found in only 12 cases out of 23; however, the changes in mineralogy that occur with baking, and the difference between sedimentary-rock magnetism and thermoremanent mag-

*The diverse sources included Brunhes and David (1903), Grenet (1945), Godard (1950), Hospers (1951), Roche (1953), Kato et al. (1954), Einarsson and Sigurgeirsson (1955), Opdyke and Runcorn (1956), Blundell (1957), Graham and Hales (1957), Creer et al. (1959), Everitt (1959, 1960), Gough and Van Niekirk (1959), and R. L. Wilson (1961).

netism, detracted somewhat from the strength of this latter evidence. A particularly forceful set of data was published in 1954 by Yoshio Kato, Akio Takagi, and Iwao Kato, who examined several magnetically reversed tabular bodies of rock which had been intruded in the molten state; these dikes of basaltic andesite cut through tuff near Sendai, Japan. Magnetization in the tuff is approximately normal but reverses within 20 centimeters of the reversely magnetized dikes. Many other clear examples known by the mid-1950's provided a formidable line of evidence for field reversals.

The evidence in early studies of reversals that occur in stratigraphically controlled sequences appears almost overwhelmingly in favor of the field-reversal hypothesis, but a crucial adjunct was lacking. Until the 1960's, virtually no detailed studies were done on the mineralogical, chemical, and physical properties of the rocks from which the evidence for field reversals had been drawn.* Thus, although the studies were detailed and careful in other respects, there remained a basis for the sustained skepticism of those who favored self-reversal. This point deserves repeated recall in assessing the evidence that follows.

In 1934, Hans Gelletich magnetically surveyed the Pilansberg dike system of South Africa, but performed no laboratory work. This swarm of dikes, the largest known, displayed reversed remanence throughout its 100-mile length. The mineralogically uniform dikes (of doleritic and syenitic rock) were injected 1,200 million years ago into their present near-vertical positions. Within the same area rocks rich in magnetite, both older and younger than the Pilansberg dikes, are normally magnetized.[27] Additional evidence for field reversals from five reversely magnetized, mid-Cenozoic dikes of basalt (tholeiite) in the north of England was tendered by J. McGarva Bruckshaw and Edwin I. Robertson in 1949.

In 1951, Alexandre Roche began a series of studies in the Massif Central of France that included basalt dikes and flows from early Oligocene through late Pleistocene age. About half of the flows were normal and half reversed; Roche was unable to detect any differences between normal and reversed rocks in the laboratory and thus concluded that the earth's field had reversed many times during Cenozoic time.[28] Roche's paleomagnetic stratigraphy was until 1960 the only correlation between a sequence of reversals of ther-

*The first were those of Domen (1960), van Zijl et al. (1962), Ade-Hall and Wilson (1963), and Cox, Doell, and Dalrymple (1963a, b).

moremanent magnetism and a previously defined stratigraphic column. Some question remained whether Roche had identified reversals more recent than the upper (late) Oligocene, since he did not study a single continuous sequence of intrusions and flows but instead compiled data from a number of scattered localities where well-dated dikes and flows cropped out. In 1957, Johannes den Boer published a report of another reversal period in the upper Tertiary rocks of the Coiron basalt plateau, to the southeast of Roche's sections.

In 1951, the work of Jan Hospers in the Tertiary lava flows of Iceland, bolstered by laboratory experiment, showed a telling periodicity of alternation between normal and reversed states, and an absence of intermediate directions. Hospers surmised that the field reversed at intervals of approximately one-half million years during the Tertiary, and that the reversal process lasted less than 10,000 years. Careful measurements in the laboratory indicated that both the normal and reversed rocks responded identically to an applied field. Hospers, as did Roche, considered the possibility of self-reversal and thoroughly rejected it.[29]

Working in the Torridonian (Precambrian) age sandstones of northwestern Scotland, Edward Irving documented 16 contiguous zones of alternating normal and reversed strata that varied in thickness from 30 to 600 meters.[30] The stability of the natural remanence, which was verified in the laboratory, provided solid evidence for field reversals, although Irving was concerned primarily with directional data for plotting ancient poles and reassembly of the drifting continents. Alan E. M. Nairn considered Irving's and Hospers's studies to be the best evidence for field reversals found by 1956.[31]

Studies by Icelanders, likely stimulated by Hospers's doctoral studies there, were published by Trausti Einarsson and Thorbjorn Sigurgeirsson in 1954 in which they examined thousands of specimens from all parts of Iceland: "This study revealed close correspondence of the magnetization with distinct formations. As a rule, the specimens taken from the same volcanic formation all have the same polarity." Rocks representing 21,000 meters of stratigraphic section were almost equally divided into normal or reversed polarity, and the dikes and the baked contact rocks, whether basalt or clay, were of the same polarization as the intrusions. They described "at least three periods during which there was reversion of the magnetic field. . . . The last one covering the Pliocene-Pleistocene boundary

. . . may be compared with the last period of inverse magnetization in France as found by Roche."[32] That and other statements in Einarsson and Sigurgeirsson's 1954 paper mark their pioneering effort at transoceanic magnetostratigraphic correlation, however crude.* (The first formal correlation of this type was done by Martin Rutten and Hans Wensink in 1960.) Einarsson's results of three years of field work, published in April 1957, presented the strongest evidence for field reversals to that date. He noted that within each magnetic group, all the rocks, of various kinds, shared the same polarity and that he had "not detected any systematic difference in volcanological or petrographical respect between normal and reverse groups."[33] A section in eastern Iceland 6 kilometers thick contained 30 different magnetic groups, and in western Iceland a 4.7-kilometer column comprised 13 magnetic groups. The total thickness of the normal group was 2,175 meters and that of the reversed group 2,250 meters; the average thickness per group was 310 meters for the normal and 375 meters for the reversed. Einarsson noted that the striking similarity of such data militated against a volcanological difference between normal and reversed groups. He revised an earlier estimate of the average length of magnetic periods, or epochs, from three million to one-half million years and then put forth several accurate and portentous conclusions: "That individual periods were of uneven duration seems also to be quite clear.† The time taken for a reversal of the field is geologically very short. The boundary sediments of erosion forms [ancient soils and evidence of weathering at the tops of lava flows] certainly do not necessitate the assumption of a longer time than 10,000 years, and even a few centuries might in most cases be sufficient." After due note of the possible incompleteness of Iceland's Tertiary magnetic record, he concluded, "It is already clear that the number of magnetic periods in the Tertiary was more than thirty."[34] That number was derived by extrapolating his data through the estimated total thickness of Icelandic lavas.

Aleksei Khramov's 1958 monograph, entitled "Paleomagnetism and Stratigraphic Correlation" in its English translation of 1960,

*The normal field of the earth, prevalent during the last 700,000 years, can in some cases impose a viscous magnetization on rocks, thus masking the true, reversed thermoremanent magnetization, if tested *only* by a field compass as was done by Einarsson. Although that sometimes happens in sedimentary rocks, it almost never does in the case of strongly magnetized basalts; thus although Einarsson did not do systematic laboratory experiments, his study merited strong consideration.

†His conclusion was in contrast to that of Khramov (1958), who believed that reversals occurred at regular intervals. The poignancy of Einarsson's accurate surmise emerges in Chapter 6.

identified him as an outspoken member of the very small community of Soviet proponents of continental drift based on paleomagnetic data. From 1955 to 1956, he appears to have been alone in his pioneering call for the paleomagnetic correlation of sediments and the erection of a paleomagnetic time scale.[35] His description of a thick sequence of Neogene and Quaternary sediments exposed in western Turkmenia included 13 polarity groups alternating between normal and reverse. He noted, "The planetary character of the large-scale variations of the field provides the possibility of a strict world-wide correlation of volcanic and sedimentary formations and the creation of a single geochronological paleomagnetic time scale valid for the whole earth."

Khramov was firmly convinced of field reversals, which he believed were "frequent and rapid." Significantly, he noted that his data indicated rhythmic variations of polarity every 400,000 years, a conclusion he reiterated in several discussions. The vast body of field data on which his conclusion was based influenced the early work on polarity time scales of Cox, Doell, and Dalrymple in the U.S. and McDougall and Tarling in Australia during the early 1960's.

Opposed to the regular, cyclic reversals posited by Khramov was A. A. Valiev's demonstration of 1960, based on study of the Margazar section of Cenozoic sedimentary rocks (continental molasse beds) of northern Fergana in central Asia, that the geomagnetic epochs are apparently of unequal length. Valiev calculated that the average duration of the normal epochs during the Miocene and Pliocene was 2.8 times as long as the reversed.*

Khramov's important work of 1958 was translated in 1960 at Canberra to provide a most valuable reference to the English-speaking world. In the introduction, he noted, "There is as yet no general work on paleomagnetism in which the results obtained abroad are summarized." Allan Cox and Richard Doell responded that year with their highly influential "Review of Paleomagnetism."

In considering the evidence for field reversal available from parts of stratigraphic sequences that showed transitional directions of magnetization, this discussion will be restricted to studies before 1960, when Cox and Doell began sustained work toward solution of the reversal question.

*Valiev's paper (1960a), which was not translated until 1961, was not as widely known nor as influential as Khramov's.

Kanichi Momose argued in 1958 that the Pliocene-age Komoro and Shigarami volcanic groups of Japan directly recorded a transitional reversal of the field as a series of intermediate directions. Momose, being firmly convinced of field reversals, stood in the minority among the Japanese paleomagnetists at that time. Francis H. Hibberd in 1961, although apparently not influential, in contrast believed that partial instability could produce the appearance of a transition, where only a smeared distribution simulating a transitional zone was actually recorded; of course Hibberd's interpretation, however aspersive of transitional data, at that date was after the fact of the cases that follow and did not rule out the possibility that Momose had in fact recorded reversely magnetized rocks. McElhinny, in 1973, provided the most complete summary of transitional studies from 1964 through 1972, and Irving exhaustively treated those available to him in 1964.

After his 1957 field work with Trausti Einarsson, Thorbjorn Sigurgeirsson measured oriented rock samples in the laboratory on a spinner magnetometer, before and after they were partially demagnetized in an alternating field. The samples covered the reversals at the beginnings of the six most recent magnetic epochs, and "special attention [was] given to places where the field observations gave an indication of a gradual transition at the boundaries."[36] Of the six reversals studied, four appeared to be gradual. Sigurgeirsson took the effects of secondary magnetization into account and plotted the results from each of these four transitions to find that they "are distributed rather regularly, indicating a definite path followed by the magnetic pole during the reversal of the magnetic field."[37] He studied the next-to-oldest transition (R_3–N_3) in detail, making 24 determinations on samples from seven localities spread over a distance of 25 kilometers. All of the results fell within 10° of a smooth curve when plotted on a graph; "The sequence of the points along the curve agrees with the position of the samples in each profile. It seems reasonable to assume that this curve represents the transition of the geomagnetic field from the reversed to the normal direction."[38] He also noted that the basalt layers, because they were free of interbedded sediments, suggested a reversal time of a few thousand years or less. In closing he emphasized that the high frequency of gradual reversals (four of six boundaries) suggested that they might well have been preserved in other basaltic regions.

Thorbjorn Sigurgeirsson, at left, and Norman Watkins collecting rock specimens in Iceland in 1964. As early as 1957 Sigurgeirsson's laboratory studies of Icelandic basalts led him to conclude that of six reversals studied, four appeared to be gradual. He pointedly remarked that the data were "distributed rather regularly, indicating a definite path followed by the magnetic pole during the reversal of the magnetic field." The Icelanders Trausti Einarsson and Ari Brynjolfsson also provided much of the strongest evidence for field reversals during the 1950's. Although Watkins's doctoral degree was from the University of London, the bulk of the research for his dissertation was done under the supervision of Allan Cox and Richard Doell in the Rock Magnetics Laboratory at Menlo Park.

In July 1957, Ari Brynjolfsson, an Icelander pursuing doctoral studies at the University of Göttingen,* published an important paper briefly describing an instrument of his own construction with which he measured the intensity and direction of magnetic moments; he also described magnetic cleaning in an alternating field, an investigation of viscous magnetization, and secular variation during the past 5,000 years in Iceland. In the report's final section he discussed the interval between clearly normally and reversely magnetized zones and drew several new and provocative conclusions: "The direction of magnetization changes gradually from reversed to normal. . . . The clockwise traces of the variation remind one of the present clockwise traces of the secular variation. . . . Perhaps the similarity indicates that the reversal of the field took place during a period of 1,000–3,000 years. It is difficult to understand how such changes could be caused by some self-reversal during the cooling process or during the time passed, as we would then expect random variations in the direction of magnetism."[39]

Brynjolfsson reported that the fine structure of those changes was determinable only after cleaning by partial demagnetization, and that the intensity of the magnetism in transitional rocks was about one-fifth that otherwise found in Tertiary basalts. In all, his evidence deserved careful consideration. The work of Trausti Einarsson, Sigurgeirsson, and Brynjolfsson on the Icelandic basalt-flow sequences provided some of the strongest evidence for field reversals available during the 1950's.†

*Brynjolfsson's doctoral dissertation was done at Göttingen under the unofficial guidance of Sigurgeirsson, as the University of Iceland did not award the doctorate in geology.

†During a three-month visit of Richard Doell and David Hopkins to Iceland in 1964, they and Pall Pallson and Thorleifur Einarsson visited and resampled the sites in which the transitions had been observed. Hopkins recalled: "The sites proved to be what Thorleifur calls 'multiflows'—packets of thin flow units of identical petrology separated from the next set above and below by weathering zones, frost breccia, or sediments. These thin flow units we had seen and later saw many times in basaltic terranes—they are simply separate cooling units of thin pahoehoe lava superposed in rapid succession during a single eruption. A single eruption from a basaltic volcano or fissure rarely lasts as long as 10 years and is commonly completed within less than a year, so Thorleifur and I felt certain that the observed transition zone represented a very rapid accumulation, and that the observed changes had taken place within less than a decade. Later that summer in Iceland, Doell sampled many multiflows . . . and found quite uniform directions of magnetization. In conversation with Doell, while still in Iceland, we concluded that what was actually observed in the Icelandic transition zones was a more or less complete decay of the main dipole field (postulated, anyhow, nowadays, for the transitions) with the consequence that at any given locality the field was dominated by the moving anomalies that cause secular variations. It was not a coincidence that 'the clockwise traces of the variation remind one,' as Brynjolfsson observed, 'of the present clockwise traces of the secular variation'" (D. M. Hopkins, written comm., xi-1-78).

Richard Doell and Thorleifur Einarsson in the field in the Esja area of Iceland in June 1964. That year Doell and Einarsson, with David Hopkins and Pall Pallson, visited and resampled the Icelandic sites in which magnetic-polarity transitions had been observed.

Edward Irving's earlier-mentioned study, published in 1957, showed the first successive reversals of polarity in a continuous sequence of "Upper Precambrian" age sedimentary strata, which he likened to the Icelandic lava-flow sequences of Hospers.[40] After discounting the likelihood of self-reversals as an explanation and presenting pertinent paleomagnetic data, he concluded that "the hypothesis of reversals of the geomagnetic field would therefore predict that there should be zones of relatively small stratigraphical thickness between those with magnetization corresponding to stable fields which have more or less random directions of magnetization. This as was shown above is what was found in this series of rocks."[41] As a pointed finale, Irving noted that "the case for special causes of reversals of magnetization is not strong when reversals of the main geomagnetic field give a simple and general explanation for the varied phenomena observed."[42] It is of considerable note that Irving was early convinced of field reversals; even though he went on mainly to pursue directionality studies toward resolution of the drift question, and did no systematic reversal studies toward a polarity-reversal scale, his choice of research did not reflect doubts over the reversal debate.*

In overview, the several lines of evidence in favor of the field-reversal hypothesis might be said to have been totally convincing by the late 1950's, in the absence of evidence to the contrary—but such there was.

The Case for Self-Reversals

In the first section of the Second Weizmann Memorial Lectures on Rock Magnetism, presented in Israel in December 1954, the Nobel laureate Patrick M. S. Blackett delineated the three hypotheses derivable from the existing evidence for the cause of reversely magnetized rocks: (1) if self-reversed rocks are so rare in nature that they can be ignored statistically, then the history of reversals of the geomagnetic field can be assumed from the normal and reversed rocks; (2) if all reversed rocks are so rendered by physical or chemical properties, then we deduce that the geomagnetic dipole has retained "approximately the same magnitude and direction [polarity] since geological history began;" (3) if some reversed rocks are due to field reversals and others to self-reversal, then the problem of differentiating them involves the task of explicating all the means responsible for self-reversal through laboratory experimentation.[43] Blackett also clearly emphasized the near-impossibility of performing definitive laboratory tests that would absolutely rule out the possibility of self-reversal; this point is a most important backdrop against which to view the events that fueled the more-than-decade-long arguments of the self-reversalists.

John W. Graham of the Department of Terrestrial Magnetism at the Carnegie Institution in Washington, D.C., published an important paper in 1949 on possible tests for the magnetic stability of sedi-

*As late as 1962, Irving, who did much significant work, mainly in directionality, was not convinced that reversal studies should enjoy research priority (A. V. Cox and R. R. Doell, oral comm., iv-78). This concern of the directionalists is early in evidence; for instance, Irving (1956b) noted in his text: "Samples from a rock series have parallel directions of magnetization but have mixed polarity; for example, of the 76 samples from the Columbia River basalts [Campbell's data, described by Runcorn (1955)], 49 are . . . with positive dips [normally magnetized], and the other 27 . . . with negative dips. . . . The only importance of the reversals in the present context is that they introduce an ambiguity in sign of the geocentric dipole, and in all calculations, *both normal and reversed rocks will be considered together without regard to sign*" (emphasis added). At no place in the summary, which opens that important early paper, did Irving mention the subject of the reversed rock data.

mentary rocks: remanent magnetization, he said, should change in direction to match the warping of the strata (this is still called the "Graham fold test"), and magnetic directions among the boulders of a conglomerate should be scattered. He also reported on a magnetized sedimentary rock horizon which he believed could be traced over hundreds of miles by means of distinctive index fossils. Relying too heavily on fossils for his evidence of synchroneity (as non-paleontologists often have) and reasoning prior to the discovery by Hospers, only two years later, that reversals probably take less than 10,000 years, Graham assumed that both normally and reversely magnetized samples came from rock beds of one well-constrained age. He thus concluded that the opposed polarities must be due to some mechanism other than reversals of the earth's field. He postulated that the reversed remanence was probably due to "stronger secular-variation foci [local disturbances of the general magnetic field] in remote epochs" when the earth was hotter and such foci were closer to the earth's surface than their present depth, estimated at 1,200 kilometers.[44] (Graham obviously had not yet adopted the idea of self-reversal in 1949; his rejection of the field-reversal hypothesis was clearly made known in his paper of 1952.) Graham was one of the first highly regarded Americans to enter paleomagnetic research, and his early work that unfortunately led to his belief in self-reversal and rejection of a reversing field affected the development of future studies, especially those of James Balsley, Arthur Buddington,[45] and others.*

Shortly after the paper of 1949, Graham was moved to inquire of Louis Néel, the future Nobel laureate, if there were possible processes by which materials could acquire magnetism opposite to that of the applied field. Néel's notable reply of 1951 proposed four mechanisms that might theoretically produce such an effect. In 1951 the theory was confirmed in part by Takesi Nagata, Syuniti Akimoto, and Seiya Uyeda,[46] who showed that groups of grains of "magnetite with impurity" from the fine-grained, quartz-rich rock (a hypersthene hornblende dacitic "pumice") ejected from the volcano Haruna, in Japan, acquired a reversed magnetism (of 0.12

*"Graham was an original, somewhat iconoclastic individual who abandoned paleomagnetism because of two mistaken ideas (self-reversal and magnetostriction) as well as Merle Tuve's opposition to his work" at the Carnegie Institution, where Tuve was Graham's superior. This quote, unattributed on request, was borne out by the commentary of several paleomagnetists. Tuve told Richard Doell in 1957 that if he wanted to remain on the staff of M.I.T. he should eschew paleomagnetism and "get into some serious geophysics, like seismology or gravity" (Doell, t. 2, s. 1, iv-17-78).

oersted intensity) when cooled from 350°C to room temperature in a weak (0.5 oersted) magnetic field; they coined the term "reverse thermo-remanent magnetism" to denote the effect.* They believed that the reversal characteristics demonstrated by the Haruna specimens in laboratory experiments were in keeping with those predicted by Néel's third mechanism, in which there are intimately mixed grains of two different types, one with a higher Curie point and lower intensity of magnetization than the other. During cooling of such a substance, one grain type (higher Curie point, lower magnetic intensity) magnetizes first, acquiring the direction of the applied field. When the temperature falls below the Curie point of the other type (with higher intensity) it becomes magnetic, but will be acted upon by not only the earth's field but also the local field created by the first, previously magnetized grains. In certain geometrical arrangements, the resulting magnetization of the second group of grains can be reversed. Therefore the two grain types are opposed in polarity when cooled, but the stronger, reversed magnetic field of the second type dominates the magnetization of the whole. Néel reported other theoretically possible interactions between grains of differing Curie points in 1955.

The laboratory experiments of the Nagata group continued,[47] and in 1957 Seiya Uyeda in a comprehensive monograph reviewed the state of knowledge concerning thermoremanent magnetism.[48] He showed that the Haruna dacite contains two distinctly different ferromagnetic minerals, one a titanomagnetite and the other a member of the ilmenite-hematite series. Contrary to their earlier belief in the interaction between the two minerals to produce reverse magnetization, "further experiments . . . on . . . separated ilmenite-hematite . . . revealed that the reverse TRM [thermoremanent magnetization] is a phenomenon peculiar to this constituent alone. This study showed that reverse TRM is a characteristic of specimens [in the] transitional range between ferromagnetic ilmenite and parasitically ferromagnetic titanohematite."[49] He also suggested that the "imperfect reverse TRM" observed in certain synthetic materials "may be related to the origin of the reverse nat-

*Richard Doell was told by Seiya Uyeda that his thesis problem under Nagata was to find self-reversing rocks; "the second one he picked up" from a cabinet "was the Haruna dacite" (Doell, t. 2, s. 2, iv-17-78). Uyeda (1978) has also published this story. The rarity of self-reversing rocks was of course not known at that date, and thus such a statistically improbable episode, early on, contributed to the force of argument by the Japanese group, who subscribed to self-reversal as a widespread mechanism.

ural remanent magnetizations of the Adirondack rocks in the U.S.A. and the Allard Lake ilmenite-hematite deposit in Canada."[50]

In 1958, Uyeda reported that ferromagnetic minerals in the dacitic pitchstone of Aso volcano in Japan exhibited the same reverse thermoremanence characteristics as those of the Haruna dacite. Although the total remanence of the rock was not reversed, "a certain part of its partial TRM has the property of self-reversal just similar to that of the Haruna rock."[51] Takao Saito discovered that the iron sand at the Sokota mines in Japan showed a partial reversed thermoremanence, although the total remanence was not reversed. The sample contained grains with "the perfect reverse TRM,"[52] which chemical and crystallographic analyses showed to be a member of the ilmenite-hematite series, a mineral first found to be self-reversing in 1954.[53]

The work of the Nagata group was highly regarded, and the question of self-reversals was widely discussed by the mid-1950's; nonetheless, only three cases of complete self-reversal, one the Haruna dacite, had been found in natural rocks by laboratory experiments at that time.

The second case of self-reversal was reported in 1954 by James R. Balsley and Arthur F. Buddington, using rocks from the Adirondack Mountains of New York. Certain rocks of Precambrian age contained hematite and ilmenite; when heated above the Curie point and then cooled in a magnetic field, they were found to be self-reversing. The accessory oxides—those minor minerals accompanying the hematite and ilmenite—were similar to those identified as responsible for reversal in the Haruna dacite. Balsley and Buddington had found the rocks not by chance, as Uyeda did, but in the course of an investigation of widespread geomagnetic anomalies found during an aerial magnetic survey by the U.S. Geological Survey in New York and several other states. The reversely magnetized igneous and metamorphic rocks surveyed were thought to range in age from Precambrian to Triassic. Their major conclusion (which confirmed an earlier surmise[54]) was that "in one group [of rocks] the accessory oxide minerals consist exclusively of members of the magnetite (Fe_3O_4)-ilmenite ($FeO{\cdot}TiO_2$)-ulvöspinel ($2FeO{\cdot}TiO_2$) series,* and the magnetic properties are normal. . . . In the second group of rocks the accessory oxide minerals are exclusively members of the hematite (Fe_2O_3)-ilmenite ($FeO{\cdot}TiO_2$)-rutile (TiO_2) se-

*In the strict sense there is no such series.

ries, and the rocks uniformly show reverse remanent magnetism and give rise to negative [aeromagnetic] anomalies."[55]

This very important conclusion was based on more than 200 oriented specimens drawn from Precambrian igneous rocks of several types and their metamorphosed equivalents (gneisses) and intensely metamorphosed sedimentary rocks of the Adirondack Mountains. In 1957, reporting on the remanent magnetism of the Russell Belt of gneisses in the northwest Adirondack Mountains, Balsley and Buddington concluded on the basis of 128 rock samples that rocks with magnetite as the predominant magnetic mineral are normally magnetized, but in a direction about 56° from the present field.* That large angle to the present field was taken as evidence that the rock magnetism was stable. Reversely magnetized rocks contained a member of the hematite-ilmenite group as the predominant mineral; the wide scattering in their magnetic directions further suggested that the mechanism of magnetization differed from that of magnetite.[56] It seemed clear that rocks bearing ilmenite-hematite minerals were not suitable for stratigraphic studies of polarity reversals.† Balsley and Buddington reported in 1958 that a layer of rock rich in hemoilmenite at Hibernia Pond, New Jersey, and an ore deposit at Allard Lake, Quebec, were also self-reversing; the latter was explicated by Charles M. Carmichael in 1959 and 1961.

*John Verhoogen recalled: "The Adirondack gneisses had an enormous effect on my thinking. At Balsley's suggestion, I spent most of a summer on W.A.E. [status] trying to figure out what was going on in the Russell gneisses. The foliation of the gneisses seemed to be one cause of the scatter in directions. Hematite is strongly anisotropic, so that any preferred orientation of its crystallographic axes could strongly affect the direction of remanence. But how was the preferred orientation related to the foliation? No one could tell. It also appeared likely that part of the remanence could have been acquired during deformation when the rocks were under stress (magnetostriction), in which case directions in the magnetite-bearing rocks would also be suspect. It was all very frustrating, and reinforced my conviction that we should understand more of the mechanisms that control acquisition of TRM before drawing conclusions about reversals or polar wander. These feelings are expressed in an NSF grant application, dated September 1958, which I append [Document No. 33]. It should be remembered that in those days no one could even explain the magnetic properties of pure hematite, which had been shown to be antiferromagnetic and should therefore have no remanence at all!" (written comm., vii-17-79).

†The cooling history of volcanic rocks is decidedly different from that of gneisses; "even if volcanic rocks contain ilmenohematite of the appropriate composition, we now know that their polarities are usually reliable" (written comm., C. S. Grommé, viii-78). Additionally, John Verhoogen has noted, "Self-reversal in ilmenohematite occurs not only in gneisses, but also in intrusive rocks (e.g. the Buck's Lake pluton of N. California investigated by Merrill) and lavas (e.g., the Haruna dacite). Self-reversal occurs only if the ilmenohematite has cooled sufficiently slowly to form exsolution lamellae of different Ti content. These exsolution lamellae rarely appear in basalts, but they may be difficult to see; they are found in many rhyolitic and dacitic lavas. There *are* basalts with partial self-reversal. Schult has described some Tertiary basalts from Germany in which the mechanism is similar to that in Gorter and Schulkes' artificial spinel. Ozima and Larson have observed partial self-reversal in submarine basalts" (written comm., J. Verhoogen, vii-17-79).

In 1954 a new uncertainty entered the picture. Naoto Kawai, Shoichi Kume, and Sadao Sasajima found that certain reversely magnetized horizontal volcanic and sedimentary rocks of Tertiary age in Japan were self-reversing; this time, however, the misbehaving mineral was magnetite, of the series $Fe_3O_4-Fe_2TiO_4$. It is important to note that whereas the studies mentioned above dealt with the hematite-ilmenite group, which are easily avoided in rock-magnetic studies, magnetite is one of the main minerals chiefly responsible for fossil magnetism in sedimentary and igneous rocks. The researchers placed rock specimens in an evacuated furnace and heated them from room temperature to 600°C; at 30-degree increments, the polarity of the specimens was measured with a sensitive magnetometer, and the "variation of the remanent magnetism with temperature was thus nearly continuously observed."[57] During continuous heating, all of the rocks suddenly switched polarity from reverse to normal at about 100°C and switched back to reverse at 350°C. Further experiments on separated ferromagnetic minerals revealed "that each single mineral grain contains three kinds of titanomagnetites having different Curie points of 120°, 410°, and 550°C, the first and the second Curie points of the three being slightly higher than the critical temperature" at which the reversals take place.[58] They also found that the phase with the highest Curie point and the phase with the lowest were common in almost all of their rock specimens; in contrast the Curie point of the middle phase varied in different specimens between 280° and 450°C. Upon heating to 800°C, the three kinds of magnetite became mixed into one homogeneous phase with a Curie point near that of the middle phase. The homogenized magnetite was also unstable in crystal structure at room temperature, and a small part of the homogenized form separated into two components, one with a higher Curie point and the other lower than the parent.[59] In summarizing, Kawai advanced still another explanation for self-reversal, different from those that Louis Néel, Takesi Nagata, and John Graham had each proposed in 1953; his mechanism provided the self-reversalists with a sort of ultimate refuge from disproof by their adversaries. When a rock forms, the acquired remanent magnetism matches the ambient field direction; the magnetically active mineral at this time is the homogeneous magnetite with an intermediate Curie point. During cooling, the two kinds of magnetite "precipitate" or differentiate from the homogeneous parent. During and after precipitation the demagnetizing field of the parent acts on the precipitates, giving rise to reversely magnetized offspring. The total remanent rock magnetism is thus the product of that of the remaining parent magnetite and the two differentiates. When the total polarization of the differentiates exceeds that of the parent, the total remanence appears reversed; when the polarization of the parent exceeds that of the differentiates, normal magnetization results. The refuge from experimental disproof lies in the fact that Kawai claimed to have demonstrated from many specimens that the rate at which differentiation took place was very slow: "It can be inferred that the magnetization of rocks may have occurred with an extremely sluggish rate which must be measured in some cases in the geological time scale"[60] However, the specter of doubt surrounding the rate of the reversal process was given a very different and sounder theoretical basis than that suggested by Kawai by John Verhoogen only two years later.[61] It appears that Kawai's claim of finding self-reversal in natural cubic oxides was open to question and thus did not have the impact that seems appropriate taken at its face value.* His publications in this subject area appear not to have been heavily cited. Furthermore, by 1958, Seiya Uyeda had shown that minerals of the hematite-ilmenite series, within a limited composition range, underwent self-reversal as a function of cooling rate.[62] However, said Verhoogen, inasmuch as experimental cooling rates in the laboratory are thousands of times faster than natural cooling, such tests are ultimately inconclusive. Moreover, because heating during experimentation often produces mineralogical changes, the original properties of the rock that brought about self-reversal may have been changed or destroyed. Thus there remained a stubborn residue of apparently unresolvable doubt.

A lecture entitled "Puzzles in the Interpretation of Paleomagnetism," written by Balsley and Buddington and delivered in Calcutta by Buddington in January 1957, is perhaps the most comprehensive review of the reversal question of that time.[63] The major puzzle appeared to be how the data obtained in studies of paleomagnetism could be so well received and so widely applied in drawing consistent interpretations without considering the facts that "some

*It appears probable that Kawai overinterpreted his thermomagnetic curves. Many of his curves (the graph J_s vs. T, or remanence vs. temperature) had a triply scalloped shape in which he identified each cusp as one of three separate Curie temperatures, representing a distinct phase. C. S. Grommé notes (written comm., viii-78) that in more than a thousand such curves the triply scalloped shape has never been seen, and he is "perfectly certain" that Kawai never observed a complete self-reversal nor had any true evidence for the magnetite-ulvöspinel solvus.

Roderic L. Wilson in Iceland during the summer of 1964. In the following year he wrote that "lavas from eleven different sites have shown a remarkably clear relation between natural magnetic polarity, the state of oxidation, and magnetic properties." He and the "oxidationists," who believed that the oxidation state of a rock could serve as a predictor of polarity, were challenged in an unpublished blind test by Cox, Doell, and Dalrymple; the "oxidationists" were not successful. Wilson was a firm believer in reversals of the earth's field and published in support of that idea repeatedly.

of the major minerals . . . have the inherent potentiality for self-reversal . . . and acquiring [magnetism] at angles to the orientation of the environing magnetic field."[64] They cited Blackett as representing the views of most geophysicists: "Although there is the potentiality for the inherent properties of minerals and mineral transformations to result in self-reversal . . . [Blackett] tends to minimize the significance of this factor in nature, at least for the time being." In contrast, Balsley and Buddington believed strongly that much more research on that potentiality was required "before it can be lightly dismissed."[65] Both men were highly regarded and influential in the earth science community, Balsley as chief of the Geophysics Branch of the U.S. Geological Survey and Buddington as a professor at Princeton University.*

Another line of evidence, which had its inception in the early 1950's, gained wide attention by the early 1960's through the efforts of Roderic L. Wilson, Patrick Blackett, James Ade-Hall, and others. The hypothesis that ferromagnetic minerals in rocks of reversed polarity are more oxidized than those of normal polarity was first

*"Buddington's primary concern was directionality and its bearing on paleogeographic reconstruction" (written comm., J. T. Gregory, i-3-78).

formulated by Balsley and Buddington in 1954, on the basis of gneisses and granites in the Adirondacks. The influence of their work during the 1950's was considerable. Subsequent work during the early 1960's on rocks from a number of far-flung localities was reported by Blackett in 1962, Wilson in 1966, Wilson and Stephen E. Haggerty in 1966 and Wilson and Norman D. Watkins in 1967; these publications provide references to the widely scattered literature. Bullard's Bakerian Lecture of 1967 sets the problem clearly and succinctly for that date.[66] The continuing search for the intrinsic factors that caused rocks to have reversed polarity included a number of studies suggesting that reversely magnetized lavas are more highly oxidized than normal ones.[67] In a summary of the evidence, Wilson focused on the Columbia Plateau basalts collected by Norman Watkins:

> Lavas from eleven different sites have shown a remarkably clear relation between natural magnetic polarity, the state of oxidation, and magnetic properties. We have been able to divide these lava specimens into five petrological categories corresponding to stages of increasing oxidation. Yet these same lavas also exhibit two of our classical tests [criteria] for field reversal! There are five baked zones with reversed magnetization, each one agreeing in polarity with the overlying lava; and one of the eleven sequences contains a set of "transition lavas" showing a gradual apparent turning over of the ancient magnetic field. . . . Yet even within this transition one can see obvious chemical changes as one progresses from normal to reversed lavas. A similar conflict has appeared in another lava sequence of Carboniferous age in Britain.[68]

The oxidation-polarity controversy erupted, in the main, after Allan Cox and Richard Doell had decided to enter the polarity studies that led to the reversal time scale. They were in contact with Wilson, Ade-Hall, and others who subscribed to a correlation between oxidation and polarity.* An unpublished blind test of the heuristic prowess of the "oxidationists" by Allan Cox, Richard Doell, and Brent Dalrymple revealed that the oxidation hypothesis was useless for predicting polarity.† Remarkably, the results of the

*Roderic Wilson was a firm believer in reversals of the earth's field; he published in support of that idea repeatedly (1962b, c, 1965, 1966b). He simultaneously believed that "the chemistry of some rocks is related to the magnetic field direction prevailing at their formation" (R. L. Wilson, 1965, p. 380).

†During a visit to Imperial College in London, Doell suggested an experiment. Cox, Doell, and Dalrymple sent Wilson and Ade-Hall almost two dozen rock samples, the polarity of which Wilson and Ade-Hall were to predict solely on the basis of rock chemistry. The "oxidationists" were not successful (Cox, t. 4, s. 2; Doell, t. 5, s. 1; G. B. Dalrymple, oral comm., i-29-80).

James M. Hall (formerly Ade-Hall) on a visit to the Rock Magnetics Project at Menlo Park in the early 1960's. Hall was convinced of the correlation between the oxidation state of a rock and its polarity.

trial were never published and were not mentioned in later publications on the subject.

In closing his Bakerian Lecture of 1967, after a brief review of the oxidation question, Edward Bullard opined, "None of the proposed explanations is able to explain all the facts, and the correlation of oxidation and magnetic polarity may be seen as one of the major unsolved problems of earth science."[69] (The way this "major unsolved problem" was evaluated by the framers of the reversal-based time scale is described in Part III.)

Partial and full reversals were also known in synthetic materials. Louis Néel's (1951) hypothesis of self-reversal was demonstrated in part by E. W. Gorter and J. A. Schulkes in 1953 on synthetic lithium chromium spinels. In these crystals, an interaction between the magnetic moments of atoms in two different positions in the crystal lattice (tetrahedral and octahedral sites) caused the respective moments to align in opposite directions; because the magnetization of each set of atoms or sublattice varies differently with temperature, the net magnetization may reverse at a lower temperature. Although such a reversal was not known in natural rocks, the laboratory evidence strengthened such a possibility and drew more attention to Néel's hypotheses.*

*According to Verhoogen (written comm., vii-17-79), Schult (1968) "found one such basalt in Germany; however, the temperature at which the sign of magnetization changes is 200°K," or about −200°F.

The question of atomic ordering with respect to possible self-reversal, raised by Néel in 1955, was sharply focused the next year by Verhoogen, whose theoretically specific model appeared, like Kawai's of 1954, to preclude forever resolution of the question by experimentation.[70] Within the self-reversal school, Verhoogen's widely known profound capabilities across a broad range of theory made him a formidable exponent. He believed that the slow migration of cations (positive ions) to preferred sites within a crystal lattice during and after cooling could produce an inversion of magnetization. He cited as examples impure magnetites containing an appropriate amount of aluminum or other elements with a suitable chemical valence. The rate of the ordering process, in which the cations move toward an equilibrium state, was not known, but he noted, "It is unlikely that reversal could occur in less than 10^5 or 10^6 years; on the other hand, ordering may be so slow that an igneous rock may be eroded away, and its magnetic constituents redeposited in sediments, before they undergo inversion; some sediments may thus have an inherent self-reversing property."[71]

Although Verhoogen's model of reversals in impure natural magnetites resulting from an ionic ordering process upon cooling was tied at the theoretical level to Gorter's empirical demonstration of that phenomenon in the compound $NiFeAlO_4$, no self-reversals of that specific type had then or have since been found to occur in magnetite, titanomagnetite, or pure hematite—the three minerals that give rise to remanent magnetism in almost all sedimentary and igneous rocks.

One must recall the very scant experimental data of the 1950's and the extent to which a person of Nagata's, Néel's, or Verhoogen's stature could influence a community that had entered a new and little-understood area of investigation. The theoretical mechanisms postulated by Néel and Verhoogen, when coupled to the few empirically demonstrated cases of total self-reversal in the laboratories (Nagata's, Balsley's, and Kawai's on natural rock; two studies showing partial reversal, one by Uyeda in 1957, and another by Saito in 1957; and Gorter's laboratory reversal of a synthetic substance) were perhaps more heavily weighted than seems justified in retrospect. For many at that time, the evidence for self-reversal provided escape from the need to suspend the constancy of the geomagnetic field, which was tantamount to an immutable universal constant; to others the few experimental cases, when examined for long and close at hand, perhaps grew disproportionately

Kenneth Creer, an early English directionalist, and Allan Cox in Kyoto, Japan, at a meeting of the International Union of Pure and Applied Physics in the summer of 1961. By that time many of the English paleomagnetists were convinced that the earth's magnetic field had reversed in the past. They did not, however, pursue reversal studies, because they were preoccupied with directional paleomagnetic research in order to prove continental drift.

against the evidence for field reversal. The problem was esoteric and treated by a small group of researchers in which the most prestigious participants were mostly in favor of explaining reversely magnetized rock bodies by self-reversal mechanisms; that situation was certainly the case in the United States but not among English researchers, who appear to have earlier accepted the inevitability of field reversals for at least certain rock groups.

It was into this ideational context of the 1950's that Allan Cox and Richard Doell entered as students. They matured to form a research program in the face of dissuasive efforts by most, but encouragement came from a rare few. Chief among these was James Balsley, of self-reversal persuasion, who played a crucial role in implementing their research in rock magnetism at the U.S. Geological Survey in Menlo Park. The efforts begun there in 1960, and also in Australia shortly afterward, finally resolved the reversal question, and produced a scale of dated reversals by the mid-1960's. But before these developments will be treated, it remains to describe a false start toward such a scale, made by Martin Rutten of Utrecht.

A Premature Start: Rutten's Reversal Scale

In 1959, Martin G. Rutten of the Mineralogisch-Geologisch Instituut, Rijksuniversiteit, Utrecht, published a little-noticed paper in a Dutch journal entitled "Paleomagnetic Reconnaissance of Mid-Italian Volcanoes." It is the first written account of an effort to define a time scale of geomagnetic polarity reversals through the use of radiometric data.

Rutten's very broad background included professorships in stratigraphy and paleontology at the University of Amsterdam (1946–1951) and another in physical geology at the State University of Utrecht (1951–1970). His earliest interests were in fossils, and upon completing his doctorate at Utrecht in 1936 he was employed by the royal Dutch petroleum group in Java and Sumatra, where he did stratigraphic studies for petroleum exploration dealing mainly with the tiny marine organisms known as orbitoidal foraminifera.[72] His many publications, which included seven books and 108 articles, ranged across a wide front encompassing paleontology, sedimentation, volcanology, evolution, coal geology, geological philosophy, and paleomagnetism.[73] In *The Geology of Western Europe*,[74] his broadest work, he clearly demonstrated that he had succeeded in trying "not to become too specialized."[75] He was apparently a divergent thinker who changed research directions frequently, was very widely read, and was likely to be attracted to a potentially useful new method in stratigraphic correlation, a subject that occupied a substantial part of his diverse career.

Rutten was formally schooled in paleontology and general geology, but not in geophysics. Thus, although his training and experience in paleontology and stratigraphy had provided him with the perspicacity to appreciate the rapidly unfolding new research tradition of the 1950's in magnetostratigraphy (as evidenced by his 1960 commentary on the work of Hospers, Einarsson, and others[76]), he lacked the technical background necessary to pursue a research program in either paleomagnetism or isotopic geochronology. "Rutten," said one of his colleagues, "was basically a paleontologist who did not understand [the technical aspects of] paleomagnetism to the extent that he could follow up on his effort of 1959."*

Rutten's historically significant paper of 1959 was not an isolated

*These views are quoted from the personal commentary of two paleomagnetists and a paleontologist-marine geologist who all had sustained close contact with Rutten (unattributed on request, vi-78).

Martin G. Rutten of the Mineralogisch-Geologisch Institut, Rijksuniversiteit, Utrecht, who died in 1970, was a highly regarded, broadly based divergent thinker who often changed research directions. He made the first effort to define a time scale of geomagnetic polarity reversals dated by radiometry; lacking a background in isotope dating and paleomagnetism, he did not follow up on his first, pioneering attempt.

one; he had early voiced belief in the desirability of utilizing polarity reversals in geologic mapping of Pleistocene deposits in Iceland and in worldwide correlation of the Pliocene-Pleistocene boundary.[77] He had still earlier demonstrated inversions in the polarity of the basalts of the Coiron plateau of France and their provisional correlation with the results obtained in 1950 by Roche;[78] a Plio-Pleistocene age was assigned to the Cône de Montredon and Pliocene for most of the flows of the Coiron plateau.[79] He was doubtless poignantly aware early on of the potentially rich service that magnetostratigraphy could provide to correlation studies in the later Tertiary. His 1960 paper, which was submitted to the journal *Geologische Rundschau* before completion of the historic one of 1959, was entitled "Paleomagnetic Dating of Younger Volcanic Series." In it he described his field methods in paleomagnetism, which, although unspecified in the paper of 1959, were very probably similarly applied to the Mid-Italian volcanoes: in his own words, "The Mid-Italian volcanoes, Vulsini, Cimino, Vico, Sabatini and Laziale were *cursorily* studied as to their paleomagnetism" (emphasis added).[80] Rutten's field method—and there is no mention of laboratory experiments—was that of Sigurgeirsson, which involved the use of a simple geologic hand compass that was laid down horizontally. "The lower edge of the northern face of the sample [5–10 centimeters across] is then held near the points of the needle." Samples that attract the south-pointing tip and repel the north-pointing tip are normally magnetized. Each sample was tested in four positions against the compass, and those not responding consistently in all four positions were rejected.[81] It is curious to note that a lengthy section in the paper of 1960 on the "requirements and pitfalls" of "this type of paleomagnetic dating" opened with a cautionary note "to be sure that . . . there is no process of self reversal within the rock" and that "the coercitive [sic] force of the cooled rock is sufficiently high to exclude practically any later magnetic induction at lower temperatures."[82] Rutten commented on a number of other points, all confirming his attention to the growing literature of paleomagnetism, his concentration on the most important papers in magnetostratigraphy, and his sharing the widespread sense of frustration (partly as a result of his own field studies in Iceland[83]) over the lack of a temporal framework for polarity reversals. He did not hold out much hope for the application of radiometric dating to magnetically determined rocks; in his paper of 1960 there was only a brief, passing mention of radiometric dating, and that was not within the context of a call for collaboration between magnetostratigraphers and geochronologists:

> A sequence of reversals found in a given series of lavas and dikes only indicates the relative age relationships within the series. It can only be used within the stratigraphic framework of other dating methods. . . . Such dating, of course, may be absolute, by radioactive methods, or relative by geologic methods, such as paleontologic, glacial or geomorphic. Notwithstanding these drawbacks, this type of paleomagnetic dating promises to be a valuable tool in the mapping of series of volcanics, where normal geologic dating is often extremely difficult because of lack of fossils.[84]

In summary, it appears that Rutten was strong in stratigraphy, especially in the use of fossils for dating rock beds (biostratigraphy), quick to sense the importance of magnetostratigraphy and very attentive to potentially applicable new developments central or peripheral to it, but was poorly equipped to pursue studies in the highly technical aspects of isotopic dating and paleomagnetism. These technical aspects were to be proved the prerequisites to success in formulating a geological time scale of magnetic polarity reversals.

The uniqueness of the Roman Volcanic Province, with its four high-potassium volcanic centers,[85] although not studied in detail by the late 1950's, had attracted much attention, since widespread ignimbrite beds (rocks of ash and other volcanic debris) are intercalated there with diverse volcanic layers and sedimentary deposits, including diatomaceous and pollen-bearing lacustrine units (lake beds). Littoral (shoreline) deposits representing high stands of sea level also intertongue with volcanic rock along central Italy's west coast. The propinquity of this fortuitous assemblage permitted Alberto C. Blanc (who aided Jack Evernden's rock-collecting activities, mentioned earlier) to define in the mid-1950's a local Italian subdivision of the Pleistocene.[86] Thus Rutten, in his search for areas which might provide controlled stratigraphic sequences containing magnetically determinable volcanic layers, came upon a 1958 report by Blanc that Garniss Curtis and Jack Evernden had derived dates on volcanoes of the Roman Volcanic Province by potassium-argon dating. In the concluding part of Blanc's address as President of the Fifth International Congress of Prehistory and Protohistory (I.C.P.P.), delivered on August 30, 1958, he had noted:

> Its application [the K-Ar dating method] has permitted Dr. Curtis and Dr. Evernden to date the volcanic group of Tolfa, north of Rome (−2,400,000 years), the moraines of the oldest glaciation so far recognized in the United States of America, the Sierran Glaciation, which are covered by a volcanic tuff (−1,010,000 years), and some tuffs erupted by the volcanic group of Bracciano, whose age varies between −230,000 and −470,000.* The method of radiogenic argon extends that of radiocarbon, whose possibilities of application are limited to the last 65,000 years. It permits the dating with exactness of prehistoric deposits which are found in stratigraphic relationship with volcanic rocks. . . . It offers particularly favorable possibilities for the application of this method to the definition of the absolute age of the oldest phases of development of humanity in Europe.[87]

Those historically important dates on the Tolfa and Bracciano volcanic groups that Blanc gave had in turn been reported earlier at that same meeting of the I.C.P.P. by Evernden, in an address that was not published; thus Rutten had to cite Blanc as the published source. (Evernden recalled, "I handed out several hundred ditto copies of those dates at that meeting,[88] but no copies have been located.) A search of Curtis and Evernden's records at Berkeley has failed to reveal exactly which K-Ar runs yielded the date of 230,000 years that Blanc cited.[89] Indeed, there are no K-Ar dates from those

*The Bracciano volcanic group belongs to the Sabatini District. Alvarez (1975) conclusively demonstrated that the "tufo rosso a scorie nere" (red tuff with black scoriae), an ignimbrite that crops out around both the Vico and Sabatini Volcanic Districts "including the portions to the south of the Sabatini (dated by Curtis and Evernden at about 470,000 years) was erupted from the Vico Volcano" and not from the Bracciano source as Blanc and Rutten both believed. (Mattias and Ventriglia (1970) first suggested but did not prove a Vico origin for this tuff.) Blanc mistakenly reported the younger date (230,000 years) from Bracciano rocks; the date was most likely derived from volcanoes north and south of Bracciano, as indicated on data cards in the potassium-argon laboratory at Berkeley.

volcanic areas of Italy that match exactly either of the dates published by Blanc, but some are quite close.

Among the Italian dates published by Evernden and Curtis in 1965, those closest to the figure of 230,000 years for the "Volcanic Group of Bracciano" were all near 270,000 years. Furthermore, the corresponding samples were not from the Bracciano Group; they appear instead to have come from Volcano Bolsena, to the north of Bracciano, and Volcano Laziale (Volcano Albano), to the south of Bracciano.* Evernden and Curtis also published five dates for the Bracciano volcano from a single horizon that were fairly close, ranging from 417,000 to 438,000 years.† This cluster agrees better with the date of 470,000 years reported by Blanc, but none of the determinations fit exactly. The log books of mass-spectrometer runs at the potassium-argon laboratory in Berkeley, where the age determinations were made, add further complications.[90] Only two dates from Italy had definitely been run in time to be reported at the I.C.P.P. meeting, which took place August 24–30, 1958: KA 302, run on July 20, 1958, and KA 304, run on April 15, 1958. A third date, KA 334 (no exact run date), was possibly also available in time for the meeting.[91]

It thus appears that only two Italian rock samples, possibly three, had been dated at the Berkeley laboratory in time for Evernden to report them to the I.C.P.P.: KA 302, a date of 2.3 million years on sanidine from the Tolfa volcanics,‡ and KA 304, a date of 431,000 years on sanidine from the "tuff with black pumices" near Volcano Bracciano (now known to have derived from the Vico volcano). The third possible date, KA 334 (from the same horizon as KA 304), was dated at 434,000 years on sanidine. There may have been other samples run and reported, but no record has been

*Data from records of the potassium-argon laboratory at Berkeley: "KA 409, Leucite, 277,000 years, *Location*: Road from Rocca di Papa to Monte Cavo, 200 meters southwest of Monte Cavo, *Volcano Laziale* [Volcano Albano], Italy, *Stratigraphy*: One of the youngest flows of the Latial Complex. Scoriaceous agglomerate with much leucite. KA 457, Leucite, 275,000 years, *Location*: Just north of Aquapendente, *Volcano Bolsena*, Italy, *Stratigraphy*: Flank flow of Volcano Bolsena. Leucite-basalt. KA 855, Leucite, 268,000 years, *Location*: Colata di lava del Divino Amore, *Volcano Laziale* [Volcano Albano], Italy, *Stratigraphy*: Leucitic lava, older than KA 409, younger than KA 348."

†The samples were KA 304 (431,000 years), KA 334 (434,000), KA 345 (438,000—a rerun of KA 334), KA 407 (417,000) and KA 408 (432,000). As noted earlier, the samples of tuff with black scoriae were collected near Bracciano but were erupted from a vent in the Vico Volcanic District to the north (Alvarez, 1975).

‡This sample, KA 302, was reported by Evernden, Savage, Curtis, and James (1964) as normally magnetized as determined by Rutten (1959); either Rutten's hand-compass determination was incorrect, Evernden's date was invalid, or there exists a polarity event, as yet unrecognized, within the presently accepted Matuyama Reversed Epoch. The possibility of self-reversal is least likely.

found to date; thus considerable uncertainty still surrounds Blanc's publication of the dates that Evernden reported at Hamburg.

Rutten remarked in his 1959 paper (he cited no reference) that "it is stated that volcanic activity began in the north and was displaced southward step by step. This area, consequently, seemed well suited to study the paleomagnetism of the early Quaternary—in particular, to date the last reversal of the earth's magnetic field from Reversed to Normal."[92] A brief description of the geography in terms of volcanic centers around Rome placed the volcano Vulsini, with its central crater lake Lago di Bolsena, northernmost. Just to the south of Vulsini lies the volcano Cimino, without a crater lake. "In conjunction to Cimino, just southward is volcano Vico in which Lago di Vico is found. Then follows volcano Sabatini with the central crater lake Lago di Bracciano. At last to the south of Rome, we find the Volcano Laziale with two lateral crater lakes, Lago di Albano and Lago di Nemi [Latin District or Alban Hills]."

Rutten was thus faced with "a problem in the geologic dating" upon discovering that all the cited volcanoes were magnetically normal except Cimino, which showed reversed magnetism.[93] Cimino was neither northernmost (oldest) nor southernmost (youngest) of the group and so raised the question whether Vulsini (the northernmost, with its crater lake Bolsena) possibly belonged to polarity period N_2.* He correctly ruled out such a possibility because "massive tuff flows from the Lago di Bolsena have spread out eastwards" to fill deep river valleys connected with the present Tiber River system, which formed after Tertiary time. He reasoned that Vulsini "must belong to the Quaternary and is then identified with the present Normal period N_1. The tuff flows from the Vico Volcano also descend into an early Tiber River and are [therefore] of more or less the same age as those originating from Vulsini."[94] He was apparently able to combine effectively his rather meager, and in part questionable, data of uncertain attribution to produce the first time scale, albeit locally based and short, through a marriage of radiometry and magnetostratigraphy.

Rutten did not present a rationale for his use of the terms Villafranchian and post-Villafranchian; more important, he incorporated them as an integral part of his time scale but made no men-

*The practice of referring to the most recent period of normal polarity as N_1, the one immediately preceding it of reversed polarity as R_1, the normal period prior to that as N_2, etc., had been initiated by Roche (1953) and followed by Einarsson (1957) in his "Magneto-Geological Mapping in Iceland with the Use of a Compass." Rutten and Wensink (1960) noted that they were "following the usage of the Icelandic scientists," who in turn followed Roche.

tion whatever of them in the text. Most likely he was influenced by Blanc's 1957 article, "On the Pleistocene Sequence of Rome," then the most complete, authoritative work treating the biostratigraphy and chronostratigraphy of the area. Blanc defined "the beginning of the eruptive activity of the Sabatini (Bracciano) volcanic group" as "one of the most notable and useful of the chronological milestones of the Pleistocene of Rome.* . . . The first eruptive materials . . . of the volcano are overlying the marine clays with [the fossil shells] *Ostrea cochlear, Pecten histrix,* and *P. comitatus* of the lower Calabrian, or of the Pliocene . . . as well as the Calabrian marine clays with *Cyprina islandica*."† After citing several more diverse lines of evidence, Blanc concluded that the Sabatini group began eruptive activity "in a climatic period colder than the present, for which the local name *Flaminian glaciation* has been proposed. . . . The Flaminian may correspond to the Mindel second cold phase or to a cold period coming somewhere during the 'Great Interglacial' of Mindel-Riss."[95] Blanc's accompanying correlation chart, which clearly placed the Flaminian glaciation after Villafranchian time, was likely the basis of Rutten's use of the terms Villafranchian and post-Villafranchian.

Rutten assigned Cimino to the ultimate reversed period (R_1), but in this matter he did not enter a discussion, thus forcing the reader to assume Rutten's sustained belief "that volcanic activity began in the north and was displaced southward step by step." It is curious that he found this assumption still tenable after concluding that Cimino's neighbors on both sides were about the same age and younger than Cimino itself.[96] The fact that Cimino belonged to the older Tuscan Province of calc-alkaline volcanism, and not the Roman Province of potassic rocks, was apparently not shown before 1966.[97]

Additional insight into Rutten's rationale behind his 1959 reversal scale may be gleaned from his 1960 paper (which was submitted for publication before the 1959 reversal-scale paper was completed) and also his 1960 paper on the Plio-Pleistocene stratigraphy of Iceland, written with Hans Wensink. In the former he extolled the 1953 paleomagnetic stratigraphic study of Roche, who placed almost the whole of the Villafranchian stage within period R_1, but

*It is also in this statement that we may find a basis for speculating that KA 409, 457, and 855 (discussed above) were erroneously attributed to the Bracciano group instead of the volcanoes to the north and south, from which they come, because they provide a less useful means of correlation than that of Bracciano—wishful thinking by Blanc, perhaps.

†The Calabrian and Villafranchian stages, at that time, were assumed to be essentially marine and continental equivalents on the basis of evidence open to question.

noted that the assignment warranted considerable further study since Roche's data were derived not from a single section but from scattered localities.* In the 1960 paper on Iceland, he acknowledged his debt to the Icelandic scientists, but noted: "As yet only preliminary studies have been made in Iceland, [which are] based on the comparisons of widely scattered local sections. Paleomagnetic dating . . . must be accompanied by geologic mapping, [or else] faulty correlations between widely separated paleomagnetic sections are apt to occur."[98]

Rutten and Wensink had earlier mapped the Elsa-Hvalfell-Armannsfell area and also the Hvalfjördur-Skorradalur area of southwestern Iceland.[99] They were able to correlate the Icelandic sequence of lavas with those of France, *first putting into practice the concept of a magnetically defined, transoceanic stratigraphic horizon.*† They concluded that the Icelandic "Graue Stufe" or "gray series," the uppermost series of plateau basalts, belonged to the "paleomagnetic periods N_2, R_1, and N_1" and contained interbedded tillites (glacier deposits) in all three periods.[100] The ideational frame Rutten brought to the Roman Volcanic Province is further explicated in his summary remark that "the paleomagnetic period N_2 is correlated with part of the Pliocene (Plaisancien-Astien), R_1 with the Villafranchian."[101]

Rutten did not include a second normal period in his scale of 1959, but did speak of the possiblity that "the paleomagnetic period N_2 of the Upper Pliocene" was represented by the "Tolfa volcanics . . . dated by Evernden as 2.4 [million years] old. These were found to present Normal earth magnetism at that date. . . . The geology of these Tolfa volcanics is, however, not well known, so this is not to be taken as certain."[102] The Tolfa date of 2.4 million years, like the others reported by Blanc in 1958, is not in perfect agreement with the date later published by Evernden and his coworkers.[103] They reported only a single date of 2.3 million years based on sanidine (KA 302) in a sample from a "deeply eroded, 'old Quaternary' volcano, 1/2 kilometer east of Tolfa, Italy, on [the] highway."[104] If KA 302 represents the date originally reported by

*Rutten practiced against his own admonitions to others in a number of particulars, including this one, since he did not provide a soundly explicated stratigraphic and structural framework for the Roman Volcanic Province. He appears instead to have exploited Evernden's radiometric dates rather hurriedly, foregoing detailed laborious field mapping as a preface.

†Watkins (1972) noted that "Rutten (1959) initiated use of K-Ar data into geomagnetic polarity history studies" but erroneously concluded that "Rutten and Wensink applied this concept to correlation between lavas of France and Iceland." Rutten's 1959 paper was not cited in Rutten and Wensink (1960) because it had not yet been written at the time of submission; Rutten and Wensink (1960) cited no *radiometric* dates.

Evernden at the I.C.P.P. meeting in Hamburg in 1958, no explanation of the discrepancy with Blanc's and Rutten's date is apparent, especially since there is no surviving copy of the unpublished report of Evernden's that Blanc quoted.[105]

In overview, it appears that Rutten rather hurriedly, and without doing the field work required for a disciplined stratigraphic framework, simply exploited a series of preliminary K-Ar dates provided by Evernden and Curtis.

Jack Evernden was originally introduced to Alberto Blanc through Kenneth Oakley, the anthropologist, of the British Museum. The early dating efforts of the Berkeley group toward increasingly younger rocks was in large measure prompted by the needs and influence of the anthropological community. It can be demonstrated that Blanc's selection of sample sites for dating by Evernden in the Roman Volcanic Province, which eventuated in the polarity scale of Rutten, was an important instance.

All of the earliest samples collected by Blanc and Evernden in Italy save one (KA 302) were collected from sites with especial anthropological and archeological significance. Most notable, and heavily collected, was the "tuff with black pumices" of Volcano Bracciano (later shown to be from Vico) in the area of Torre in Pietra, which was known to be older than the Cassian and Flaminian glaciations and younger than fossil-dated lake beds and fluviatile (streambed) sediments of the important Paleolithic site of Torre di Pagliaccetto, near Torre, and other sites. Blanc's then most recent publication in 1957, written prior to Evernden's arrival in the summer of 1957, was concerned mainly with the Pleistocene sequence of Rome, including "the chronologic position of the local paleolithic horizons in reference to the main paleoecologic Quaternary events" with emphasis on the Lower Paleolithic sites at Torre and the collection of paleontologic evidence in the area northwest of Rome.[106] He significantly noted in the abstract of that paper that the paleontologic and stratigraphic studies at Torre had been supported by the Wenner-Gren Foundation for Anthropological Research since 1956. Blanc was thus highly motivated to produce results from that site and so was inclined to direct Evernden there. Significant potassium-argon dates constituted a formidable contribution, especially at that time, toward guaranteeing continued research support from any foundation.

It is most curious that at no time before or after publication of his historic 1959 polarity-reversal scale did Rutten ever communicate with Curtis or Evernden.[107] Why a man of his demonstrated interests and initial accomplishment made no effort to play a further role in bringing radiometric and magnetostratigraphic data together remains open to question. It is a likely surmise that it was due to his growing recognition that he personally lacked, and could not gain access to, the capabilities in radiometry and rock magnetism prerequisite to such an effort.* The sense that Rutten thought himself an outsider among members of both those communities was confirmed in conversations with a number of scientists who knew him.

It is significant that Allan Cox, Richard Doell, and Brent Dalrymple in the historic first polarity-reversal scale of June 15, 1963, used the three dates reported by Rutten in 1959; they were especially dependent on Rutten's data at that time because all six of the dates they reported from magnetically determined Californian rock samples were older than 0.98 million years.[108] Rutten's two youngest dates, determined by Evernden and Curtis on the Italian volcanics, were "from 230,000 to 470,000 years old." It is also curious that inasmuch as Rutten attributed the "Sabatini Volcano" dates to "Evernden and Curtis (cf. Blanc, 1958)," Cox, Doell, and Dalrymple did not verify those dates at Evernden and Curtis's laboratory;[109] it is all the more curious since Dalrymple was still in residence at Berkeley, and Cox and Doell were only 50 miles distant in Menlo Park. That unlikely situation was largely engendered by interpersonal friction between Evernden at Berkeley and Cox, Doell, and Dalrymple of the Rock Magnetics Project at Menlo Park (a story detailed in Chapter 5). In 1959, Cox and Doell perceived Rutten's opportune attempt to combine conveniently available potassium-argon dates with his questionably derived polarity determinations as a "light" attack on the problem of polarity reversals. They nevertheless "began to feel somewhat competitive at that time" about their research on reversals.[110] Thus Rutten's paper provided further impetus to their already formulated plans for the research program that was to eventuate in the reversal time scale, which in turn unexpectedly became the master key in confirming seafloor spreading in 1966.

*Recall that Tjeerd Van Andel of Stanford University (a Dutchman who was a graduate student under Rutten) said that no mass-spectrometer dating capability was acquired by the Netherlands until the mid-1960's (oral comm., vi-9-78).

Chapter 5

Major Participants, Mentors, and Implementors

TIME SCALES of geomagnetic polarity reversals, after Rutten's rather premature one of 1959, were contributed by a small group of earth scientists who, unlike Rutten, were themselves practiced in both radiometry and paleomagnetism. The polarity-reversal scale evolved from 1963 on as a construct of rock magnetic and isotopic data gleaned from the earth's crust in small increments. Its earliest versions grew from the attempt to demonstrate that on a global scale, contemporaneous rocks possessed the same magnetic polarity; such a demonstration was repeatedly cited as would-be proof for the reality of field reversals. The virtues and potential applicability of such a scale were well known and discussed long before the advent of potassium-argon dating of very young rocks; thus, as might be expected, the earliest efforts by paleomagnetists at establishing synchroneity among widely separated magnetic horizons involved the counsel of Quaternary stratigraphers and paleontologists. A similar collaboration was documented for the case of the young-rock radiometrists at Berkeley in Chapter 2. During the 1950's it slowly became apparent that reversals of the earth's magnetic field occurred with a frequency beyond the limits of discrimination afforded by biostratigraphy; but before useful isotopic dates were available, the paleomagnetists were much dependent on, and in some cases in sustained close communication with stratigraphically oriented researchers in later Cenozoic rocks.

In the late 1950's, after a gestation period several years long, the enigma of reversely magnetized rocks grew to overshadow other questions in paleomagnetism for some, but not for most. The first research program specifically aimed at resolving the field reversal versus self-reversal question was conceived at Berkeley in the late 1950's and later undertaken at the U.S. Geological Survey's research center in Menlo Park, California, by the young geophysicists Allan Cox and Richard Doell, who astutely invited to collaboration Brent Dalrymple, a recently graduated potassium-argon geochronologist. All three were trained at Berkeley and came to contribute the majority of the data during the pioneering stages of the polarity-reversal scale. The Australian Ian McDougall, who had acquired dating skill during his 1960–61 postdoctoral year at Berkeley, also contributed substantially to the evolution of the scale by his collaboration at the Australian National University in Canberra with paleomagnetists Donald H. Tarling, Francois H. Chamalaun, and others. In addition to the major efforts at formulating a scale of reversals in Menlo Park and Canberra, C. Sherman Grommé and Richard Hay, at Berkeley, in 1963 importantly demonstrated the first normal-polarity data point lying within the R_1 period, reported as 1.8 million years by Evernden and Curtis in 1965. Evernden, Savage, Curtis, and James in 1964 matched some polarity data with K-Ar dates, but did not alter the boundaries of the scale then extant.

Doell

The first to enter paleomagnetism among those mentioned was Richard R. Doell; his graduate thesis opened rock-magnetics research at the University of California at Berkeley.[1] Born in California in 1923, schooled in Carpinteria and later in Santa Barbara, he graduated from high school in 1940 with a long-term interest in mathematics and drafting, subjects that were central to his later professional contributions. A portent of his remarkable instrument design and building success lay in his hobby of building and flying model airplanes. An early interest in aeronautical engineering, prompted by a family friend who was an airline executive, led him to enroll at the University of California at Los Angeles as a mathematics major; he intended to transfer later to the California Institute of Technology, in Pasadena. A lack of dedication to studies, abetted by a too-rich freshman social life in a fraternity, resulted in poor academic performance, which was ended by the start of World War II. After failure to gain admission to the Naval Air Force due to an asthma and sinus condition, he was drafted into the Army early in 1942 and sent to the University of Oklahoma to study civil engineering, which occupied most of his military service until discharge in November 1945. The decision to enroll at U.C. Berkeley as a physics major, he recalled, sprang in large measure from his close friendship with two service comrades who were returning there, and because "Berkeley was a state school." It would be inexpensive, he could live on the benefits of the G.I. Bill, and "it was simple to transfer my papers from U.C.L.A."[2] He had no clearly focused academic or career goals but was "more interested in having a good time." His scholarship suffered, and he decided to leave school until a clear career plan emerged to focus his studies. Departure from Berkeley was followed by marriage and a move to Los Angeles, where he worked as a draftsman at a construction company. The termination of the brief marriage and job was followed by relocation to Santa Barbara. There, he recalled, "browsing through the newspaper, I saw an ad for a company doing geophysical exploration. I didn't even know what the name meant, but the qualifications required somebody who had drafting ability and knew a little bit about physics and math." He was employed by the United Geophysical Company as part of a field exploration team conducting operations in the Santa Barbara Channel on several boats working out of Santa Barbara. "Until that point," said Doell, "I had never been interested in geology," nor had he ever studied geology. It was on a camping trip to Yosemite with Duncan Robertson, his project chief, that he first "thought about what was in a rock." A tenure of two and a half years with United Geophysical, two years of which was spent in the Persian Gulf near the island of Bahrain, allowed him to learn much about seismic exploration at first hand. After the Persian Gulf project ended, he was sent to Calgary to calculate the thickness of glacial till deposits from seismic-reflection (sonar) records; he much enjoyed the work and decided to pursue a degree in geology in order to advance within United Geophysical. His former military service friend, William Burke, prevailed upon him to return to Berkeley. In the summer of 1950 he decided to pursue a B.A. degree in geology at Berkeley; the idea of graduate study had not yet occurred to him. Doell was interviewed at Berkeley by Perry Byerly, who examined his undistinguished academic transcript and told him, "I guess you want my advice, young man? That is: you have a good job; keep it." Byerly's advice did not deter Doell. He declared a geophysics rather than a geology major, since he had already completed much of the required physics and mathematics. Because he was on a leave of absence from United Geophysical, they employed him while he was an undergraduate during the summer of 1951 in Santa Paula and again during the summer of 1952, when he decided to enter graduate school.

During his junior year at Berkeley he met and married Ruth Gertrude Jones, "a very eager and good student" in biochemistry who went on to acquire a Ph.D.; living with Ruth was a positive influence on his scholarship, and he received good grades.

Perry Byerly and John Verhoogen, the only two geophysicists in the department, were the most influential in Doell's undergraduate training. His interest in geology led him to enroll in a number of elective geology courses beyond those required of geophysics majors, and as a result he was exposed in classwork to almost the entire faculty. It was in Verhoogen's Geophysics 122 class that he elected to write a senior paper treating the development of an instrument for recording the magnetic properties of rocks; that choice was engendered by his continued interest in the instrument's possible application upon his return to United Geophysical. John Graham's early papers on paleomagnetism and others on geophysical logging techniques also contributed to his early interest in rock magnetism. He noted that United Geophysical "did magnetic

Seven years after receiving his doctorate, and by then employed at the U.S. Geological Survey in Menlo Park, Richard Doell, at left, visited his former mentor John Verhoogen at Berkeley in December 1962.

and gravity surveys as well as seismic;" thus his paper involved "magnetic properties as a correlation tool in oil wells."[3]

Verhoogen recalled: "In 1952–1953, I believe, I gave a seminar on the subject of paleomagnetism which was attended by Dick Doell, who had a great deal of geological experience and was looking for a thesis topic that would not be too remote from geology."[4] Doell recalled that at about that time, John Graham, shortly after completing his degree at Johns Hopkins, gave a talk at Berkeley and afterwards conversed with him. Doell, in turn, visited Graham at the Department of Terrestrial Magnetism of the Carnegie Institute, in Washington, D.C., where he inspected the laboratory equipment and talked about the work of Patrick Blackett's group in directionality studies at Imperial College, London. Conversations and readings at Berkeley, in the seminar and out, touched on the work of Graham, Blackett, Runcorn and Clegg, and Balsley and Buddington, and the possibility of sampling the vast time span represented in the rocks of the Grand Canyon to resolve the polar wandering and continental drift questions. Verhoogen spoke at length about this period.

I was interested mainly in physical properties of minerals and rocks at that time, in that seminar in 1952. We talked about magnetic properties of rocks. . . . We read through the papers of Néel, the French physicist, who had first developed the theory of the acquisition of thermal remanence, and talked about Néel's ideas on the mechanism for self-reversal of magnetization. It was at that time that Doell decided to build a magnetometer here and to start doing what is now called paleomagnetism. He built his instrument in the fall of 1953, while I was away on sabbatical at Cambridge [University]. I saw the work that Edward Irving had been doing on the Torridonian [Precambrian] of Scotland, and I became much impressed by the fact that rocks were reliable carriers of magnetization. [On another occasion he told me, "I hadn't the faintest idea of how you can build a magnetometer."*] It was [Doell who] suggested building a spinner magnetometer; what little money was needed was surreptitiously diverted, as I recall, from an O.N.R. [Office of Naval Research] grant [for a study of] "metamorphic reactions." In those days the National Science Foundation was just starting, and most of the money for research in geophysics came from O.N.R., and I had a grant to do research on a project which had nothing to do with paleomagnetism at all, and suddenly, we found our-

*Verhoogen, oral comm., ix-10-78. Verhoogen had a widely known bent for the theoretical and a near-aversion to the instrumental and experimental. Doell recalled vividly that after he and William Fyfe had "badgered" Verhoogen for months to visit the paleomagnetics laboratory, he finally arrived. "He looked in, looked around, and said, 'Very nice, let's go get some coffee.' He was absolutely uninterested in equipment . . . only in results" (Doell, t. 2, t. 1, iv-17-78).

selves with equipment which had nothing to do with the purpose of the grant. But they never caught on to that, and I never told them. The spinner magnetometer was built with these surreptitiously derived funds, and it was only years later that we got N.S.F. money for the work.*

During 1953, before gaining access to O.N.R. funds, Doell began to design an instrument incorporating a vibrating coil, but finally found the signal-to-noise ratio untenable. He noted that the problem lay with the moving detector, which must be operated in field-free space, isolated from all magnetic influence except that of the sample being measured. After a frustrating period, he became convinced that the design could not be made functional. During this trying interval, Verhoogen became increasingly aware of Doell's need for funding and diverted the O.N.R. funds to the project. A second spinner magnetometer for Doell's geological work was built under contract by machinists in the Engineering Department from plans secured by John Verhoogen. The design was identical to that of the machine being constructed at Berkeley's Lawrence Radiation Laboratory for Ronald Mason of Scripps Institution of Oceanography and Imperial College at London. Doell recalled,

The problem was that it had some highly tuned filters in it, commercial filters and amplifiers, and in order for the thing to work it had to have a very constant speed. It was driven by air jets and had a little air turbine in it—there seemed to be no way it could stay fast enough. It had two separate electronic circuits, and it compared the signals from the reference system on one end of it and the rock on the other. In order for those to be meaningful there couldn't be any phase shifts [between the two circuits]. These were highly tuned circuits [in] which you needed to keep the noise down. Any time the speed changed a bit, the phases shifted like mad; so we had a very bad system at first.†

Upon completion, the magnetometer was installed in a small lab outside of Verhoogen's office in Bacon Hall, but it was soon moved,

*Verhoogen, t. 2, s. 1, viii-11-77. Verhoogen's early interest in paleomagnetism, Doell's decision to build a spinner magnetometer, and the diversion of funds from the O.N.R. grant are recounted in Document No. 13, a letter of i-21-69 from Verhoogen to Edward Irving. The first N.S.F. support for the magnetics studies came in 1959 in response to a research proposal (Document No. 33) entitled "Physical Chemistry and Magnetic Properties of Fe-Ti Oxide Minerals," submitted on ix-15-58. Verhoogen noted (t. 2, s. 1, viii-11-77) that "before 1959, we had been operating mostly on a $5,000 grant from the American Petroleum Institute." At about that time in 1953, another air-drive spinner magnetometer was begun in U.C. Berkeley's Lawrence Radiation Laboratory, funded by Scripps Institution of Oceanography, for Ronald Mason of Scripps. Because it never performed well, Doell never tried to gain access to it and did not include it in his plans (Doell, t. 1, s. 2, iv-17-78). Mason's later work in marine magnetic surveys is treated in Chapter 8.

†See Doell's further comments on this point on pp. 181–82.

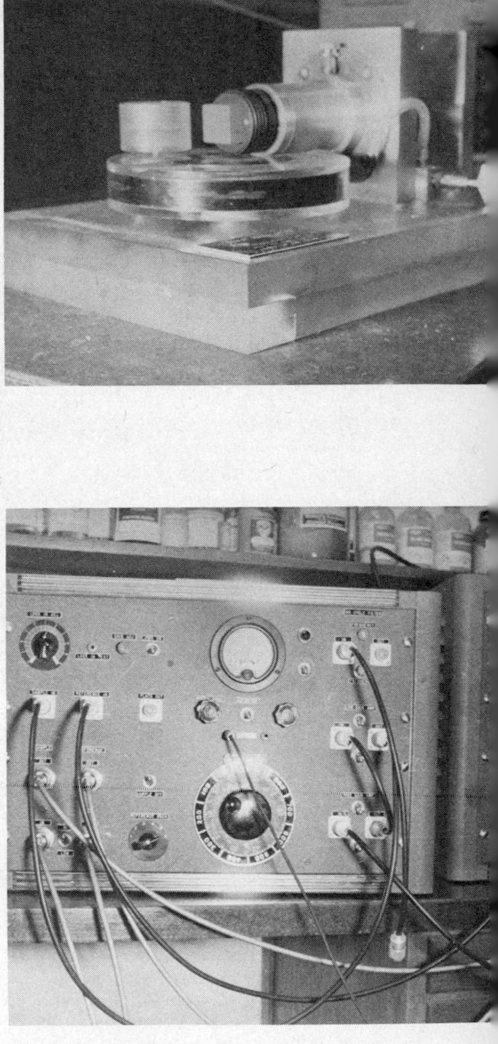

Three views of the first spinner magnetometer built from Ronald Mason's plans by the Engineering Department at Berkeley. At lower right, the original electronics panel used in conjunction with the spinner shown above.

because of its noisy operation, to a room across the hall from that used by Joseph Lipson on the third floor of Le Conte Hall. The physics building laboratory was well supplied with power and accoutrements that enabled Doell to assemble "a good setup with field-free space."[5] The spinner that Doell acquired with the O.N.R. funds was not sensitive enough to measure many of the rock specimens he collected from the Grand Canyon; thus toward the end of his thesis work he gained access to a spinner magnetometer at Standard Oil and Gas Company in Tulsa, which was "pretty much of an exact duplicate of the slow-speed one used by John Graham at Carnegie."

Discussions with Verhoogen led Doell to apply unsuccessfully for a grant from the Bishop Museum in Honolulu in order to study the basalts of Hawaii as a doctoral thesis problem.* "The geology of Hawaii had been pretty well written up. The ages of the rocks were considered to go back quite far into the Tertiary; there were lots of good sections . . . so it seemed like an ideal place to do in probably better detail than Hospers had done, mostly with the Scottish [basalts]."[6]

Doell's orientation in the early and mid-1950's, during his graduate school years and afterward, was much in keeping with the interests of paleomagnetists in general—essentially directionality studies and attempts to explicate the nature of rock magnetism and its reliability. The question of reversal was very new and not then regarded as a suitable dissertation problem, especially for geophysicists with strong geological inclinations and training like Doell. His attention had been earlier directed to the Grand Canyon as a logical site from which to derive directionality data applicable to the polar wander and continental drift questions, in which he had done considerable reading in undergraduate courses.† He noted that his interest in Hawaii "was related to continental drift." Jan Hospers had concluded that there was no discernible drift since as far back as the Eocene, and thus Doell's early collections in California were made with an eye to testing Hospers's conclusion. His thesis was divided into two major sections, one treating the Grand

*Allan Cox later similarly applied to the Bishop Museum and was refused funding.

†At the N.S.F. Rock Magnetism Conference of viii-7–9-54, "Keith Runcorn wanted to go and sample the Grand Canyon. . . . He applied to Louis B. Slichter at the Institute of Geophysics to give him some money for field work, and I think Runcorn was very surprised when he learned from Slichter that the Institute had already given me money to send Doell to the Grand Canyon to collect and that Doell, in his thesis, had already determined several poles from the Precambrian for the Grand Canyon section" (Verhoogen, t. 3, s. 1, viii-16-77; Doell, 1955b).

Richard Doell coring basalts on the Hawaiian Islands.

Canyon sequence and the other the blue sandstone and siltstone of the Miocene Neroly formation near Livermore, Martinez, and Santa Cruz, California.[7] Doell noted that the Neroly problem "figured as the good half of my thesis," which revealed that the rocks got their magnetism from a chemical coating acquired long after deposition. Chemical remanent magnetism was at that time a new, little understood process, and chemical experiments came to occupy much of his time.* He recalled that during his graduate studies, "there were a lot of things on reversal, but that was a problem

*Doell's sustained interest in chemical magnetic remanence during his tenure at the University of Toronto is evidenced in a letter to John Verhoogen of xi-10-55 (Document No. 14C).

that I was *not* particularly into."[8] Doell built the first demagnetizing equipment at Berkeley in the course of his thesis work, but "it only went up to 10 oersteds." (He also collected from some lava flows in the Sierra Nevada, but the samples were lightning-struck and beyond the limits of his weak demagnetizer.) "We at first thought that the magnetism in the Neroly [formation of Miocene age near Berkeley] was reversed, because we calibrated the thing exactly backwards for a while, and that was pretty exciting, but it turned out when we recalibrated the instrument that it was normal." The Neroly problem was originally selected because of Doell's familiarity with the geology and also to determine if John Graham's fold test was valid.[9] During the years of his thesis work he recalled the coffee conversations and seminars centered on the work of the English directionalists and the exciting new reversed rock data being contributed during the early 1950's by Graham, Balsley, and Buddington.

Doell had returned to Berkeley to pursue a bachelor's degree in

Doell in the Grand Canyon collecting rocks for his doctoral dissertation in the summer of 1954.

order to advance his career in oil exploration, but his superior academic performance had led some on the faculty, notably Perry Byerly and John Verhoogen, to invite him to graduate study. He entered graduate school as an alternative to resuming work with United Geophysical, which was anxious to have him return. His decision to stay in research and seek a university position upon graduation was arrived at slowly during his graduate studies.

In the spring of 1955, shortly before completing his Ph.D. in June, Doell began job hunting in a very limited geophysics market; after an unsuccessful interview at Pennsylvania State University, he responded to an advertisement for a geophysics faculty position at the University of Toronto, received a telegraphed offer from J. Tuzo Wilson (who enters this history in Chapter 7), and accepted it. Within that week he was tendered a second job offer by Robert Shrock of the Massachusetts Institute of Technology, after delivering an invited lecture there. He preferred the position at M.I.T., but, Ruth Doell recalled, "the people at Toronto would not let him out of his commitment."[10] Shrock assured Doell that he would hold the M.I.T. position open for him until he completed the one-year contract at Toronto.

Upon Doell's arrival at Toronto in August 1955, he knew that his stay would be short and thus he "helped mainly in putting together a paleomagnetics laboratory with Tuzo Wilson's graduate students. There was work in rubidium-strontium and also potassium-argon [dating] being conducted in the department; [Richard D.] Russell was trying to find out how Reynolds's glass machine was doing." His main interest during that year was centered in "a restudy of the varved clays of the east."[11]

Doell revised his Ph.D. thesis for submittal to the American Geophysical Union for publication, completing the work by October 1955. In a letter to Verhoogen, he noted that "Professor Wilson and Professor Watson are quite interested in rock magnetism (no doubt due to a recent talk by Professor Blackett on the subject), to the extent that we are now well along on the construction of a magnetometer (very similar to the one at Cal)."[12] It is noteworthy that in that first letter to his mentor after departure from Berkeley, two and one-half of the four pages were occupied by a discussion of magnetometers, gravimeters, and new developments in the field of instruments; much of the discussion was centered on the instrumental strengths and weaknesses of the department at Toronto. That emphasis was quite in keeping with his early development and strongest interests, and with the talent praised by several interviewees. "Doell's greatest genius is conceiving of and building instruments—the best I've seen," said one, and another, "Doell is a genius with instruments." Correspondence with Verhoogen during the Toronto year also included discussions of Kittel's relation of ferromagnetism and antiferromagnetism in hematites, the significance of the directional grouping of the Grand Canyon rocks, reconciling paleobiogeographic and paleomagnetic data, and a heavy teaching load that detracted from his efforts at research and "getting the spinner magnetometer operating properly."

The Toronto group "appeared to be getting committed" to doing some studies on glaciers on Ellesmere Island, Canada, during the 1957–58 International Geophysical Year. In preparation, a group of about a dozen under the leadership of John Jacobs embarked for a stay of more than two months on the Salmon Glacier in British Columbia, close to the Alaskan panhandle, during the summer of 1956. Doell was asked to go along to set up the seismic work and train people for the work on Ellesmere. He wrote Verhoogen: "Of course I know nothing about seismic shooting on ice, but as you once advised, I shall get a book! What interests me most is the possibility of taking along the Worden gravimeter. . . . I think that we shall be able to [map] the ice-rock surface fairly accurately . . . (using the seismic stuff for control at a few points). The density contrast is wonderful!"

His main interest in the gravimeter apparently stemmed not from questions of pure gravimetry or the machine's application to seismology and structure but rather from how these questions applied to rock magnetism, as evidenced by his "plan to try a method of getting the polarization of the basement rocks."* Such a method "had to assume no polarization in the seds overlying the basement rocks—where the 'sed' [sedimentary rocks or sediments] are ice this should be a greatly more valid assumption."[13] Doell's seismic work on the ice that summer of 1956 and a second shorter stay during the following summer before returning to M.I.T., said his wife Ruth, "did not turn him away from paleomagnetism but only led him to recognize the dullness of seismology work."[14] The resulting paper, "Seismic Depth Study of the Salmon Glacier, British Columbia," was not published until February 1963.

He left Toronto to commence the academic year at M.I.T. as an

*In principle, if magnetic anomaly data are known for a region that has been mapped gravimetrically, the direction of polarization can be determined.

assistant professor in the fall of 1956, and "tried streamlining what had been a two-year course at Berkeley into a one-year geophysics course at M.I.T. and set up a graduate course in paleomagnetism."[15] During the second year, he conducted "a graduate seminar on the reversal problem; I had all the students reading Néel." Doell recalled:

That year we did reversal problems, self-reversal and field reversal and the controversy. . . . Having two markedly different theories, as it were, to explain one phenomenon was an awfully fun thing to teach and talk about. It was for that course that I'd been sort of collecting data that [in 1960] got in the paper Allan [Cox] and I did.[16] I had been abstracting and going over all of the literature and compiling [paleomagnetic] poles and trying to sort out what was good and bad. While I was at M.I.T. in 1958, I gave a paper at the A.G.U. [American Geophysical Union] summarizing what I was learning from my compiling.[17]

Doell was invited to teach the summer session at Berkeley in 1958; it proved to be a most eventful reunion with Allan Cox. Doell continued: "Allan and I discovered we'd been doing essentially the same thing . . . so we decided to collaborate." Cox and Doell's "Review of Paleomagnetism," published in 1960, was the most comprehensive treatment of the subject to that date and did much to bring the potential usefulness of paleomagnetism to the attention of the geological community and to "make the subject more respectable."* Its publication by two "newcomers" aroused some ill feeling among paleomagnetists; especially irked were members of the directionalist group, including Edward Irving, at Cox and Doell's emphasis on reversals and their characterization of the directional data as equivocal.[18]

The review paper grew from a friendship begun during 1955 in the department of geology at Berkeley, when Cox was a senior undergraduate and Doell was completing his Ph.D. dissertation. Toward the end of the academic year, John Verhoogen introduced the two; Cox recalled,

Dick Doell was just finishing up, and he showed me around the lab and got me interested in the equipment. . . . He and I went out on a field trip . . . to Table Mountain over by Jamestown [California] and camped out. We collected some samples from Table Mountain, brought them back, and measured them. That was my first sampling trip. So I could see the way

*James Balsley had originally been invited to write the review paper on paleomagnetism for the Geological Society of America but declined due to his time-consuming administrative duties in the U.S. Geological Survey.

the techniques worked. Dick and I overlapped enough to have a good personal contact at Berkeley. Then he went on to Toronto and then eventually to M.I.T., but we maintained contact all the time.[19]

Doell did not find the social scene at M.I.T. to his liking and made few friends while there. He also came to feel that his dedication to teaching, which he considered an important function, was not likely to contribute to his professional advancement; furthermore, the teaching duties did not allow him to pursue enough research. His sense that a professorship placed instruction above research appeared in conflict with the "cavalier attitude of the M.I.T. staff about teaching duties."[20] Late in the summer of 1958, after much talk and planning collaborative research in paleomagnetism with Cox, Doell told his wife that upon completion of his third year (1958–59) at M.I.T., he would like to move to Menlo Park to begin a rock-magnetics project at the U.S. Geological Survey. That incipient plan had grown in part from Cox's long-term part-time employment by the Survey and encouragement from the chief of the Geophysics Branch, James Balsley. Since early in his graduate studies, Doell had also been encouraged in his paleomagnetic research by Balsley. Doell shared Cox's apartment in Berkeley; they spent most of their time collaborating on the "review of Paleomagnetism" paper, "recalculating Fisher statistics on an old Marchant calculator and backpacking.* I don't think the reversal problem was *the* central problem at all. There were several problems we wanted to do; Allan had been working on the stratigraphic approach, trying to get dated lava flows from paleontology." They were interested, as were most other paleomagnetists at that date, in the "general continental-drift thing."[21]

Doell's third year at M.I.T. was begun but not completed; in November 1958 a malignant melanoma was discovered on both of his legs. Surgical removal of the lymph tract of his right leg resulted in moderate edema. The operation was particularly depressing to him because he was usually very active physically. The original prognosis was not encouraging, and he and Ruth decided to return quickly to California, where both their families lived.†

The idea of returning to California to work with Allan Cox had grown in Doell's mind long before the discovery of his illness. In a

*Their recalculations led to the discovery of mistakes on the part of more than one well-reputed geophysicist, but they "feared to tell people that they'd made mistakes."

†Document No. 15, letter of xi-24-58 from Cox to Balsley, describes Doell's surgery and Cox's paleomagnetic work and status with the Survey while still a student at Berkeley.

letter to Balsley in 1958, Cox, who was employed by the Survey and in sustained communication with Balsley by that time, explained his role in bringing Doell to write Balsley about a position at Menlo Park with the Survey:

> Dick stayed with me while teaching at the University [U.C. Berkeley] this summer. We have long had the vague wish to work in the same vicinity (—for a while we considered the possibility of going to M.I.T.). In line with this, we had a good time this summer designing an ideal magnetic lab we would like to build somewhere sometime (total cost: $12–15,000, plus a year of our time). As we were speculating, it occurred to us that it might conceivably be possible to actually do something like this at Menlo Park—you had mentioned the possibility of some laboratory facilities here. Even if we worked only part time on magnetic work, we could, I think, get a pretty good lab going in a year on fairly limited funds. Dick, as you know, would be hard to beat as a person for this sort of work. [Doell had met Balsley as a first-year graduate student at Berkeley and had kept up contact with him at M.I.T., where Balsley had given seminars, and at meetings of the American Geophysical Union.] A thing which once bothered me about the possibility of going to M.I.T. was that Dick's and my interests were too similar. Being together over the summer, we find this no longer true. Besides, the *dangers of isolation* now seem more important to both of us.[22]

That letter sets out clearly their complementary relationship, their plan for collaboration in rock magnetism, their desire to work in the Survey, and their crucial dependence on Balsley's approval of such a plan. The letter also noted that Louis Pakiser, then chief of the Ground Survey Section of the Geophysical Branch, had favorably impressed Doell with the suitability of the U.S. Geological Survey as a place to do research and thus had furthered his "resolve to sound the Survey out about the possibility of working in Menlo Park."[23] In eagerness, Cox assured Balsley that if Doell, because of his illness, were "unable to complete some project in magnetism," Cox would "of course accept the responsibility for doing so."*

Cox

Allan V. Cox was born on December 17, 1926 in Santa Ana, California to working-class parents who desired that he attend a Lutheran parochial school in the sixth through eighth grades in the

*Document No. 16, letter of ix-30-58 from Cox to Balsley, also touches on work in Idaho with Harold Malde sampling for magnetic determinations in the Eocene rocks of Washington, and

city of Orange, some five miles from his home; that "unrewarding" experience led to his "drifting away from religion." He continued on to high school in Orange to remain with his circle of friends, but transferred to the local high school in Santa Ana in the tenth grade. One of his early "great intellectual excitements . . . came when the teacher of chemistry had to go into the Navy . . . and had only six weeks to spend teaching chemistry." The class thus worked "three or four hours a day, five days a week on chemistry." Cox recalled that "just plunging into something with enthusiasm and working hard at it was a big experience for me. I decided I wanted to become a chemist. That's what I majored in when I went to college." Having had that teacher was the crucial factor in his first choice of a career; "He was an inspired teacher—really. I think it was more the personal contact than being attracted to the subject." Cox came from a home in which the garage was "always full of electronics chassis" accumulated by his father who, though not educated in electronics, was given to repairing radios and doing electrical wiring for friends. His mother graduated from high school in the same year he did—she had attended night school while raising a family and doing creative writing. Both parents were well disposed toward education, and permitted him to build a chemistry laboratory in the basement of their home. He spent much time in the library reading both fiction and nonfiction, which included "most of the books of Freud that were in the library." He "always read a lot" and bought books with his meager income.[24] His teachers were aware of his interests and ability and recommended that he attend college. "That's where I got that idea, not from home at all," he recalled. Singing the hymns of Bach at the Lutheran church with his sister, Lois, was a prelude to his interest in classical music, which led to his constructing an early high-fidelity music system while in high school.

Immediately after graduation from high school at the age of seventeen, Cox enrolled at Berkeley during the summer session of 1944. After that first quarter he joined the Merchant Marine for three years, where he was trained in radio telegraphy and to a lesser extent in electronics. Much free time aboard ship provided the opportunity to read across a broad front including philosophy, anthropology, science, and whatever else appealed to him.[25] Upon discharge in 1948 he reenrolled as a chemistry major at Berkeley,

---

the remark that Cox and Doell had "embarked on a review of paleomagnetism" for eventual publication in the Geological Society of America *Bulletin*.

where he felt "sort of naive"; although he liked the content of chemistry intellectually, he had not formulated any definite career plans. In addition to chemistry, he had considered several other undergraduate majors, foremost among which was philosophy. "I spent a lot of time trying to read philosophy but saw that it didn't seem quite logical and tight enough to be satisfying to me at that age."[26] He recalled, "I think one reason that I switched to earth science later was that I saw the kind of life-style you could have in earth science, a combination of field work plus lab work. That was so much more attractive than a career working in a lab all the time. That was one of the reasons I shifted." During those undergraduate years at Berkeley, he was apprehensive about the rise of McCarthyism, which "tended to demoralize" him as a student in the sense that he "couldn't imagine functioning very well in a society that would end up being the way that McCarthy seemed to be going." His politics were "fairly far toward the left" and he felt a feeling of "alienation . . . from what was going on around me. . . . I think at that time I felt I was probably on some 'Un-American' list that would have prevented me from functioning very well as a scientist." His falling morale was reflected in his grades. At about that time he befriended Ken DeWitt, his laboratory partner in his quantitative analysis course, who had worked the previous summer as a field assistant with the U.S. Geological Survey.[27] In the winter of 1950 DeWitt arranged a dinner meeting for a small circle of friends, which included Clyde Wahrhaftig, who met and was favorably impressed by Cox. Through that contact Wahrhaftig offered Cox employment as his field assistant in Alaska during the summer of 1950.[28] That summer of work under Wahrhaftig in Alaska was a most crucial turning point in Cox's career; the poignancy of that experience is clear in his recollections:

That was a group of very high morale people who worked in the Alaska Range [Troy L. Péwé, David M. Hopkins, Thor N. V. Karlstrom, Henry Coulter, John R. Williams, Ernie Muller, and Clyde Wahrhaftig[29]] and in all parts of Alaska. It opened up a whole new world for me. It was a chance to work really closely with scientists who were first-rate people on an exciting project that I could understand. At Berkeley, being a student was a very impersonal thing in those days. Being in the chemistry department as I was, one didn't know professors very well. The typical relationship with your advisor would be that you'd leave your study list, he'd sign it, and you'd pick it up again from his secretary. So it was a very impersonal kind of intellectual atmosphere, at least in the circles that I knew prior to com-

Allan Cox examining rock glaciers at the head of Virginia Creek, Alaska, in August 1950.

ing into geology—whereas in Alaska, I had the chance to work with Clyde Wahrhaftig, who was a great intellect and a very stimulating person. That began opening up the world of geology to me. I became aware of a group of people working on important problems who are intellectually first-rate. They were friendly, warm people and they still constitute the largest group of my friends.[30]

Wahrhaftig was highly impressed with Cox's performance in handling horses and general resourcefulness and also regarded him as a fine intellectual companion, versed in Proust and Bach, who could also "cook superbly," a singularly important skill among

Clyde Wahrhaftig, now a professor at Berkeley, was the major influence in starting and shaping Allan Cox's early career in geology. Cox worked for three field seasons with the Alaskan group that included Wahrhaftig. The group was greatly concerned with glacial correlation and instilled in Cox a sense of the power of a new, widely applicable means of correlation. While Cox was still an undergraduate, he collaborated with Wahrhaftig on a paper that stimulated research in a field quiescent for decades. The photograph was taken in 1952.

isolated field geologists!* Cox recalled, "I didn't think geology was a very deep subject. It seemed very qualitative." He found that his newly ignited interest in geology grew greatly after that first summer in Alaska, which prompted him to read in the subject for pleasure. He continued to "conceive" of himself as a chemistry student but found increasing difficulty in applying himself to studies, with the result that his grades suffered: "the 1950–51 academic year was a real disaster." The difficulty grew from his investment in reading geology instead of the required chemistry for his coursework and also from "worry about the political scene."† In the spring of 1951, he felt he was undergoing some sort of crisis, so instead of taking his final examinations he went for a month-long solitary hike in the Sierra. "A bunch of incomplete grades" resulted in the loss of further deferment, and he was drafted.[31]

Before reporting for active duty in the fall of 1951, he worked

*Document No. 18B is the U.S. Geological Survey 1950 field season report by Clyde Wahrhaftig to P. Lewis Killeen, chief of the Alaska Section in Washington, D.C., xii-4-50. Cox was recommended for future work with the Survey; he was rated as excellent in all categories save that of heavy work due to a knee injury early in that field season.

†Among Cox's geological readings, he regarded *Principles of Geology* by Gilluly, Waters, and Woodford as an influential and profound introduction to the subject.

again for Clyde Wahrhaftig.[32] Unlike the first summer, spent largely in isolation with a party of five, the second, centered in McKinley Park at the base camp, involved contact with many geologists and the opportunity to attend a geologic conference. During the first summer he had become interested in the study of rock glaciers in the Alaska Range, in which Wahrhaftig had begun an outstanding study. He recalled, "I was aware then of the challenge of trying to explain rock glaciers and understand them, but by the second summer I really got into it and had my own ideas. That's one of the most interesting and exciting scientific projects I ever worked on. By the end of the second summer I think I'd decided that I should really go into geology. At least I knew I didn't want to continue in chemistry. But when I got back from Alaska, I had to go into the Army." He was inducted on October 10, 1951.*

During his Army service from October 1951 to October 1953, he was assigned to radar microwave school and later to Signal Corps duty at Fort Monmouth, N.J., to be trained in the operation and maintenance of microwave communication equipment.[33] During that time he corresponded with Clyde Wahrhaftig, who sent him a variety of books on geological topics, including the regional geology of the area where he was stationed. Wahrhaftig's warm personal approach, both in furthering Cox's interest in the subject and in filling his long-felt need for a learning experience unlike the detached, impersonal one he had known as a chemistry student, was vital in his decision to "go back to school and work hard as an earth science major" by the time he was discharged. Cox philosophically recalled: "The Army was the drabbest sort of nonintellectual experience . . . but still I got a lot out of the Army. The main thing was to see that life could be very drab. . . . I think it was a good cure for the loss of morale I'd felt as a student. I saw that if one wanted to be despondent and not do much with his life, it was possible, but I discovered that the bottom of the barrel really wasn't that interesting!"

He was thus well primed on discharge to return to his studies at Berkeley. Like Richard Doell, Cox found it "more convenient to go into geophysics than geology" because of a rich physics and mathematics background that included honors math courses; he was

*Cox, t. 1, s. 2, v-2-78 (quoted). Document No. 18E is a letter from Clyde Wahrhaftig to the Selective Service System, Local Board No. 47, in Berkeley, ix-51. He asked the board to reconsider Cox's draft classification, noting his "outstanding geologic work" and "considerable promise." Wahrhaftig noted Cox's change to a geology major and hope that Cox would be available for work with the Survey again in the summer of 1952.

"only one course short of either a physics B.S. degree or a chemistry or a math B.S. degree." In conversation, Cox repeatedly emphasized the contrast between the cold impersonality of the chemistry and mathematics learning environments and the "very personal kind of department" of geology, in which Perry Byerly would punctuate his lectures in geophysics with stories about Beno Gutenberg and other classical figures he had known very well. After a very successful first midterm exam, Byerly sent a note asking Cox to come by to discuss career plans; it was out of these discussions, while an undergraduate, that he first considered a future in geophysics. Cox was careful to note, "It would be nice if I could say that I liked the intellectual framework of geophysics more than chemistry, but I'm afraid at that age, my reasons were mainly just personal."[34] Here, as in the earlier case of Cox's personal contact with Wahrhaftig, the human element displaced intellectual factors in forming career plans.

Clyde Wahrhaftig's interest in rock glaciers had led him to participate in a committee effort within the Survey to codify their terminology; he had also published a paper on frost-moved rubbles by the time of Cox's first field season. Wahrhaftig's enthusiasm for geomorphology was contagious, and Cox soon found himself drawn into the research on rock glaciers. Wahrhaftig considered Cox an "intellectual giant" who was "way ahead" of him and much closer to his own broad range of philosophical and artistic interests than anyone else in the field party. While Cox was in the Army, Wahrhaftig worked on a paper during 1951 and 1952 summarizing information on rock glaciers and doing some field studies. He was in the process of revising a final draft in the spring of 1954, just before leaving for the third field season with Cox in Alaska. Wahrhaftig recalled,

> Cox contributed a lot of original ideas to my thinking about rock glaciers in the field and a lot of observations. At the end of that summer I suggested that, if he didn't mind, I wanted to make him a coauthor of that paper. Cox said that he'd not be a coauthor of anything unless he had actually participated in the writing and the research. His contribution to that paper is probably what got me the job at Berkeley. He insisted on doing something that had not even occurred to me, which was looking into the physical theory of glacier flow, which I was not aware of until he found those papers on ice-deformation theory. Basically I contributed the qualitative geologic observations, except for the statistics, but the physical theory is largely Cox's contribution.[35]

Upon return to undergraduate studies in the fall of 1954, Cox would often work at the Survey on weekends without pay, gathering data and determining glacier dimensions from stereoscopic aerial photographs of the Alaska Range. Cox had read Harold Jeffrey's estimation of the viscosity of a lava flow using the flow velocities, the slope of the bed, and the lava thickness, from which the stress could be derived and the viscosity then calculated. "Cox also had the brilliant intuition that the more rapidly moving upper part demonstrated that the behavior of rock glaciers is similar to ordinary glaciers," said Wahrhaftig.[36] The rock glacier paper, written by Cox while still an undergraduate, is now regarded as a classic treatment of a recondite subject.[37] It stimulated much research in a field that had been quiescent for decades and represented a significant growth stage in Cox's career.

The influence of Wahrhaftig and others working in Alaska was important in forming interests and ordering priorities among the many problem areas that a young geologist such as Cox might choose to pursue. Wahrhaftig regarded himself mainly as a Pleistocene geologist, concerned with correlating glacial histories, especially in the Alaska Range, as were other geologists of the Alaskan group with whom Cox worked for three field seasons.* Long-term uncertainties in such correlation, and the frustrations they engendered, occupied a central place in the thinking of the Alaskan group, and much of their conversation was concerned with correlation. This exposure had a profound effect on Cox, instilling and heightening in him a sense of the value of a better, more widely applicable means of stratigraphic subdivision and correlation of the Pleistocene.

The first time I became aware of a dating problem . . . on a very small scale was in working with rock glaciers. We needed to get criteria for telling active from inactive ones. There were some that hadn't moved for a long time and others that were still moving. We observed that the ones that weren't moving were covered with lichens, so we knew they had not been active during the time the lichen was growing. It then got to be a question of how long does it take a lichen to get going on a new surface. That was the first time I was concerned with trying to date things—when did the rock glaciers become inactive, how fast did they move. . . . Through that

*Letters of iv-10 and iv-17-51 from Wahrhaftig to Troy Péwé show the deep concern of the Alaskan group with glacial chronology and correlation. Wahrhaftig, in the second letter, concluded, "As you see, we have a terrific problem in correlation to solve" (Documents No. 18F and 18G).

concern I then became more aware of the problem of dating young rocks, recent rocks, and of dating Pleistocene glacial advances. Troy Péwé was working on that, so I became aware of the importance of dating glaciations through our contacts with him.[38]

Cox also learned of other stratigraphic and dating efforts in the Pleistocene and through his long-term contact with David Hopkins. Wahrhaftig, an old friend of Hopkins since their Survey association in 1942 in Seattle, had written to him in 1952 about Cox, then stationed in New Jersey. Hopkins's main interest was Pleistocene geology and archeology, and he had "struggled with dating problems all of his life"; he invited Cox to his home in Washington, D.C., for weekend stays.[39] Hopkins was at that time at work on his doctoral dissertation at Harvard on the Quaternary geology of the Seward Peninsula of Alaska, including glacial, periglacial, and volcanic areas. The visits to Hopkins's home, which provided relief from an "intellectually barren Army experience," included talks far into the night on a broad range of topics. Cox recalled, "I'm sure that Dave had a much deeper influence on me than I can even remember. He maintained the contact and was always interested in me when I was in school. . . . Through the interest of Wahrhaftig and Hopkins in Pleistocene dating, I became interested in that very early in school."[40] During that interval the Alaskan group, including Wahrhaftig, Hopkins, Péwé, Coulter, Karlstrom, Williams, Muller, and others, were "aiming for a glacial map of Alaska." Under Péwé's leadership, they eventually published a circular entitled *Multiple Glaciations in Alaska—A Progress Report* in 1953.* Cox had also been influenced by reading Elliot Blackwelder's work on glacial advances on the east side of the Sierra Nevada and had, immediately before his Alaskan field work of 1954, "spent the whole summer hiking . . . and visiting the classic localities that Blackwelder had described, . . . trying to recognize his evidence for four glacial advances on the east side of the Sierra."[41] Early on as a graduate student, he read the two volumes of J. Kaye Charlesworth on Quaternary geology that were mainly concerned with global cor-

*Recall earlier commentary on the preference for research problems in old rocks by isotopic geochronologists and the prestige that seemingly accrued with increasing rock age. Among the authors of the multiple glaciations paper, "there was a bit of ego involvement in that certain people in the group felt it very important that they have an older sequence than anybody else. There was a lot of jockeying about for the oldest and longest sequence of glacial deposits to be recognized" (Wahrhaftig, t. 1, s. 2, ii-18-78). David Hopkins also noted that until recently, Quaternary geologists traditionally were regarded by most in the geologic community as working on surficial, less important problems that did not merit the priorities accorded studies in older rocks (D. M. Hopkins, oral comm., iii-13-78).

relations. The plethora of deposits with their romantic local names appeared like "something out of the Middle Ages"; he viewed Dave Hopkins as very much in the middle of what he recognized as a "horrible problem." "I think by the time I had become interested in reversals, almost as a kind of hobby, I'd developed and maintained a serious interest in the correlation of Quaternary deposits."[42] A sustained interest in the methodology of correlation was demonstrated repeatedly during later stages in Cox's career.

Upon entry to graduate school in September 1955 Cox was undecided about a dissertation subject, but was somewhat attracted to gravity because of his belief that it was the major driving force in tectonic processes; he was not encouraged to pursue such a study by the faculty. John Verhoogen and others on the faculty were early of the opinion that Cox would be a likely successor to Richard Doell in the paleomagnetism laboratory. Cox much admired Perry Byerly and was stimulated by his lectures, however he viewed seismology as a major tool not only in oil exploration but also in detecting underground nuclear blasts in foreign countries.* He thus regarded it as "mainstream" and "Big Science," and found, too, that it was not in keeping with his desire to pursue studies in "something that was slightly offbeat" that he could have to himself.

For a seminar on volcanology given by Howel Williams, Cox wrote a paper on sources of volcanic heat; he also treated volcanic morphology and stratigraphy, which later proved very useful when he began to use volcanoes as "magnetic tape recorders."[43]

A most important backdrop against which to view Cox's selection of a thesis was the question of continental drift. In the U.S., the drift question was not often openly raised in seminars, symposia, and conferences, and when discussed found few outspoken advocates during the 1950's. The departments of geology and paleontology at Berkeley provided a slight exception in the person of John Verhoogen, who has already been described as holding intellectual sway among a most notable faculty of earth scientists. Verhoogen, graduated from Brussels in 1933 as an Ingenieur des Mines and trained in 1934 at Liège by Paul Fourmarier as an Ingenieur-Geologue, was an early advocate of drift who recalled that

*In contrast, Jack Evernden left his professorship at Berkeley for AFTAC (Air Force Technical Applications Center) because he wanted to do something for world peace (oral comm., J. F. Evernden, ix-10-80). Researchers there had been attempting for several years to devise a method of discriminating seismic waves generated by nuclear explosions. Within a year Evernden had drawn upon his training as a seismologist to provide a solution. John Verhoogen remarked, "With Evernden's contribution it became possible to police nuclear explosions. It [will] condition all treaties . . . regarding nuclear explosions" (oral comm., iii-19-82).

"just because Fourmarier was opposed to it, I became favorably disposed to it, just on the general principle that you don't believe what your professors tell you." Shortly after joining the Berkeley faculty, he spoke favorably on drift at a Paleontology Department seminar in 1948 that included off-campus scientists; in contrast to most geophysicists at the time (including Harold Jeffreys of Cambridge, the most influential opponent to drift), Verhoogen saw "no physical objections to it" and thought that "the lack of a mechanism did not preclude it."[44] His formal research had by 1944 led him to advocacy of convection within the earth's mantle, an idea that was prompted by his consideration of David Griggs's presentation on mountain building at the 1939 meeting of the International Union of Geodesy and Geophysics; also while working in the Congo (now Zaire), he had read Arthur Eddington's *Internal Constitution of Stars*, in which the 1924 theorem of Von Zeipel was discussed. The theorem held that a rotating body such as a star or planet with internal heat sources could not be in a state of hydrostatic equilibrium, and thus some kind of flow must occur; it there-

Allan Cox at Buckeye Hot Springs, Mono County, California, on the eastern flank of the Sierra, in early September 1955, at the time of his return to graduate school at Berkeley.

Allan Cox and Nancy, at a cabin on Cody Creek, Alaska, in July 1954.

fore supported the theory of convection currents, which were held to be the cause of continental drift.

The paleontological community was particularly opposed, especially those with interest in the later Mesozoic and Cenozoic record of North America. Their data, drawn from a continent undergoing essentially east-west movement during that interval, reflected little of the ecological change that would have resulted from a great north-south continental excursion.

The question of drift was one of the major topics raised at meetings of the "Berkeley Geology Club." The club was started mainly by Allan Cox and Mark Christensen* "partly as a reaction against the Geology Department seminars, which were very technical and dry and weren't concerned with what the students considered the

*Former Chancellor of the University of California at Santa Cruz, now Professor of Earth Resources, U.C. Berkeley.

big ideas in geology."[45] They usually met at a place where they could get beer and invited people, such as Philip B. King of the Geological Survey and Adolph Knopf, whom they thought would be "concerned with larger ideas." During the first series of meetings at which student papers were delivered, they decided to take a vote on the question of drift, and "the ayes had it." The advocates of drift were a strong majority. Some of the faculty were "outraged" on hearing of this.* Cox recalled that the departmental seminars, in contrast to those held by the students, were generally narrow; but Verhoogen tended to provide the exceptions, as on his return from an I.U.G.G. meeting in Rome in 1954, when he gave

> one of the most lucid talks I've ever heard on the larger issues in geophysics. Everything seemed connected. It was a beautiful exposition. He was very much aware of the issue of polar wandering. . . . He was favorable to the idea of convection in the mantle, which he regarded as a key idea.† . . . For my own research he suggested several topics. One was to produce chemical remanent magnetization in the laboratory. [This topic was likely prompted by the earlier work of Doell, who discovered chemical remanent magnetization for the first time from field evidence.] Verhoogen got an exciting letter from Phil Du Bois in which he reported that Ernie Deutsch, and I think also Ted Irving, had picked up paleomag evidence [from the trap (basalt) rocks of Decca] for the northward drift of India. But at that time, there weren't good Eocene [paleomagnetic] pole positions from elsewhere in the world. Their pole position for India fell in the South Atlantic, somewhere off Florida. The simplest interpretation was northward drift of India.[46]

This finding excited Verhoogen greatly, and he suggested that Cox try to obtain a good North American Eocene pole position

*Letter from Cox to Doell of i-13-58 is Document No. 19. Cox recalled: "The Geology Club has been a great deal of fun. The reaction of FJT [Francis J. Turner] and the faculty was that they didn't care what we did, as long as we didn't *involve* them in the proceedings. . . . Several of the faculty were P.O.'ed because they weren't invited to the Continental Drift Symposium. One of the reasons the Club has caught on, I suspect, is its slightly revolutionary flavor" (Cox, t. 2, s. 1, v-4-78; Christensen, t. 1, s. 1). Brent Dalrymple noted that "one of the rules of the Geology Club was that no faculty were allowed, a rule still in effect when I was there [1963]" (written comm., x-16-78).

†It was earlier shown that Turner's bringing John Reynolds to Berkeley from Chicago, for proposed conventional work in mass spectroscopy, unexpectedly eventuated in the innovation of the young-rock K-Ar dating program. Analogously, Verhoogen recalled that "my concerns about convection were directed toward resolution of questions about the generation of magma and transfer of heat to the surface and temperature distribution in the mantle, but *not* directed toward resolution of the drift question" (Verhoogen, t. 6, s. 2, ix-20-77). Nevertheless, his studies and opinion about convection went far during the 1950's and 1960's to help open discussion about drift. Verhoogen's open-mindedness, near-avant garde posture in drift and other questions, and intellectual leadership extending beyond the Berkeley campus provided a model of scholarship for both Cox and Doell. Verhoogen's influence on both appeared repeatedly in nu-

for comparison.[47] Howard D. Gower, who had returned to Berkeley for graduate study after employment with the Survey in the mid-1950's, told Cox of the lower to middle Eocene Crescent volcanics of the Olympic Peninsula in Washington. Cox spent about a week sampling the volcanic rocks along Lake Crescent, but was discouraged to find them highly altered. On Gower's recommendation, he contacted Parke D. Snavely for information about the Siletz River volcanics of Oregon, which Snavely had mapped and named in 1948.[48] Snavely showed Cox a section of lower to middle Eocene rocks that included volcanics with interbedded sediments and sedimentary rocks intruded by numerous volcanic sills (tabular bodies paralleling the sedimentary layers). Snavely was thoroughly familiar with the area, and he believed that the Siletz River volcanic region, which he interpreted as an oceanic-island sequence, was much more stable than the Olympic Peninsula that Cox had fruitlessly sampled earlier.[49] Cox recalled: "I picked the Siletz River [rocks] because they were marine, and in order to get good tight age control it would be good to have some marine fossils. The Siletz River [rocks] were paleontologically dated, and they were basalts."[50] The Eocene magnetic field direction recorded in the Siletz River volcanics was found by Cox to be about 60° clockwise from present north.

Snavely had no substantial reason to believe the rocks themselves had rotated, although he had observed horizontal shearing in the field, which indicated that faulting had disturbed the rocks. Snavely remembered that "the thinking at that time was on a smaller scale; there was no reason to expect large-scale rotation."[51] Cox recalled:

> As luck would have it, the pole position for the Siletz River agrees fairly well with the pole from the Deccan traps, but the two poles are located where they are for completely different reasons. The Siletz River basalts, we now know, *were* rotated around a local vertical axis; they were somehow twisted around as they were accreted in North America or shortly afterwards, whereas the Deccan traps were carried northward with India. But that early result led me to become less enthusiastic about continental drift. It seemed like a curious coincidence, that two of the first Eocene studies in the world should give very close pole positions out in the South Atlantic and that neither India nor Oregon had drifted. The simplest explanation, and the wrong one, is that this represented polar wander and

---

merous subtle ways extending to their phraseology, idiom, and even gestures—a not uncommon trait among favored students leaving a mentor they admire.

not any evidence for drift at all. My impressions of Verhoogen's reaction to all of this, I think, was that he agreed with my interpretation that it didn't supply strong support for continental drift. And, if that were true, it meant that polar wander was a very rapid phenomenon, and you had to be very careful about interpreting paleomag data in terms of continental drift. It robbed the Deccan trap result of a lot of its significance. We now know it was just a fluke, because the coast of Oregon might well have rotated in the other direction, in which case the Deccan pole result wouldn't have agreed with it.* [It appeared that] it was going to be very, very hard to prove anything about continental drift from pole positions, and that was one reason I decided not to pursue that field [directionality studies], and to work on reversals instead.[52]

It was, in fact, Snavely's belief in the stability of the Siletz River block that led him originally to suggest it to Cox for paleomagnetic directionality studies. Had Cox gotten early results more in keeping with those being produced by the English, which suggested drift, he might well have been diverted from pursuing reversal studies. These had from first encounter seemed to him "like an amazing phenomenon." He wondered why "everyone wasn't running around and talking about reversals."

Cox's new research direction did not come from John Verhoogen but "through personal contacts with Jim Balsley in the Geological Survey, whom I'd gotten to know through Wahrhaftig," by way of Balsley's assistant chief, Mary C. Rabbitt.† In his initial excitement about the reversal problem, Cox had been close to Verhoogen in his thinking about the causes of reversed rocks, but came to have more doubts about self-reversal because of the influence of Hospers's papers.[53] Balsley's work on reversely magnetized rocks was well known in the Survey; it was natural, then, that Howard A. Powers, who had started work in the Snake River Plain of Idaho in 1954,[54] sent several basalt samples to Balsley for measurement, some of which were normal and others reversed. Because Balsley's administrative duties in the Survey and continued collaboration with Arthur Buddington kept him from personal pursuit of the

*Contrary to others, Edward Irving believed correctly in 1960 that the Siletz River block had rotated (Irving, oral comm., xi-30-78; A. V. Cox, oral comm., ix-5-78).

†An undated letter of 1956 (prior to June) from Cox to Rabbitt notes earlier conversation concerning Rabbitt's suggestion that Cox be employed by the Survey on a W.A.E. basis. Cox's intended summer work in the basalts of Oregon and Washington, additional collecting in California, Verhoogen's "especial" interest in studying possible mineralogical mechanisms of reversal, and Cox's plan to collect during two summers are outlined. Cox inquired "if the Geophysical Branch is interested in any of the program" he presented (Document No. 19A). In a letter to Cox of vii-26-56, Rabbitt confirmed Balsley's intent to "go ahead with the appointment" of Cox to a position in the Survey (Document No. 19B).

problem, he asked Allan Cox to "look into the possibility of self-reversal in those rocks." At the start of that study, Cox was excited about self-reversal and "expected it to be the explanation for the reversely magnetized rocks."[55] Cox found the Snake River Plain a potentially fruitful area of inquiry for his dissertation, because Garniss Curtis was providing isotopic dates from ash beds from the region for Howard Powers. Powers had made it known to the Berkeley group that the Snake River Plain had many virtues as a study area in rock magnetics. Powers was working with Harold E. Malde, who had joined him in the spring of 1955; Powers worked mainly on eruptive materials and processes while Malde's interests were centered on stratigraphic problems. Malde had been influenced by Pierre Bout of the Claremont Graduate School, who had worked on the volcanic stratigraphy of the Auvergne in France; Malde had also investigated the magnetostratigraphy of Iceland and was familiar with the work of Roche, Hospers, Khramov, and others and had corresponded with Hisashi Kuno concerning reversely magnetized later Cenozoic basalt flows of Japan.[56] Malde and Powers had taken a dip needle into the field for preliminary magnetic determinations on several of the basalt flows.[57]

During the summer of 1957, Cox collected oriented samples. One of his major objectives was to test whether the magnetic properties would be reproducibly uniform among a number of widely separated localities (over 10–20 miles) within the same lava flow; he made collections from two different lava flows.[58] One of the flows, for which Malde had established stratigraphic continuity, stretched over 20 miles. The observations and experiences in the Snake River Plain went far in building Cox's confidence in attacking the reversal problem.

Cox spoke at length about his thinking:

Pat Blackett was strongly committed to self-reversal,* and he was such a powerful and impressive man. He came by Menlo Park, and I saw him over in Japan at a meeting. Verhoogen [then] favored self-reversal. Blackett . . . said that if there was any statistical correlation between petrology and

*Blackett may have been strongly committed to self-reversal by the late 1950's, but he seemingly was undecided about the question in December 1954: "Accepting then, that some reversed rocks do indicate a reversed earth's field but that some do not, we have to face the problem of distinguishing one sort of rock from the other. In some cases this is a very difficult matter. . . . There are, however, a number of valuable deductions from the experimental facts of rock magnetism which do not depend on [differentiating field from self-reversed rocks]. The most important of these deductions is concerned with . . . evidence . . . to test . . . the highly controversial hypotheses of Continental Drift and Polar Wandering" (p. 8, 1956). Here too is a clarion call for the mounting of directionalist studies.

polarity, then that would be enough to show that self-reversal was the causal mechanism. In fact, that same correlation, in the gross, shows up in the Snake River Plain, but I didn't ever believe that was the explanation because of some detailed observations I had made during my graduate work. One observation was the presence of both oxidized and nonoxidized rocks within one flow that was entirely reversed. . . . So for me, the fact that the flow formed at one time as one cooling unit, was all reversed, and yet had oxidized and nonoxidized basalt within the same flow, was very convincing. That was a crucial observation and one, by the way, that I didn't ever publish. The Snake River rocks turned out not to be suitable for precise dating with the techniques then available. . . .

The other observation that was very important was finding baked zones in the Snake River basalt. The importance of these is that the minerals present in the baked zone are completely different from those in the basalt. The basalt had magnetite or titanomagnetite; the baked zone contained hematite. It strains the imagination to conceive of a mechanism that would cause the hematite to always undergo self-reversal when the magnetite undergoes self-reversal. . . . The [polarity of the] Snake River Plain baked zones agreed with the polarity of the overlying flow. That was crucial when I was working on my thesis.

In addition there was stratigraphic control; all the rocks that were younger than a certain horizon were normal, and all the rocks that were older than a certain horizon were reversed. The stratigraphy had been beautifully worked out by Malde and to some extent by Powers, but the detailed work up there was done by Malde, who was a young guy working in the Snake River Plain.* I would stay with him when I was up there during the summers of 1957 and 1958, so I got to know the stratigraphy extremely well and have a lot of confidence in it. The crucial observation for me was the perfect correlation of the baked zones. We didn't *ever* find a reversed baked zone in a normal flow or vice versa; they always correlated. . . . After I'd completed my thesis and was working for the Geological Survey, I was still interested in self-reversal, and spent a year or two seeing if there might be some correlation between petrology and polarity. And there would be, in a certain sense. In the Snake River Plain, for instance, there was the same correlation that R. L. Wilson later reported. A large fraction of the reversed flows were more oxidized, and a large fraction of the normal flows were not, but there were exceptions. It seemed to

*Malde, Cox, and several others in this book are related by a curious set of ties. Allan Cox and Kirk Bryan, Jr. (now a professor of meteorology at Princeton University) served as Wahrhaftig's field assistants in Alaska in the summer of 1950. Bryan's father, a professor of geology at Harvard, had strongly influenced the careers of Wahrhaftig, Hopkins, and Malde while they were students there; all three later became recipients of the Kirk Bryan (Memorial) Award in Geomorphology of the Geological Society of America (Wahrhaftig in 1967, Hopkins in 1968, and Malde in 1970). Each of the three influenced Cox in stratigraphic matters; Wahrhaftig and especially Hopkins consulted also with Doell and Dalrymple during their research.

Harold E. Malde in his office at the U.S. Geological Survey in Denver in February 1978. Allan Cox recalled: "The stratigraphy [of the Snake River Plain] had been beautifully worked out by Malde and to some extent by [Howard] Powers. . . . I would stay with [Malde] when I was up there during the summers of 1957 and 1958, so I got to know the stratigraphy extremely well and have a lot of confidence in it. . . . We didn't *ever* find a reversed baked zone in a normal flow or vice versa; they always correlated."

me that the exceptions were the really important observations, and for Blackett the exceptions were unimportant.[59]

Cox vividly recalled that among his many influential experiences with John Verhoogen, one of the most important occurred on the occasion of the 1957 meeting of the International Union of Geodesy and Geophysics in Toronto; "that was my *big* intellectual experience in graduate school." In an unprecedented act, Verhoogen sent Cox, still a graduate student, to represent the Department of Geology and Geophysics at the meeting.

That was unheard-of in those days at Berkeley. . . . Verhoogen, I understand, was criticized and had to defend himself. . . . We'd been reading the papers of [Julius] Bartels, [Emile] Thellier, [Sydney] Chapman, and [Edward] Bullard, the list of all the great people who were almost like gods in my mind at that time. I remember going up there and actually meeting all those people. It was so exciting to me because it made science become real.

John Verhoogen was bold in his support of those he regarded as promising. In 1952–53 he gave a seminar on the magnetic properties of minerals that was attended by Richard Doell. Verhoogen recalled that Doell was looking for a thesis topic and suggested building a spinner magnetometer to do a paleomagnetics thesis. Verhoogen thereupon surreptitiously diverted what little money was needed from an Office of Naval Research grant on metamorphic reactions to build the magnetometer. In 1957, in an unprecedented act, Verhoogen sent Allan Cox, while still a graduate student, to represent the Department of Geology and Geophysics at the meeting of the International Union of Geodesy and Geophysics in Toronto. "That was unheard of at Berkeley," Cox recalled; "Verhoogen . . . was criticized and had to defend himself."

The technique of magnetic cleaning using alternating fields, which is very crucial in paleomagnetic work, was being discovered or rediscovered by a lot of groups separately. I'd been building my own apparatus at Berkeley. Verhoogen always referred to my AC demagnetization cleaning apparatus as "black magic" because I could start with really messy-looking data, and have it all become orderly and neat. I had read a clue in an article and just developed my own apparatus. It turned out that two or three groups around the world developed similar apparatus independently. I remember hearing Bullard, one of the greats in the field, and Bartels I think and maybe [Takesi] Nagata, getting in a big discussion about AC demagnetization. They were all independently rediscovering demagnetization. There are a lot of little details, but they were arguing how an AC demagnetization apparatus would work on the basis of first principles in physics. I'd figured out a lot of those things myself. At the I.U.G.G. I saw that if these great minds were grappling on a very lofty plane, with what was a fairly pedestrian lab problem—that there might be some hope for me! I knew the answer to the question they were discussing, but didn't say anything because I was rather shy. Bartels was so pleased that data, so painfully collected over decades by patient people, was now being brought to bear on the larger long-term problems of the magnetic field. There were talks on self-reversal, continental drift, and all the exciting ideas. I'd also met Ted Irving, and he was working on drift. I remember his saying that I should work on drift because that's where the action was. At that time I was committed to self-reversal.

A very important event took place when I got back to Berkeley. Verhoogen had given his brilliant summary of the 1954 I.U.G.G. [meeting] to the department, and when I got back I also had to report to the various seminars in our department on what I thought were the important ideas that had been presented at the 1957 I.U.G.G. I worked very hard in making my presentations. In those days at Berkeley we didn't learn about spherical harmonics as a mathematical tool for analyzing a field; this wasn't taught in courses there. I had to dig all of that out myself so I could review work on changes of the present field. It was then that I recognized the importance of the overall decrease in dipole moment. I had to do an enormous amount of self-education in order to give a decent summary. I think Verhoogen anticipated all that. He probably did this just to draw me out; to put me on the spot in a sense.

I think going to that meeting was crucial; hearing the real giants arguing about a lab technique that I knew something about made me realize that I wasn't too far from becoming a member of the club. It suddenly wasn't inconceivable that I could do the kind of work that they were doing. As a student, always seeing published work, which is so impressive and intimidating, you can't even imagine doing something similar yourself.[60]

During those graduate student years, 1955 to 1959, the paleomagnetics laboratory in the physics building was located next door to the room where Joseph Lipson was conducting potassium-argon research with John Reynolds. Cox (and Doell) interacted frequently with Lipson. Cox recalled, "I knew [Lipson] well. The dating of young volcanic rocks was one of the major enterprises going on at Berkeley when I was a grad student. It was something we all followed with a lot of interest.* I had Curtis and Evernden as teach-

*In a proposal submitted in 1958 to the U.S. National Committee for the International Geophysical Year, Cox planned as the major objective of his proposed research "to study the remanent magnetization of volcanic rocks of late Pliocene and Pleistocene age in order to obtain information about the earth's magnetic field during the past million years." In a brief but pointed review of the literature, he cited especially the work of the Icelanders. The proposal

ers, in courses that weren't related to dating at all. They were very good teachers. [K-Ar dating] wasn't the sort of thing that was taught in a course over there at that time, it was just done."[61]

Cox and Doell were unique among paleomagnetists in that they were early exposed to and continually informed of unfolding developments in the pioneering young-rock dating program at the potassium-argon laboratory from its inception onward. Cox and Doell's attempts to find willing collaborators among geochronologists (most of whom were not early impressed with the importance of polarity-reversal studies) constitute crucial future episodes addressed in subsequent sections.

Balsley, the Bungalow, and the Camp Followers

By the summer of 1958, Cox and Doell had laid plans for a long-term collaborative effort at the U.S. Geological Survey at Menlo Park, California. Cox had been employed on a part-time basis for several years by the Survey, had made a commitment to full-time future work with them, and had been in sustained communication with James Balsley, chief of the Geophysics Branch from 1953 to 1960. Balsley had long known Doell and was then apprised of both his professional efforts in paleomagnetism and his recent severe illness, which led Doell by the fall of 1958 to return to California from M.I.T. That September, Cox wrote to Balsley explaining Doell's circumstances and his desire to take up research in rock magnetism at Menlo Park before completing the academic year at M.I.T.; Doell then telephoned Balsley to inquire directly.[62] Balsley encouraged Doell to begin the rock-magnetics project with Cox at his earliest convenience.[63]

Balsley's role in the development of the Rock Magnetics Project at Menlo Park was extremely important; his unique combination of endowments permitted him to understand the importance of the research program proposed by Cox and Doell, to be much in favor of its execution, and to act with insight and expedition toward its implementation. Balsley's absence from the Survey would surely have encumbered and could have precluded development of the project.

James R. Balsley had acquired a B.S. in geology and geophysics

---

touched on a number of important reasons for pursuing polarity-reversal studies; in particular, "a new approach to the problem has been opened up by refinements in the potassium-argon age dating method which enable the dating of potassium-rich volcanic rocks as young as 10,000 years" (Document No. 19B-1).

James R. Balsley pioneered in the development and application of the airborne magnetometer, and early pressed for a solution to the problem of reversely magnetized rocks. As chief of the Geophysics Branch of the U.S. Geological Survey, he implemented the establishment of the Rock Magnetics Project at Menlo Park. Allan Cox thought that "Balsley spanned both worlds; he saw the need for both the theoretical work and efforts in application."

from Cal Tech in 1938 and then entered the Harvard Graduate School to pursue a Ph.D. in geophysics with David Griggs. His work with the Survey, begun in the summer of 1939 and followed for a number of years, was concerned wtih chromite, vanadium, and titaniferous magnetite deposits.[64] The Adirondack Mountains were a major source area for magnetite and thus received much of his attention (detailed in Chapter 4). By the summer of 1943 his ground-level magnetic study included the iron-ore deposits of the Upper Peninsula of Michigan.

Balsley played a pioneering role in the development and application of the airborne magnetometer.[65] David Griggs, at that time Chief Scientist for the U.S. Air Force, was cognizant of Balsley's mineral mapping efforts and was also familiar with the Navy's magnetic submarine detector. Griggs therefore invited Balsley to participate with the Office of Naval Research and Bell Telephone Laboratories in the development of the airborne magnetometer for use in geophysical prospecting. In the summer of 1947, the magnetometer was flown in the Upper Peninsula of Michigan, where Balsley first encountered the negative magnetic anomalies that led him into extended research, eventuating in a number of publications with Arthur Buddington of Princeton University and others.[66] By the summer of 1948, Balsley had taken specimens to the Department of Terrestrial Magnetism of the Carnegie Institution and begun work with Thomas Murphy, Alvin McNish, and John Graham and found that the specimens had strong negative remanent magnetization. E. Harry Vestine and Alvin McNish were particularly interested in the magnetic field of the earth and believed that Balsley's airborne instrument could be of service to them.[67] Balsley's early discovery that the remanent magnetization was very much greater than the induced magnetization, in the case of the Michigan dikes, and that negative anomalies were present where positive ones were expected, led him to research that placed him at the center of the self-reversal versus field reversal question. He was a pioneer in rock-magnetic studies in several respects and knew and spoke with all of the prominent figures in the field, including Takesi Nagata, Seiya Uyeda, John Verhoogen, the Carnegie group, Louis Néel, and others. He was actively encouraging both Doell's and Cox's efforts in paleomagnetism long before they began the Rock Magnetics Project at Menlo Park. As Cox saw it, "Balsley spanned both worlds; he saw the need for both the theoretical work and efforts in application."[68] Doell recalled, "Balsley provided very strong support of the Rock Magnetics Project in funding and other ways. We were never strapped for anything at any time, because of Balsley's efforts. He thought that the Rock Magnetics Laboratory studies would lend themselves to aeromagnetic studies. Balsley was a strong self-reversal advocate."

Although Balsley jokingly mentioned his wish that Cox and Doell's experiments would show that self-reversal was the real cause of reversed rock groups, Doell said:

> He essentially gave us free rein. . . . Jim would obviously approve of what we were trying to do. I don't remember him ever trying to talk us into doing any particular thing, or trying to talk us out of doing anything we wanted to do. . . . There was no direction in that sense at all. . . . We didn't always get exactly everything we asked for, but that was because Jim was much more sensible than we were. He'd been doing this long enough to realize that we wouldn't get everything done we'd hoped to. . . . I don't think he had more funds than he knew what to do with, and I don't think all of the projects were supported as generously as ours. Jim thought that paleomagnetism and magnetic studies in general would be a great boon in interpreting the aeromagnetic anomalies and [related projects] for which the Survey and the Geophysics Branch were very heavily committed. In fact, his laboratory in Washington [D.C.] was geared to doing a lot of magnetic properties studies for aeromagnetic interpretation.[69]

Balsley also advised Doell and Cox to hire technicians to do the time-consuming technical chores and reoriented them to certain aspects of the project that resulted in great economies in time. Doell recollected that "with Jim's help, we managed to change a lot of ideas in the Survey about getting more technical help than a lot of other professionals had been used to."

In comparing the level of support and the research context provided by the Survey with those generally encountered elsewhere, Doell remarked, "I don't think you could hope to get this kind of stuff at a university, or even at many research labs."[70] The Rock Magnetics Project staff was apparently not made to feel that their research was to be oriented in terms of satisfying the larger needs of the Survey. They were left to their own theoretical concerns, allowed to formulate and execute experiments on their own terms, and supported at a level that was clearly seen as privileged by many from other projects within the Survey.*

Shortly before arriving in Menlo Park to assemble the Rock Magnetics Laboratory, Cox wrote to Balsley with a proposed plan for a nonmagnetic room in which to conduct their experiments, "stressing that it should be made out of wood." Cox recalled Balsley's short response, a note saying "Dear Allan, I approve of your plans for building a redwood bungalow for two. PS: Please leave room for a third some day, if I can get out from under all this administrative work."† After taking up residence, Cox asked Balsley what he wanted him to work on, to which Balsley rejoined, "Well, I hired you because I thought you knew what you wanted to work on." Said Cox, "That's the last time I ever asked anybody in the Survey what he wanted me to work on. It's an ideal place to work."[71]

In January 1959, after Cox and Doell had laid plans for the Rock Magnetics Project, Doell underwent further surgery for the removal of part of his lymphatic system, which required a long recuperation. To save money and time, he and Cox purchased electronics parts, including Heathkits, for use in building electrical components for the lab.‡ The parts were purchased in California and sent to Massachusetts for Doell to assemble while convalescing; they thus had a number of instruments, including a Wheat-

*The "privileged" status the Rock Magnetics Project seemingly enjoyed may have contributed to certain frictions (discussed below) that arose in the course of Cox and Doell's attempts to secure radiometric dates on rock samples with known polarities. Bearing directly on this point, Appendix B contains a "Parody on Privilege," and Appendix A, a "Note of Concern."

†In Document No. 16-1, Balsley notes difficulties in attempting to secure space for "our 'magnetic laboratory' in Menlo Park."

‡A letter of ii-10-59 from Balsley to Gordon Bath details equipment purchases and the decision to have some "instrument work done by Kenneth Harper at Cambridge [Mass.] where Dick Doell can supervise the work" (Document No. 16-2).

The tarpaper-shack home of the Rock Magnetics Project. Allan Cox remembered that "there was a man . . . who was the chief administrative officer of the Survey at Menlo Park. I told him I'd like to use one of those empty shacks. He said, 'Well, they aren't official government buildings.' I said, 'Yeah, but Art Lachenbruch has a little lab off in the corner of one; would it be possible for me to put some equipment in the other one?' He had a key to the building in his hand and said, 'Well, it's not an official building.' He threw the key on the ground and turned around and walked off. I studied that gesture for a long time, picked up the key, and it fit the door and so I moved in." "The shack," in expanded form, still serves to house the Rock Magnetics Laboratory at Menlo Park.

stone bridge, voltmeters, an oscilloscope, some resistance substitution boxes, and vacuum-tube voltmeters, by the time of Doell's arrival in Menlo Park in March 1959.[72] While Doell was in Massachusetts, Cox, who was still in part-time status at the Survey and close to completing his Ph.D. thesis, took the first step toward acquiring their "redwood bungalow."

The first thing we did was get a place to put our lab. I remember very clearly, before Dick got out here, I decided I wanted to work in the tarpaper shacks, which weren't recognized as official buildings on the part of the Geological Survey. [These buildings are still designated "temporary."] . . . I told [Glendon J. Mowitt, chief administrator of the Survey] I'd like to use one of those empty shacks. He said, "Well, they aren't official government buildings." I said, "Yeah, but Art Lachenbruch* has a little lab off in the corner of one; would it be possible for me to put some equipment in the other one?" He had the key to the building in his hand and said, "Well, it's not an official building." He threw the key on the ground and turned around and walked off. I studied that gesture for a long time, picked up the key, and it fit the door and so I moved in.[73]

And thus was the initial space acquired that serves still to house much of the Rock Magnetics Laboratory at Menlo Park.

Upon occupancy, Cox wrote Doell explaining his progress in obtaining electrical components and machine tools and the installation of four 50-amp power lines. He detailed the use of space and especially noted that "one corner will be kept free of wiring so we can put our field-free region there." The 20-by-100-foot building was described as "grubby . . . but the most spacious lab I've ever been in." Gordon Bath (an administrator) informed Cox that the Survey was well endowed with funds that year and invited requests for equipment. Cox excitedly wrote Doell, "Thus emboldened, I gave [Bath] the following list, with priority in the order indicated: (1) low level amplifier, ca. $400; (2) fluxgate heads for field-free space, ca. $1,500; (3) recording analytical balance, $3,600."† He also discussed a number of questions relating to equipment and his plans to complete his thesis and begin work in residence at Menlo Park by June 1959—which he did.

Doell recalled,

Balsley used to come out [to Menlo Park] temporarily for . . . months . . . and helped us to get rolling. At that time, research projects with their own supporting facilities, like electronic equipment or instrument makers, were essentially unknown. All that was done rather centrally. The chief of the administrative division of the Survey [Glendon Mowitt] was a very strong person. . . . He believed very strongly in the separation of things: a paleomagnetism project should do paleomagnetism; map makers would only make maps; and machinists all had to be grouped in one place . . . under the control of machinists. It was a management argument and it was very strong in the Survey.[74]

*Arthur Lachenbruch is a National Academician who has specialized in heat-flow studies.

†The letter (of i-13-59; Document No. 19) contains Verhoogen's concerns about the completion of Cox's thesis, remarks on Creer's and Nairn's magnetic results from South Africa supporting Alex Du Toit's idea of the movement of Africa, and Walter Munk's inquiry of Cox concerning the evaluation of paleomagnetic data in support of polar wander.

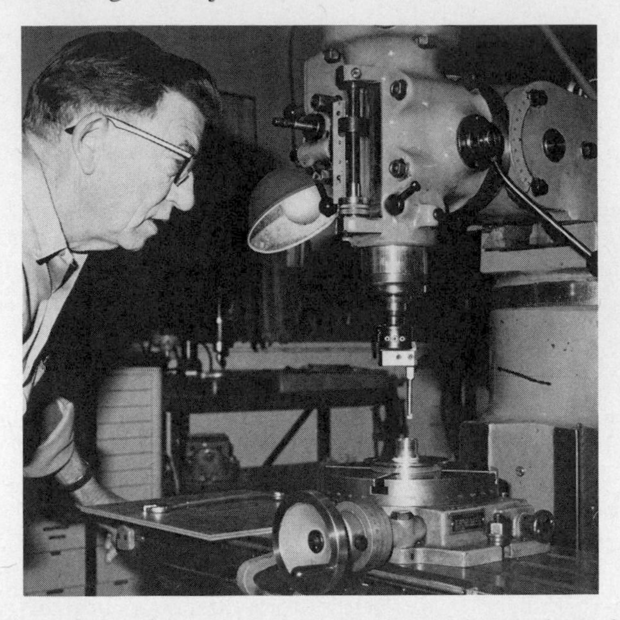

A recent photo of Major Lillard at the vertical milling machine in the Rock Magnetics Laboratory, Menlo Park. The exclusive use of technical staff and machines within the Rock Magnetics Project was arranged by James Balsley; it was crucial to their success.

Balsley circumvented the encumbrance of such a policy: he arranged the temporary assignment of Major Lillard, a master instrument builder and technician from Washington, D.C., to the Rock Magnetics Project.* Lillard's role was crucial in equipping and organizing the laboratory; the desirability of keeping him beyond his temporary duty assignment of six months was early recognized by all.† "Balsley finally talked Mowitt into the advisability of letting Major stay on." Doell had requested Lillard by name, for the reasons that Nathaniel Sherrill, a technician in the Rock Magnetics

*Letter of iii-13-59 from Balsley to Bath requesting that Lillard be assigned to Menlo Park for "a couple of years . . . helping Doell and Cox set up the magnetic laboratory." Balsley wrote in the hope that Bath "may be able to start some wheels rolling before coming out here," and concluded that "we may now have the opportunity to build a first-class magnetic laboratory, and can do it in a year or two with the help of Lillard" (Document No. 19-1).

†In a letter of viii-27-59 to Glendon Mowitt, Balsley emphasized that Lillard was being detailed to Menlo Park to assist Doell and Cox in establishing "a research laboratory concentrating for the next several years in paleomagnetic investigations." Fearful that Lillard might be put to use instead in a geophysics instrument shop, and thus have his usefulness to the Rock Magnetics Project reduced, Balsley forcefully requested "that any agreement regarding the relationship of Major Lillard to the proposed central shop in Menlo Park be delayed until my return" (Document No. 19-2).

group recalled: Lillard "could build anything anybody could design and build it better than almost anybody; he was an old world craftsman in the finest sense of the word and had great physical endurance."[75] Cox and Doell knew that Lillard had built a spinner magnetometer at Silver Spring, Maryland, for Balsley and that he was fully familiar with the laboratory components they required. Doell continued:

> We managed to get a lathe and eventually a milling machine and some drill presses. But each of those pieces of equipment had to have Mowitt's permission, and he fought over each one. I remember going to Packard of Varian Corporation and getting documentation . . . that it was cheaper for each research group to have its own machine tools, the idea being that equipment was a lot cheaper than personnel. That difficulty went on for a long time until we finally got the mass spectrometer for our own lab [in 1964].*

In the episode of the mass spectrometer, recounted below, in which Mowitt was not involved, the difficulties reached almost destructive proportions.

Long before occupying the laboratory, the design and building of equipment for it were largely conditioned by decisions regarding research plans formulated by Cox and Doell. They were explicit in noting that at the outset, research problems were selected primarily to yield publishable results most rapidly. Simple, reliable equipment that could rapidly be made functional and was not costly was thus sought. Their decision imposed fundamental limitations on the scope of problems they would be equipped to address. There are two types of magnetometers for studying rocks: one is a spinner, in which the rock is rotated near an electrical coil; the other is an astatic magnetometer in which the rock is positioned near a small pair of balanced magnets hung by a fine fiber. The English paleomagnetists generally employed the latter type, and Americans used spinners. Neither Doell nor Cox had any substantial experience with astatic magnetometers. Doell explained how they proceeded.

> There are basically two kinds of spinners. The one that I developed at Berkeley was the first one of [the] type in which the rock was spun on one

*Doell, t. 4, s. 2, iv-26-78. Cox and Doell shared the same sense of frustration in the face of bureaucratic encumbrance; Balsley's role was especially vital in circumventing many difficulties. In a letter to the acting chief of the Theoretical Geophysical Branch, Cox blended wit and acerbity in a poignant volley against bureaucratic inanity; much of his character is revealed (letter of v-18-66 is Document No. 20-1).

end of the shaft and a little reference magnet and coil system was on the other; then these two signals were fed into a transformer. Almost immediately the size and angle of the reference signal could be changed until it matched that of the sample signal in the little transformer. There are very weak signals at this point, and only after the signals are mixed can they be run through filters and amplifiers and some sort of a display scheme. The other type of system treats the reference and the sample signal separately and filters each of them to get rid of noise before you do the comparison. The second type of magnetometer is inherently more sensitive; you can measure very much weaker rocks, but the problems involved are getting a very stable frequency so the filters can be highly tuned. [It is very difficult to get an accurate instrument that is also highly sensitive to weak signals. In contrast, in the simple type in which the signals are mixed first, highly accurate results are more easily obtainable, but sensitivity is sacrificed since there is much "noise" in the system.] So the thing we started right away was to build the simple, accurate, but not very sensitive type of equipment. One of our first priorities was also a demagnetizer tumbler* arrangement and a coil. Those were things that were pretty new at that time. . . . There was only a handful of people doing measurements, and everybody pretty much knew what each other's equipment was like and how it operated. As with most labs before, we thought we would have one piece of measuring equipment that was supposed to do all things. We were hoping to develop a big enough program that would warrant two instruments: a simple, not-too-sensitive one first that we could get going very rapidly, and a sensitive, more time-consuming one later. As it turned out, we eventually built three of the simple machines, which were going at the same time.[76]

The simple machines were used because of their reliability, and because the strongly magnetized basalts Cox and Doell measured in their research in polarity reversals did not require the more sensitive type.

The first monthly project report (for April 21 to May 20, 1959) bore the title "Remanent Magnetization of Rocks Project"; that title persisted until January 1960, when it was changed to "Rock Magnetism Project." Only Doell and Cox were listed as project personnel until the report of June 21 to July 20, 1959, in which Major Lillard appeared as a transfer from Silver Spring, and Thomas

*Demagnetization is the reduction of remanent magnetism aimed at selectively reducing only the softer or less stable components. Such techniques may involve alternating-field demagnetization, thermal demagnetization, and chemical demagnetization. Alternating-field techniques were first regularly done by As and Zijderveld (1958), who demagnetized each specimen three times along three axes in an alternating field. In 1959, Kenneth Creer began tumbling specimens about two axes at right angles, perpendicular to the axis of the demagnetizer coil. Doell and Cox (1967) used a more complicated three-axis tumbler.

Lagerquist was noted as "temporary for the summer."[77] During the first few months, much time was spent in acquiring wooden furniture and equipment, and Doell as project chief conferred with William Vaughn and others at the electronics laboratory of the Survey in Denver and also with Louis Pakiser. In Washington, D.C., he talked with James Balsley and Robert Moxham and also discussed magnetometers with Walter Wrigley, Win Markey, and John Haovorka. Cox and Doell were still at work on the "Review of Paleomagnetism," but Doell had also begun the design of a spinner magnetometer and demagnetization equipment. The first semiannual progress report of May 31, 1959, confirms the remarks of Cox and Doell that the concept for the Rock Magnetics Laboratory arose through discussions between them during June and July 1958.[78] The equipment list for future installation included two spinner-type magnetometers,* an AC degausser (demagnetizer), a saturation magnetometer, a thermal degausser, and an astatic magnetometer. Their research aims were directed at study of the earth's magnetic field in geologic time, or paleomagnetism, and the relationships between remanent magnetic properties and rocks. They intended that their first magnetic measurements would be "in the form of a continuation of studies already undertaken by Cox and Doell" in Pleistocene basalts of Idaho (recall earlier discussion of Snake River Plain results), Eocene basalts of Oregon (the Siletz River volcanics data that led Cox to eschew directionality studies), and Precambrian and Paleozoic sediments of Arizona.[79]

The feverish work pace within the project during the latter half of the first year is set out in detail in the monthly project reports for that period.† The semiannual progress report of November 30, 1959, cited Lillard's usefulness in organizing and operating the laboratory, the status and proposed use of equipment, the construction of a modified Schmidt net with 1/10-degree accuracy for the graphical reduction of experimental data, a magnetic survey of the Menlo Park grounds, the completion of two final and two preliminary manuscripts, including a long and, ultimately, very influential

*One, a motor spinner, was to have a sensitivity of 10^{-5} to 10^{-6} emu per cubic centimeter (emu/cm^3); the other, a turbine spinner, was to have a sensitivity of 10^{-7} to 10^{-8} emu/cm^3. "The motor spinners ended up with a sensitivity of 10^{-3} to 10^{-4} emu/cm^3 and the turbine spinner *never* worked" (written comm., Grommé, iii-19-81).

†Monthly Project Reports and Semiannual Progress Reports for the Rock Magnetics Project are on file in the Isotope Geology Branch office, U.S. Geological Survey, Denver, Colo. Duplicates of most are also on file in the Office of the History of Science and Technology Project, Bancroft Library, University of California at Berkeley, under the Project in Geomagnetic-Reversal Time Scale History.

Allan Cox and Major Lillard in the Rock Magnetics Laboratory, about 1963, following the successful operation of a newly built magnetometer.

review article for *Advances in Geophysics* titled "Paleomagnetism," and a restatement of their research aims expressed earlier in the first semiannual progress report.[80] The progress in the laboratory was widely known and attracted many notable figures; the monthly report for October–November 1959 alone noted as visitors Seiya Uyeda, John A. Jacobs, John Belshé, Ronald W. Girdler, and Gordon J. F. MacDonald. Most of the world figures in paleomagnetism visited the laboratory within its first year of operation. A typical monthly report, such as that of September 20, 1960, mentions six "manuscripts in preparation, with estimated dates of transmittal to the Branch" between September and December of that year. In addition to a seemingly herculean schedule at the Survey, Cox and Doell held various honorary and other appointments in the geophysics department at Stanford University, arranged and supervised seminars, and gave a two-term course in 1963. They also helped build up a small but fairly complete laboratory, and helped supervise graduate student research.[81] Their effort was undoubtedly a significant influence in the award to Stanford of a three-year research grant, under the U.S.-Japan Cooperative Program, in which Doell, Cox, and Dalrymple were listed as "principal investigators." They submitted that proposal, in September 1962, through the U.S. Geological Survey and Stanford jointly. The proposal, detailed shortly, summarized the aims and needs of the project at Menlo Park; it resulted in a grant of $120,000 in June 1963, which greatly increased the productivity of the project.[82]

Cox, Doell, and Dalrymple repeatedly emphasized that the Pliocene-Pleistocene boundary question and the shortcomings of biostratigraphy and physical stratigraphy in long-range correlation were major if not the foremost scientific problems to which a magnetostratigraphic time scale could apply. Cox recalled:

> The possible use of such a scale provided much of the basis for the Survey supporting it for a number of years: it was built in from the very beginning; that was one of the things that we hoped to do and eventually did. We tried tying in with known paleontology (the formal stages) because that's what existed. This had been done by the Icelanders before; they talked about N-1, R-1, the most recent reversals. You could begin to see the correlations already. You could clearly correlate between the Snake River Plain and Iceland . . . but the trouble was that *you couldn't do it with any precision*, because the formal stages themselves weren't very tightly defined and frequently not recognizable, for one reason or another, over great distances. Even in Italy [where the type sections defined the formal stages], they had been unable to correlate with any certainty between the continental deposits and marine deposits at the Plio-Pleistocene boundary. That was the trouble that biostratigraphers got into when they tried to assert that there was correlation when it hadn't been established. We started by taking the biostratigraphic approach [and frequently sought the counsel of Survey stratigraphers such as David Hopkins and Jack Wolfe], but we were very keen to get potassium-argon dating going because I became increasingly aware that the tools of paleontology just weren't sharp enough to establish the worldwide correlation of a reversal. It wasn't that we went all out using paleontology up to a certain day and then on that day reviewed the situation and decided that paleontology wasn't adequate and then decided to switch our efforts. That wasn't the situation at all. At all times we would have been happy to get potassium-argon dates, and we continued, as long as I was working on reversals, to try to tie the dating in with paleontology.[83]

The important work of Curtis and Evernden during the late 1950's was known to all both at Berkeley and the Survey; Cox and

Doell were especially well informed because of their contact with Joseph Lipson and the proximity of the paleomagnetics laboratory to that of the radiometrists at Berkeley. But it was clear that the virtues of rock magnetics did not give its practitioners priority in the minds of the overburdened young-rock potassium-argon geochronologists. The geologic community, by the late 1950's, was clamoring for dates in solution of classical, highly regarded problems. Thus two young men engaged at the Survey in little-known and questionable rock-magnetics work were by the time of the formation of their project in 1959, and afterwards into the early 1960's, clearly in search of potassium-argon dating capability, in whatever form it might become accessible. Cox recalled that

> the Berkeley group was having great success in dating young rocks, but they weren't interested in this problem of polarity reversals, and so there weren't enough really young things dated to begin to put the reversal story together. *At that stage, we still weren't able to get people who knew how to do dating to do it.* In our first round of work we simply began looking at the paleomagnetism of rocks that had already been dated for other purposes. The first formation we looked at . . . was the Bishop Tuff. I had sampled the tuff already when I was in graduate school, because that was being dated by Curtis and Evernden. I did a very thorough [magnetic] study of the Bishop Tuff and found . . . it was normal. *We became camp followers to the people doing the potassium-argon dating for their own reasons.**

Brent Dalrymple recalled, "Cox and Doell tried to get Evernden and Curtis interested in working on the reversal problem, but they were busy with other things."[84]

Cox was informed of the potassium-argon dating work of Brent Dalrymple by Clyde Wahrhaftig. During his first graduate school year at Berkeley (1959–60), Dalrymple had accompanied Wahrhaftig in the field and discussed the selection of a thesis problem. Wahrhaftig subsequently served as a member of Dalrymple's doctoral committee. On July 15, 1961, Dalrymple, Wahrhaftig, and Robert Fleck, then serving as Dalrymple's summer field assistant, drove to Westgard Pass and Eureka Valley, California, in order to collect basalts that Dalrymple intended to date in order to time the uplift of the Sierra Nevada (which many considered an overly am-

*Cox's remarks here (t. 3, s. 2, v-9-78 [emphasis added]) are reminiscent of those of Arthur Maxwell (oral comm., viii-17-76), who was relegated to the use of small, irregularly spaced intervals of time aboard oceanographic research vessels during the early 1950's when he was conducting little-appreciated pioneering studies of heat flow through the ocean floor. In contrast, explosion seismology then enjoyed high priority and was treated with deference.

Richard Doell, Edward Roth, and Allan Cox coring basalt in 1961 in the Berkeley Hills. Depicted is an early system for coring built by Cox at Berkeley. Cox recently recalled that the system had been used in collecting samples from the Snake River Plains for his thesis: "It was assembled from off-the-shelf commercial parts including a diamond drill, a 4-horsepower Briggs-and-Stratton engine, and a flexible shaft of the type used in settling concrete. The drill rig was heavy and awkward to pack, but it worked" (written comm., Cox, i-19-81).

bitious thesis project at the time). From there the party proceeded to the Berkeley Geology Department summer field camp run by Mark Christensen, Charles Gilbert, Donald Weaver, and Charles Corbato.[85] Cox and Doell had been collecting basalts on the eastern flank of the Sierra and were visiting the Berkeley camp. It was there that Cox and Doell first met Dalrymple, who remembered:

It was Allan Cox, really, who got me interested in the reversal time scale, just because he did spend a great deal of time discussing it with me. I presume that was an attempt to get me interested in it, but Allan's also the kind of person who has always been very interested in students. He explained paleomagnetism and reversals and what they were doing and why the reversal time scale was important. We spent quite a few hours that night sitting around the campfire drinking beer, just the two of us discussing that particular problem. Then over the next day or two we took some

G. Brent Dalrymple at Sanostee Mesa, New Mexico, Navajo Reservation, in the summer of 1958. While an undergraduate at Occidental College, he worked for a mining exploration company and used the magnetometer shown in searching for titanium ore.

field trips together to sites of lava flows that I was going to work on. . . . The entire Berkeley field camp faculty group was assembled, and Dick and Allan were demonstrating their first working model of a field fluxgate magnetometer. We agreed at that time to keep in touch with each other, so that if there were any volcanic rocks I was dating that they could sample, they would [measure their magnetism], and, if there was any way that I could possibly contribute some radiometric ages to their problem, I would do that. Although I was concerned with my own thesis work at that time, we made a kind of first tentative contract and informal research agreement. It wasn't an agreement to collaborate on anything, but was an agreement to keep in touch, so that whenever it was mutually convenient to put the data together, that would be done.[86]

Through that eventful first encounter in 1961, Dalrymple learned of the incipient polarity-reversal studies program and the possibly crucial way his newly acquired dating skill might aid a collaborative effort to formulate a globally applicable correlation tool. Dalrymple was destined to derive the first young-rock potassium-argon date specifically for polarity-reversal studies* and to join the Rock Magnetics Project. Dalrymple's role in constructing a temporal framework for magnetostratigraphy by potassium-argon radiometry was entered by a circuitous route. The story touches on familiar places and people, among whom Ian McDougall, of the Australian National University, figures prominently. Dalrymple's and McDougall's roles in this history took remarkable symmetric turns, and their differences as well as their similarities are richly illuminative of the human side of the polarity-reversal scale endeavor.

Dalrymple and the N.S.F. Challenge

Like both Doell and Cox, G. Brent Dalrymple did not first major in geology in college but changed to it from physics, to which he had been attracted because of his early interest and strength in mathematics. Born in Alhambra, California on May 9, 1937, he was educated in Bell, graduated from high school there in 1955, and went on to Occidental College because a cousin "said it was a nice place." In an introductory geology course in which he enrolled to meet course requirements outside his physics major, he was "greatly fascinated by the lectures of Joseph Birman and after two or three months listening to him lecture about geology, went in to see him about changing to a geology major." His decision to enter graduate school at Berkeley did not rest on "any marvelous or penetrating reason." He had met and married Sharon Tramel, an education major, at Occidental; both wanted to remain in California, and Sharon had received several offers to teach in the San Francisco Bay area. Although he "did not know much about the Berkeley faculty" and had no idea of what or under whom to study, he came to Berkeley toward the close of his senior year, where

*The sample was collected by Dalrymple from Owens Valley, California, in the Big Pine quadrangle during a field trip with Cox and Doell in the summer of 1961 (Berkeley laboratory data card for the whole-rock obsidian sample is Document No. 21).

Charles Gilbert evaluated his transcript and informed him of his acceptance to graduate status. At the close of his first academic year in 1960, he was offered a teaching post in the six-week summer field course in the White Mountains of California, east of the Sierra. He taught the course with Garniss Curtis, Jack Evernden, and Donald Weaver.* Dalrymple's first contact with Curtis and Evernden proved a rewarding experience in several ways. He recalled that "they treated me as an equal instructor; we were all there for the same purpose doing the same job, and I really felt comfortable about it." The sustained social interaction in a close living and teaching environment brought Dalrymple closer to Curtis and Evernden, and "coming off that summer, an alliance with them and seeking them out to do a dissertation under them would have been the most natural sort of thing." The three had discussed potassium-argon dating, which struck Dalrymple as a useful thing to know, but he did not, by summer's end, intend to do a thesis in it. His early inclination, largely as a result of his undergraduate training, was to pursue geomorphologic and petrologic problems, which Birman, his "role model" at Occidental, had done, but he was attracted increasingly to radiometry because "it was fairly obvious it was going to be important in the earth sciences, and it was a new and interesting thing." Dalrymple continued:

> I'm kind of a knob-twirler at heart, and that part of it appealed to me—a chance to twirl knobs and pound on some rocks. When I asked Curtis and Evernden if I could learn some more about it by working in the laboratory, I didn't really have any thoughts, at that moment, about pursuing that in any detailed way. I just wanted to know something about it; and at that time you couldn't take a course in it. I suppose the nearest thing was John Reynolds's mass-spectroscopy course, which was basically on the design and performance of the mass spectrometer. There were no textbooks on the subject either! I went into potassium-argon dating because it was a fun thing to do. . . . almost none of the decisions I've ever made throughout my life have had any kind of a deep penetrating meaning. They just seemed and felt like the right thing to do.[87]

Dalrymple was greatly influenced by Joseph Birman, for whom he had worked as a field assistant in an attempt to clarify the glacial stratigraphy in the Sierra Nevada.† Birman's teaching prowess and

*Donald W. Weaver, who specialized in Cenozoic, foraminiferal biostratigraphy, is now a professor of geology at the University of California at Santa Barbara.

†Birman had served as Clyde Wahrhaftig's field assistant; Wahrhaftig was Dalrymple's thesis adviser at Berkeley.

enthusiasm, and the fact that he gave his students considerable responsibility, place him with Garniss Curtis as one of the two most important influences in Dalrymple's career. At the opening of the 1960–61 academic year, Curtis, who takes great interest in his students, arranged for Dalrymple to take instruction in the potassium-argon laboratory from John Obradovich, who was by that time a senior graduate student, dating glauconites under Evernden and Curtis. Dalrymple's first day at the lab was an occasion of curious historical symmetry.

> I went in the K-Ar laboratory one morning and met John Obradovich and a fellow named Ian McDougall, who was also in there for the very first time. We started out learning about vacuum and argon-extraction systems on the same minute of the same hour of the same day; we worked along in parallel, for the time that Ian was there, in Reynolds's room, which contained Geology's argon-extraction stuff on the first floor of Le Conte Hall.[88]

During that year Dalrymple and McDougall became friends on both an academic and social basis. The parallelism between them was sustained as Dalrymple went on to provide the young-rock dating capability for the Rock Magnetics Project of Doell and Cox, and McDougall, a year after his return to Australia from his post-doctoral year at Berkeley, assumed leadership in the Australian assault on the polarity-reversal scale.

Dalrymple's earlier field work had included a number of disagreeable summers in the desert; he was thus "looking for a place that had some pleasant surroundings" in which to do a thesis that would combine his deep interest in geomorphology, instilled by Birman, with his new-found fascination with potassium-argon radiometry. After conversations with Wahrhaftig, Curtis, and a number of others, he decided to examine the history of uplift and erosion of the Sierra through K-Ar dating of volcanic rocks, which covered various erosion surfaces at different times; it was, in fact, a project nearly balanced between his two interests.[89]

Following Dalrymple's initial contact with Cox and Doell in the summer of 1961 at the Berkeley field camp, Dalrymple communicated with them on about a monthly basis. In the spring of 1962, Dalrymple was tentatively offered a job in Zürich, Switzerland, at the Eidgenossische Technische Hochschule by Agusto Gansser, who intended that a young-rock K-Ar radiometrist trained at Berkeley set up a laboratory at the Hochschule in order to define the Cenozoic plutons (bodies solidified from molten rock at depth)

Brent Dalrymple, at left, doing thesis field work at the Hetch-Hetchy Reservoir in the summer of 1961 with his undergraduate assistant Robert Fleck (now a geochronologist with the U.S. Geological Survey). Dalrymple recalled: "I met Allan [Cox] and Dick [Doell] for the first time at the Berkeley field camp about a month after this picture was taken. It was Allan Cox really, who got me interested in [the reversal time scale]. . . . We spent quite a few hours that night sitting around the campfire drinking beer, just the two of us discussing that particular problem. . . . He explained paleomagnetism . . . and why the reversal scale was important."

of the Southern Alps.[90] The offer of two years of work had been tendered via Jack Evernden, who encouraged Dalrymple to accept it; Evernden had loosely contracted to send somebody there with a Reynolds mass spectrometer. Brent and Sharon Dalrymple were delighted at the prospect of a two-year stay in Switzerland and agreed to go. Over the next six months Evernden and Dalrymple continued to discuss the offer frequently, but Dalrymple never received word from Gansser, or saw any written communication to Evernden, concerning a contract. By Christmas time of 1962, Dalrymple's uneasiness had mounted greatly in spite of Evernden's reassurances that the job in Zürich would be forthcoming. Dalrymple recalled:

> It was at the western meeting of the American Geophysical Union, held at Stanford University, between Christmas and New Year in 1962: we were sitting out in the parking lot in Dick Doell's Volkswagen bus, and he offered me a job at the Geological Survey. . . . [Until that time,] with the potential job in Switzerland coming in, there was no percentage in trying to get heavily involved in a reversal time-scale experiment at all but I hadn't received anything in writing myself nor seen anything in writing. So when Dick made a very concrete offer to come down and work with them on the reversal time scale and take a permanent job with the Geological Survey, I accepted it.[91]

Dalrymple noted that additional factors guiding that decision included the exciting quality of the reversal experiment, a chance to remain in the Bay area, and the prospect of spending some years working with Doell and Cox, two "extremely likable people who were easy to get along with."

The acceptance of the offer to work at the Survey engendered some ill will between Dalrymple and Evernden. It was later exacerbated when Dalrymple collaborated with Cox and Doell, providing dates for the historic first reversal scale of June 1963, which appeared in *Nature*;* he had derived those dates in the laboratory of Evernden and Curtis while still a student at Berkeley. Evernden was disturbed about the use, without his consent, of data derived in his laboratory. Evernden was, as has earlier been shown, a brilliant, persevering worker who was without equal "in terms of sheer energy and ability to accomplish a lot in a short period of time." Dalrymple noted that "I don't know anyone who can do better at that than Jack Evernden. He was the real driving force behind the Berkeley lab for years and was typically there [before 7] o'clock in the morning and [often] wouldn't leave till midnight. He did that day after day after day."[92] Evernden was in fact a founding father of the young-rock dating program at Berkeley, had a profound commitment to the research program, appeared nearly compulsively driven toward its success, and was thus most likely to be deeply concerned about the activities and products of the laboratory and its staff. Dalrymple's idea of what he could do with data

*This ill will is the reason alluded to in Chapter 4 that Cox, Doell, and Dalrymple used Evernden's dates at third hand, as reported by Blanc and cited by Rutten, rather than verifying them from Evernden's laboratory records and specimens.

that, as he recalled, "was derived by my own effort and on my own time, in that laboratory," was not congruent with Evernden's.[93]

The U.S. Geological Survey's potassium-argon capabilities began in 1959, as recounted in Chapter 3, when Cyrus Creasey and Paul Bateman built a mineral-separation laboratory and gas-extraction line in Menlo Park. Two Berkeley-trained radiometrists had arrived in Menlo Park before the end of 1961—Ronald Kistler, in June 1960, and John Obradovich, in December 1961. For almost two years, starting October 6, 1960, argon gas samples extracted at Menlo Park were "bootlegged" to Berkeley for analysis on the Reynolds mass spectrometer. In 1960 the Isotope Geology Branch was organized under Samuel Goldich, who transferred the argon-extraction line at Menlo Park to his jurisdiction and purchased a Nier-type mass spectrometer in 1961. It is important to recall that the requests for rock dates far exceeded the capabilities of the potassium-argon laboratory at Menlo Park (and the one in Denver) from its inception.[94] The limited capability and great demand engendered competition in securing dates, and serious questions and frictions arose regarding priorities. Cox and Doell grew increasingly aware that their need to secure a large number of dates in a short time could not be met by the Survey alone.

An important opportunity to expand that capability arose in 1962 as an outgrowth of President Kennedy's success in funding a cooperative program with Japan for the scientific investigation of the Pacific Ocean.* William Benson of the National Science Foundation, a friend of Doell's who had earlier played a role in securing funds for his paleomagnetic research at Toronto, asked Doell about the possibility of a cooperative project with the Japanese in paleomagnetism. By that date Verhoogen, Cox, Doell, and other Americans had long been in contact with Takesi Nagata, Seiya Uyeda, and others in Japan. The organization of a conference was the preface to the project, and a great deal of transoceanic correspondence (soon to be cited) ensued before the First U.S.–Japan Seminar on Paleomagnetism, which was held on December 6 and 7, 1962, at the University of Tokyo. It was attended by Balsley, Doell, Verhoogen, Victor Vacquier, Robert Hargraves, Arman Frederickson, Robert Oetjen, LeRoy Sharon, Stanislaw Vincenz, and Roy Hanson; Japan was represented by Takesi Nagata, Syuniti Akimoto,

*"It is most likely that the President's name had only ritual significance. The real reasons were all at a much lower bureaucratic level, probably within N.S.F." (unattributed on request).

Naoto Kawai, Hisashi Kuno, Minoru Ozima, N. Watanabe, and M. Yamanaka.

The seminar was centered on a review and state-of-the-art summary of paleomagnetic research, in both the United States and Japan, which set forth what the group regarded as the main research areas and how much emphasis was to be accorded each.* The concluding section of the summary report listed the most important subjects for investigation in the cooperative program. The first topic mentioned was "paleomagnetic studies of Miocene to Recent rocks of the Pacific Ocean and surrounding areas":

> Especially important are studies of Pleistocene and Pliocene rocks. . . . It will be necessary for complete success in this program to have all rocks studied well dated by the K-Ar method. From these studies, the following knowledge should be gained: (1) the bounding ages of the different polarity epochs, (2) morphology of the geomagnetic field during each epoch, (3) relative amount of variation between epochs and between areas.† These studies should be most important of those undertaken under the cooperative program.

After archeomagnetic studies, paleomagnetic studies of Cretaceous rocks, and theoretical studies of rock magnetism, "*Age Dating*, because of the extreme importance in these studies of dating problems, . . . is included as a separate item. It is very clear that the success of the program depends on a very large amount of work in this field."

Prior to the seminar, Doell had written to Nagata, who was then visiting at the University of Pittsburgh; in that letter of September 24, 1962, he stressed the heavy reliance of paleomagnetists on "geologists interested in Pliocene-Pleistocene stratigraphy and correlation, and on isotope age dating methods—principally the K-Ar method." Doell's intention to use the potassium-argon capability of the Isotope Geology Branch at Menlo Park for the project was implicit in his statement that he had suggested to E. Dale Jackson, a Survey geologist and administrator, that Ronald Kistler be sent to Japan to discuss potassium-argon dating with Japanese scientists (David Hopkins, in the area of Pliocene-Pleistocene stratigraphy, and Meyer Rubin, in carbon-14 dating, were also included). Doell

*The "Report of the U.S.-Japan Seminar on Paleomagnetism," prepared by Nagata and Doell, Drafting Committee, is Document No. 22.

†These three goals show plainly that the Rock Magnetics Group accepted the existence of polarity epochs as such and was working toward their temporal definition.

added that those "latter three persons . . . are very competent in their respective fields, and moreover, are already interested in and working with us on the problems concerned." *

Other correspondence and interviews all bear out the primary concerns of Cox and Doell with the need to acquire potassium-argon dating capability for their objectives. Doell wrote Nagata: "Among the many paleomagnetic studies that might benefit from U.S.–Japan cooperation, we chose that of geomagnetic field reversal and secular variation as ones we were most interested in and most capable of carrying out. These studies, as you know, are the ones that we've been working on for some time, and our proposals to the N.S.F. *were essentially for an intensification of the work we are already doing*." †

It was not until June 25, 1963 that the National Science Foundation awarded $120,000 for partial support of the project "Paleomagnetic Studies of Selected Miocene through Recent and Historic Rocks of the Eastern Pacific Basin Area" to the U.S. Geological Survey.[95] Almost simultaneously with the N.S.F. award, Brent Dalrymple, who had been listed as a principal investigator on the research proposal, joined the Rock Magnetics Project at Menlo Park.[96] Dalrymple had been offered the Survey job in December 1962; it was apparent to Cox and Doell by that time that the proposal, which had been submitted in September 1962, was virtually assured of success because of a number of factors. These included the apparent eagerness of the National Science Foundation to comply with the President's wishes for a U.S.–Japan program, and the involvement and support of internationally recognized geophysicists in both the United States and Japan. Rock magnetists made up a rather small, intimate community, who were fully sympathetic to their research aims.

*The letter is Document No. 23. With evident concern, Doell remarked in closing: "Dr. [Arman] Frederickson plans to submit a proposal [for studies under the U.S.-Japan Committee]; his studies concern self-reversal in connection with reversals of the geomagnetic field. I have not seen it but from the subject matter I am somewhat apprehensive that his studies might duplicate or be competitive with ours, and maybe even more competitive with what we hope will be an important part of the Japanese program."

†Doell's letter of x-5-62 (emphasis added) is Document No. 23A. Doell noted, alluding to his earlier remarks on ix-24-62, that "I hope, Tak, that from the above [explanation of prior arrangements and research intentions] you will better appreciate . . . my apprehension concerning Prof. Frederickson's proposal, since in my 'frame of reference' this did seem to appear as a competitive proposal." Cox and Doell were additionally concerned with the possibility that Rutten might extend his polarity-scale work of 1959.

The "final" version of two proposals to the U.S.-Japan Committee for Scientific Cooperation in Paleomagnetism is Document No. 24. A cover memorandum notes routing from E. D. Jackson to R. M. Moxham on ix-7-62.

During late 1962 and early 1963, before the arrival of Dalrymple in June, the Nier mass spectrometer was not yet operating in Menlo Park; the demand for dates was far in excess of the few samples being "bootlegged" to Berkeley for analysis, which virtually precluded the possibility of dating samples for the Rock Magnetics Project. There was some resentment in Menlo Park regarding the privileged status of the project, which enjoyed the indulgence of administrators with "clout" who viewed Cox and Doell as "the shining lights of the Survey."[97] The late Robert M. Moxham, who became chief of the Theoretical Geophysics Branch in 1962, recalled that "Cox and Doell were first-order scientists and thus deserved funding and support. I tended to lean over backwards and fund Cox and Doell at the expense of the others."[98]

The technical staff and resources afforded the Rock Magnetics Project was in marked contrast to that allocated to the Isotope Geology Branch at Menlo Park. John Obradovich recalled, "When Ron Kistler and I built the isotope lab there, we were the machinists; we were the guys who did the dirty work and built everything there. Cox and Doell were able to sell their concept of what they wanted to do; people would listen, and they got the money. . . . We [geochronologists] were good enough to do the analyses but not allowed to decide on the merits of the problem."[99] By the time of Dalrymple's arrival, the Isotope Laboratory's relationship with several segments of the staff at Menlo Park had become strained over the questions of priorities in dating rocks. The Rock Magnetics Project was viewed by Kistler and Obradovich, as it was by most others, as not entitled to special privileges. The strain continued to grow after Dalrymple's arrival. It is important, in this context, to note again that in 1962, paleomagnetics was not widely understood nor was it regarded by most geologists as deserving of special priorities.*

Long before the formal award of the N.S.F. grant in June 1963, an agreement had been worked out in "extensive conversations" between Doell and Goldich.[100] Robert Moxham, in a letter for the record, noted that Goldich agreed on the urgent need for his lab to undertake potassium-argon work: "Doell has in mind hiring, at GS-11, Brent Dalrymple. . . . It is estimated that to fully set up a K-Ar project in our lab will include, in addition to salary, about $20,000. Doell indicated he would be willing to cover this cost with

*See Appendix B for a delightful parody of the work of the Rock Magnetics Project at Menlo Park.

no increase in funds, if absolutely necessary."* Moxham later noted that "Cox and Doell prevailed upon me to get the budget to hire Dalrymple. There was no escaping the logic of their arguments. I had no reluctance scientifically—only budgetary."[101] The agreement between Doell and Goldich was, from the first, to place Dalrymple in the Rock Magnetics Project of the Theoretical Geophysics Branch, and not in the Isotope Geology Branch. It was agreed that Dalrymple was to use the laboratory of the Isotope Geology Branch, which was run by Kistler and Obradovich. That presumption was so strong that Dalrymple, before coming on the payroll, commuted to Menlo Park from Berkeley with Obradovich for three weeks in order to familiarize himself with the laboratory and aid in preparing for his own residence there. Upon arrival at Menlo Park, Dalrymple helped in the laboratory in the construction of spikes, examined thin sections for their suitability in dating for the Rock Magnetics Project, and prepared his doctoral thesis for publication.†

Within a few months, a great number of samples of magnetically determined rocks had accumulated, as Dalrymple was refused access to the use of the Isotope Laboratory because of "various excuses" offered by Kistler or Obradovich. A mounting sense of frustration prevailed in the magnetics laboratory. The historic first polarity scale by Cox, Doell, and Dalrymple had appeared by June 1963 in *Nature* and had been followed on October 5 by the scale of McDougall and Tarling.[102] The sense of urgency that gripped the Rock Magnetics Project could only be allayed by rapidly dating the great number of accumulated specimens in order to improve and extend the scale and publish the new findings. (The development of the polarity scale *per se* is treated in Chapter 6.) Doell had made extensive collections in 1960 and 1961, and Allan Cox and David Hopkins had earlier returned specimens from the Pribilof Islands of Alaska. Dalrymple had already examined thin sections of those rocks and selected specimens for dating while still at Berkeley, and

*The three-year budget for the Rock Magnetics Project, submitted by Doell, Cox, and Dalrymple on v-3-63 to the Office of International Science Activities, U.S.-Japan Cooperative Science Programs, stipulates $60,000 for isotope analyses for age determinations (100 per year) and analytical laboratory costs of $13,800; thus more than half of the entire N.S.F. grant was devoted to obtaining dates, as worked out in the agreement between Doell and Goldich (Document No. 27).

†The monthly report of the Rock Magnetics Project for the month ending vii-20-63 noted that G. Brent Dalrymple joined the project and "would be responsible for our K-Ar dating studies. . . . Particularly important is the K-Ar dating program for Miocene to Recent rocks of paleomagnetic interest that has been worked out in conjunction with the Isotope Geology Branch" (Document No. 27A).

there were "literally several hundred potential samples" waiting to be dated. Marvin Lanphere, trained by Gerald Wasserburg at Cal Tech, and assigned as a geochronologist to the Alaskan Geology Branch, also was not able to gain access to the Isotope Laboratory in order to meet the dating needs of his branch.[103] Both Lanphere and Dalrymple were told by Goldich that the laboratory was in Kistler's and Obradovich's hands; they in turn invoked Goldich as the authority who had denied access to Dalrymple and Lanphere. During that first year in residence, Dalrymple was able to run only about 20 samples, and those only after a "forced agreement" extracted by Doell that Dalrymple be allowed the use of the laboratory for two weeks. The experiments proved useless due to improperly prepared spikes that had been supplied to Dalrymple.[104]

A most telling series of letters from July 5, 1963 to January 1, 1964, most of them between Goldich and Doell, reveals much of the basis for the disagreement, which according to Dalrymple "cost the Rock Magnetics Project a full year's delay in acquiring dating capability."[105] About two weeks after Dalrymple's arrival, Goldich wrote to Doell and Cox that "we have made considerable progress since our discussions, which led to a cooperative investigation wherein we shall provide K-Ar age determinations for your paleomagnetic studies." Goldich also noted that (1) a new mass spectrometer was to arrive in Denver similar to the Menlo Park instrument; (2) Harold Mehnert was being trained in Menlo Park to operate the Denver machine, and Goldich hoped that Dalrymple would work with him in the future; (3) it would be desirable for Dalrymple to spend some time with Mehnert in Denver to "get him off to a good start"; (4) problems in Denver required that Mehnert would need Dalrymple's help in Denver; (5) the extraction train and mass-spectrometer parts were being charged to the Rock Magnetics Project account.[106]

The letter of July 5, 1963, was the last one of the cited series that was devoid of some note, however subtle, of the growing discord between the Rock Magnetics Project and the Isotope Geology Branch. On October 23, 1963, Goldich wrote:

It is my understanding that this is a cooperative effort. *We are neither a service Branch,* nor an inactive group. We have a program of our own which is also significant, worthwhile, and of great interest to our people and to many Survey geologists. For this reason I agreed to set up a laboratory in Denver to avoid obvious problems and difficulties that would arise if we tried to do this work in our Menlo Park Laboratory. . . . During the tooling up in Denver, possibly some arrangements can be made in Menlo Park for the analysis of critical samples; however, this should be an equitable arrangement and satisfactory to Ron Kistler. *Joint publication* is not only desirable but I feel it is imperative in fairness to the great contribution our Branch will be making. Ron Kistler has a great investment of time and hard work in our Menlo Park Laboratory. I should like to mention that our investment in the mass spectrometer and auxiliary equipment comes close to $50,000. . . . I hope we can obtain reliable ages in quantity for the paleomagnetic project.[107]

Doell responded on October 29, 1963, with a four-page typewritten letter that elucidated a number of formerly enigmatic, important particulars:

We realize, of course, the additional problems you undertook in order to do the isotope work for our project. However, during the early discussions (Western AGU meeting at Stanford), we considered several other possibilities, including that of our doing all our work, and you indicated that the best way for us to proceed was to hire Dalrymple on our project (to select and prepare samples, coordinate efforts, etc.) and that the isotope work should be done in your Branch—we supplying the funds that would be needed to set up additional facilities. These decisions were formalized less than two weeks later (Moxham's memo of January 8, 1963).

We hoped and expected that these funds would allow you to proceed with our work in a manner that would not affect the fine and significant work of the Isotope Geology Branch.

We also understood at that time that you thought we would be getting reliable isotope data by this fall, and possibly even by last July if everything went well. We certainly could not expect more, and with a formal "go ahead" a short time later, we were looking forward to having the program rolling before now. However, as we know from bitter experience in getting our own laboratory going, it is often difficult to realize such objectives. We are a bit puzzled, though, by your suggestion that additional funds would permit an acceleration of the program—one that we had hoped would be in operation by now.

As you point out in your letter, rapid developments in the field (See McDougall and Tarling in Oct. 5, 1963 issue of *Nature* and cf. our *Science* article!!), which we anticipated almost a year ago, do require as rapid a realization of our plans as feasible.* In this connection, we hope that the work load at Menlo Park will permit Brent to get a few critical runs done here while the Denver facilities are being completed. After recent talks with Ron Kistler, I feel that he fully realizes our problem and will do all he

*The race for the polarity-reversal scale was well under way by this time, and a sense of competition pervaded the project.

can to schedule this work, hopefully in a manner that won't disrupt the other programs Ron has.

Doell continued in a discussion of Dalrymple's contribution in making the Menlo Park isotope laboratory functional, and equipment purchases by the Rock Magnetics Project for the isotope laboratory; he also noted other contributions by his group in aiding the isotope group. Doell then turned to what he had earlier been loath to discuss:

I have been avoiding our basic problem—that of authorship—but I wanted first to make clear our feelings on the specific points you raised about additional problems our work has caused, the "accelerated program," and the recognition of contributions to the Menlo Isotope Lab.

I feel that the main reason for our "authorship problem" stems from the different manner in which we approach "service work" (a poor term in this connection, perhaps, but I can't think of a better one). I feel that this aspect of our project and your Branch is not too different and I should like to spend a few words explaining our position on "service work," since it seems almost antipodal to yours.

We encourage others who are interested in, or have need of, rock magnetic data to undertake a "do it yourself" program. We have helped personnel in the Regional Geophysics Branch and Special Projects Branch to build and set up equipment patterned after ours in Washington and Denver, and we have trained their personnel in the proper use of this equipment. We also encourage people to use our own laboratory here at Menlo Park, asking only that their interest be great enough to entice them to learn the techniques, interpretations, etc., so they can do the work themselves.

We certainly recognize that this is a different approach from yours—requiring that all isotope studies in the Survey be done in the Isotope Branch—but I suspect that both approaches have merit. Otherwise one of us would no doubt have to change. In any case, I feel we should recognize differences such as these and seek a way in which we can get the job done and bring credit to both programs.

Getting down to the specific matter of authorships, it now appears that our isotope work that involves geomagnetic phenomena will be pretty well supervised by Brent. He selects (or sadly, more often rejects) material suitable for reliable determinations, he determines how it is to be prepared (whole rock, feldspar, etc.), he submits samples for, and analyzes data from, K analyses here at Menlo, and he submits (or will) material to Denver to have the Ar extracted and measured by methods that he will have specified. Meanwhile, the magnetic work is undertaken on samples Allan Cox and I select, the methods used are those we specify and that we have developed and trained our personnel to carry out. (It also now seems likely that Brent will train the Denver isotope people in the extraction methods.) Finally, it seems very unlikely that any persons other than Brent, Allan, or myself will interpret these data or write the reports describing their scientific significance. *We feel that there is an obvious and clear-cut demarcation between what constitutes authorship and what warrants acknowledgment in these particular studies involving geomagnetic phenomena.*

It seems that you feel that the isotope analysis work is, in general, of publishable significance in its own right regardless of other aspects. I sincerely respect this view and understand the need your personnel have for authorship. However, as stated above, the fulfillment of this need on the geomagnetic phenomenon work is quite inappropriate, and we must seek other means to satisfy the need.[108]

Doell also, in a long paragraph, indirectly suggested that the Isotope Branch periodically publish a list of all isotopic work done "as a possible means of satisfying the publication needs of Isotope Geology Branch personnel." Doell closed noting that "as a matter of record, I should like to point out [that] to date we have not received a single K-Ar date specifically obtained for our program."

On October 31, 1963, Kistler responded to Doell's letter of the 29th and noted that (1) the Isotope Branch had, in fact, not produced any of the 100 per year promised dates for the Rock Magnetics Project; (2) the decision "decided at the branch level of administration" for the dating to be done in Denver "stands"; (3) Kistler and Obradovich felt that they should "have a certain amount of time and freedom to work on some problems that require use of the equipment" they had built; (4) the Menlo Park isotope lab had on hand 66 samples from projects in Denver, Washington, and Menlo Park to determine in addition to a request from Marvin Lanphere to run 39 samples for the Alaskan Project; and (5) 150 of the 550 spikes that Dalrymple had worked on went to the Magnetics Project, and that the electronic technician employed "by our two projects spent at least 90 percent of his time in [the magnetics] lab."[109]

On November 4, 1963, Goldich responded to Doell with acerbic brevity:

Thanks for your long memorandum. Rather than dwell on the inaccuracies and omissions I should much prefer to say that I am sorry that the progress in our Branch has not been better, and I hope that we can have the laboratory in Denver running soon. I feel this is important in view of the competition you are facing in the area of your research. Please be assured that I am doing everything I can. Once we have the apparatus in operation

I think we should stop and review our cooperative arrangements. I shall retain a Philadelphia lawyer to cope with you at times. Best regards and of course our best wishes for the success of your project.

Sincerely yours,
S. S. Goldich, Chief
Isotope Geology Branch

P.S. Your suggestions for running my Branch are not really appreciated.[110]

The grant from N.S.F. would have permitted the Rock Magnetics Project to build their own potassium-argon laboratory, but the Survey had had poor experiences with mass spectrometers at the Hawaiian Volcano Observatory and elsewhere; Goldich thus wanted all isotope dating concentrated in the Isotope Geology Branch. The need to secure independent K-Ar dating capability became increasingly apparent as time wore on and the needs of the Rock Magnetics Project went unmet. Dissatisfaction grew in many quarters of the Survey with the way the Isotope Branch exercised authority in meeting requests for dates, and late in 1963 an *ad hoc* committee was formed to investigate the isotope operation. Doell had not been invited to participate in the deliberations by the administrators who had formed the committee, and he interpreted that fact as reflecting a lack of confidence in his leadership. He was prompted to tender his resignation as Deputy Assistant Chief Geologist for Experimental Geology at Menlo Park, but the resignation was never effected.[111] The reasons Doell was excluded were never made clear; apparently Doell himself, a central figure in the controversy with the Isotope Geology Branch, was probably regarded as unsuitable for membership on a committee working to solve the problem.

The Rock Magnetics group's sense of frustration culminated in a letter from Doell to Robert Moxham in late December 1963.[112] Doell addressed the major aspects of the problematic situation, emphasizing the obligation to complete the research that the Survey had incurred in accepting the N.S.F. grant. Doell stipulated:

As principal investigators we have now decided that the only possible way we can proceed is to begin construction of complete K-Ar facilities in the shortest possible time. . . . We see no other way in which we can (and over the next two years) fulfill our obligations to N.S.F. . . . I feel very strongly that in all fairness to us and to N.S.F., we should be allowed this freedom of action [specifically, to build gas extraction lines and a mass spectrometer within the Rock Magnetics Project for the project's own use] to meet our obligations or we should terminate our studies at once.[113]

At that time both Robert Moxham, chief of the Theoretical Geophysics Branch, and Samuel Goldich, chief of the Isotope Geology Branch, were under the jurisdiction of Wayne R. Hall, Assistant Chief Geologist for Geochemistry and Geophysics. Doell recalled,

> It finally got to the point that I told Wayne Hall and Bob Moxham that I would have to leave the Survey; Brent [Dalrymple] was of the same opinion, [as was Cox,] and it looked like they were going to let us leave. I was crying my woes to Paul Bateman one day, and he said, "Don't do anything; let me talk to Wayne Hall. I've known him for a long time." . . . I'm sure it was as a result of Paul Bateman's talking to Wayne Hall that he finally decided to give us permission to come back and present our case.[114]

Dalrymple recalled, "Dick Doell and I flew back to Washington to an arranged meeting in Bob Moxham's office with Dick, myself, Bob Moxham, Wayne Hall, and Sam Goldich to get the question of laboratory facilities straightened out."[115] Dalrymple also sent a letter sketching the costs involved in building a mass spectrometer for the project.[116] Doell described the meeting:

> We asked that we be given permission to set up our own isotope dating facility in terms of building and operating a mass spectrometer and extraction lines. These were the big holdups in the Isotope Geology Branch. I'm sure they hadn't been able to get these things going because of a lot of other very important commitments. By then in the [Magnetics] Project with Brent Dalrymple, our machinist Major Lillard, and our electronics man Nat Sherrill, all working with us, we had what we felt were more than ample facilities and personnel to do this. The only thing that had been holding it up before had been the general policy at that time, and I think a reasonable one in terms of past history, that all isotope studies had to be concentrated within the Branch where there was a lot of expertise.* I'd flatly told them before that we were going to leave the Survey, because we didn't feel that we could carry out the obligations of the grant, and that we would write the National Science Foundation why we were doing it. I guess we had a pretty powerful lever. They weren't about to let us go through with that one, once they realized that we probably would.[117]

*Recall that Goldich had been called to form the Isotope Geology Branch in response to the Survey's diminished reputation in isotope dating during the 1950's. Clyde Wahrhaftig (oral comm., i-17-80) noted that the branch was also formed to control activities involving radioactive materials. Doell and other informants praised the job Goldich did in organizing the branch and bringing it to a state of excellence. In keeping with earlier remarks on the tendency within the Survey to centralize, Wayne Hall (oral comm., x-28-80) remembered that a conflict arose once over organization between Ralph Cannon, who was doing lead-isotope dating in Denver, and Goldich, who insisted that each type of specialty be segregated. Cannon's request that each research project group be diversely composed to enhance functionability was like the request of Doell and Cox.

Brent Dalrymple and the Reynolds-type glass argon mass spectrometer, at Menlo Park, about 1971.

Doell's request was granted. He and Brent Dalrymple returned home and started building tables in Doell's garage to make the extraction lines. "Our whole idea," said Doell, "was to duplicate exactly the facilities at Berkeley, . . . to duplicate the Reynolds-type machine."[118] Dalrymple noted,

Between the time we began construction of the lab and the mass spectrometer and the time we had the mass spectrometer operating was six months. Dick Doell and I built the extraction-line tables on the weekends and at night. The Bayard-Alpert valves, or the "Grandt" valves as they were later known, and the source and collector parts we bought from Cliff Grandt at Berkeley. We had the valve drivers and the brackets built at a local machine shop to drawings that I provided the machinist. The rest of it was built out of glass tubing. We bought the extraction-line bottles from

Morley Corbett, who blew them at Berkeley; at the same time, we ordered a mass-spectrometer tube from Corbett. We paid about $3,500 for that tube with the conductive coating on the inside, and he took responsibility for providing the entire mass-spectrometer analyzer. It was basically Berkeley equipment, copies of the ones in the Geology Department.[119]

Within one month of their return from the Washington meeting, one gas-extraction line was in operation, and a second was under construction along with the Reynolds machine.[120] By May, construction of the mass spectrometer was in its final stages, and data had been gathered to calibrate the new argon 38 bulb tracer system.* Dalrymple and Lanphere also separated about 1,000 grams of muscovite from a Cretaceous granite from Death Valley to distribute worldwide for use as an interlaboratory argon standard—the first one produced since the early 1950's.† On July 29, 1964, Dalrymple, justifiably excited, sent Robert Moxham news that they had produced a beam in the mass spectrometer and had gotten a good peak shape and background condition; he concluded that after baking and calibration the instrument would be ready for routine operation. In the August 20 monthly report he noted that the dating of the Sierra Nevada and Japanese samples for the reversal time scale had begun.[121] Thus did the Rock Magnetics Project acquire its own crucially necessary potassium-argon dating capability, after a year-long delay in which a substantial backlog of magnetically determined rocks from a great number of localities went undated. It is tempting to ask, in retrospect, the extent to which the delay in dating forestalled the confirmation of seafloor spreading.

During 1962 and early 1963, while Cox and Doell sought dates at Menlo Park in fruitless desperation, Ian McDougall had ready access to the potassium-argon facility built by Jack Evernden at the Australian National University, but McDougall was not interested in polarity-reversal studies—yet.

McDougall and Tarling

Ian McDougall has been mentioned before: in 1960, as part of the Australian National University's entry into potassium-argon dating

*Bulb tracer systems, which are used to mete argon gas, generally consist of a gas pipette of specific volume coupled to a reservoir (bulb) that can be evacuated, checked for leaks, and filled with tracer argon such as argon 38. Bulb systems are usually more precise than manifold tracer systems, which consist of a large number of individual tracer tubes joined by capillary tubes to a manifold, thus allowing a large number of tracers to be prepared simultaneously.

†These developments were reported in the monthly report for iv-21 to v-20-64 (Document No. 20H). Dalrymple noted, "The new Ar-38 bulb and the muscovite standard were a *joint* proj-

of young rocks, John Jaeger turned McDougall from research in petrology to geochronology through a carefully planned postdoctoral year in the potassium-argon laboratory at Berkeley. At the beginning of the 1960–61 academic year, he began instruction in K-Ar dating at Curtis and Evernden's laboratory on the same day as Brent Dalrymple. McDougall left Berkeley in the spring of 1961, at the close of his postdoctoral year; in October 1963, he and Don Tarling published the first Australian polarity-reversal scale, less than four months after the publication of the first scale by Cox, Doell, and Dalrymple. Once having begun such studies, McDougall assumed leadership in the Australian effort to define a polarity-reversal scale and collaborated first with Tarling and later with François H. Chamalaun, and to a lesser extent with H. L. Allsopp and Hans Wensink. In what quickly became a race for the polarity-reversal scale, McDougall provided the major impetus and continuity of effort in Australia to gather data on the reversal behavior of the earth's magnetic field. He had departed Berkeley with the newly acquired skill to date young rocks, but he had not acquired along with that skill the sense of the profound significance of polarity-reversal studies that he might have gotten from repeated contact with Cox and Doell, nor did he begin collaboration immediately with a paleomagnetist upon his return to Canberra. The reasons behind his attitude lay largely in his professional background.

Ian McDougall, one of five children, was born in Hobart, Tasmania, in 1935, and, like his father, he attended a Quaker school for thirteen consecutive years. He entered the University of Tasmania in 1953. The Australian educational system, which is closer to the English or the European rather than the American model, required that he specialize in the last year of secondary school. An external examination was set by the University of Tasmania, based on five subjects selected by the examinee; among his subjects McDougall chose chemistry, physics, mathematics, and geology. He recalled: "It was unusual to do geology as a final-year subject in high school, and that still remains the case in Australia. I had to make special arrangements to do it because it wasn't really taught at the school I was at. I had to do much of it at a technical college at

ect by Lanphere and me. In fact, Marv Lanphere helped me a great deal in building the lab for the Rock Magnetics Project. It was quite clear that the new lab was also his only hope of access to an operating lab as long as Ron Kistler and John Obradovich remained in charge of the main lab" (written comm., xi-8-78).

night school."[122] The University of Tasmania was in Hobart, thus it was natural for him to pursue his B.Sc. Honors degree there, which involved a fourth year of study with emphasis on research toward a thesis in geologic mapping.

The mapping included an area about 20 miles north of Hobart consisting of rocks typical of the Gondwana sequence that contained dolerite sheets in place. "At Hobart one would have had the ideas of continental drift as a matter of course—we had Carey there." S. Warren Carey provided an "inspiration" for McDougall; Max Banks (in biostratigraphy) and Allan Spry "gave one the basic geological training. Carey was more the man who made one think about the really broad aspects of earth science." McDougall was trained in a department like Berkeley's that strongly emphasized "a field study base, in which the staff would put you in an area where mapping was required" in order to fulfill the requirements for a B.Sc. Honors thesis problem. He was most attracted to petrologic study in hard rocks, and had become increasingly interested in the thick dolerite sheet composing Mount Wellington near Hobart. Allan Spry, the Senior Lecturer in petrology, mentioned that in an older thesis he had found a description of a much more acidic or silicic rock associated with dolerite than anyone had seen up to that time. McDougall traveled with Spry and Carey to the area and found what are now called granophyres.* With the encouragement of both Spry and Carey, McDougall began to consider study of the Tasmanian dolerites, which cover about half of the island of Tasmania, as a subject for his doctoral dissertation.

> Undoubtedly there was some connection [of the dolerites] with continental drift because of the widespread occurrence of similar types of rocks in all the Gondwana continents [India, Africa, Australia, South America, and Antarctica] of about the same age, as far as one knew at that stage. I think it may have had some bearing on my selection of the dolerite problem, but my immediate interest was understanding what happens chemically and mineralogically, after the emplacement and during the cooling, to one of these several-hundred-meter-thick sheets. I was also interested in other Gondwanaland dolerite rocks and thought that they must reflect a presumably major tensional event related to the breakup of Gondwanaland.[123]

Unlike his counterparts at most other Australian universities, in

*Very silicic differentiate of dolerite; a porphyritic extrusive rock with a micrographic, holocrystalline groundmass.

Ian McDougall, at left, and François Chamalaun operating the Reynolds-type mass spectrometer in the isotope dating laboratory at the Australian National University. McDougall's start in reversal studies was prompted by the work of the Rock Magnetics Group at Menlo Park. Once in his own laboratory he provided leadership and continuity of effort among the Australians and engendered a spirit of competition with the Californians that appears to have heightened productivity in both research centers.

which superior students were retained, Carey encouraged McDougall to go outside Hobart to pursue graduate study.

McDougall's decision to enter graduate school at the Australian National University in 1957 "arose from contacts that Carey had with J. C. Jaeger." Jaeger had a close connection with Tasmania, having come from the mathematics department there. As an undergraduate McDougall had been encouraged to attend the A.N.U., and Carey was "instrumental" in introducing McDougall to Jaeger.

Before taking up his graduate scholarship at the A.N.U. under Germaine Joplin, a petrologist trained at the universities of Sydney and Cambridge, McDougall and Joplin corresponded extensively. He conducted two months of field work in preparation for his petrologic studies of the Tasmanian dolerites; his overall aim was understanding the Red Hill dolerite dike, with emphasis on genesis, not on alteration.[124]

McDougall noted that John Jaeger too had been strongly influenced by Warren Carey; they were both on the staff at Tasmania before Jaeger was called to head the Department of Geophysics at the A.N.U. According to McDougall, "Jaeger's appointments were always in fringe fields, since he thought that the major breakthroughs would come in the interdisciplinary fields. Edward Irving was also an important intellectual influence along with Jaeger—we had a broader vision than most students because we had come through Carey, who was a broad generalist, and then exposure to Irving and Jaeger."

When Garniss Curtis visited the A.N.U. in 1958, McDougall heard him and was "much impressed by Curtis's approach and techniques." McDougall did not at that time consider entering the field of radiometry; as he was completing his Ph.D. thesis, he applied for a scholarship to Cambridge to continue petrologic studies.

In late 1959, before McDougall had completed his thesis, Jaeger informed him that he was already making arrangements to get into potassium-argon dating and to take over the Department of Radiochemistry in the same school, which meant acquiring the Metropolitan-Vickers mass spectrometer and John Richards. This step gave Jaeger the base to expand seriously into isotope geology. (How Jaeger acquired K-Ar dating capability for the A.N.U. and how Evernden assembled and operated the laboratory there were detailed in Chapter 3.) The commentary on Jaeger's force of character from a number of informants leaves little doubt that the young McDougall was strongly prevailed upon to take up a year of study under Curtis and Evernden at Berkeley. McDougall recalled: "I felt very diffident about entering K-Ar radiometry when Jaeger first approached me. I had a great deal of diffidence about whether I had sufficient math, chem, and physics background in order to enter K-Ar radiometry. I know I had this feeling right through until I went to Berkeley, and I think it probably persisted there."[125] In January 1960 he applied for a Commonwealth Scientific and Industrial Research Organization Scholarship; by April 1960, at the time McDougall submitted his Ph.D. thesis, arrangements had been made

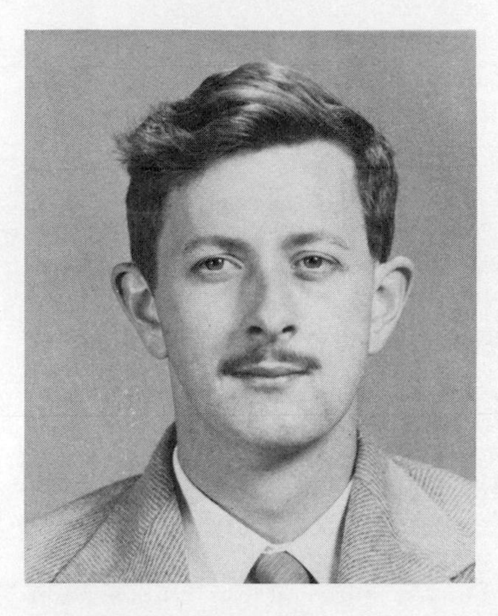

Ian McDougall had this passport photograph taken in 1960 just before leaving Australia to begin his postdoctoral fellowship at Berkeley.

through Curtis to spend the postdoctoral year in Berkeley, and the scholarship had been awarded.

On the way to Berkeley in August 1960, he spent "three or four days in Hawaii" in order to visit Kilauea Volcano, since he was quite interested in lava lakes and had earlier corresponded with the staff at the Hawaiian Volcano Observatory. Although he knew he would be learning potassium-argon radiometry in the coming year at Berkeley, he had "no thought of dating volcanic rocks on that first visit to Hawaii." McDougall recalled: "John Obradovich taught me the ropes and gave me my first lessons in glassblowing and running an extraction line." Early in McDougall's year at Berkeley, John Reynolds gave a course of lectures on mass spectrometry and isotope studies, which "was a very valuable introduction . . . and the only formal teaching" that he had at Berkeley. The remainder of his education there "was learning by doing and watching people like Obradovich. . . . Curtis was very kind and considerate." After an early period of learning techniques and running samples under the general direction of Obradovich, he came to feel that he might start examining samples of the Tasmanian dolerites that he had earlier sent to Curtis. During the next year, the samples were dated successfully. According to McDougall, "The successful dating of the Tasmanian dolerites opened the possibility of dating other dolerites; before leaving Berkeley I wrote to South Africa to get Karroo dolerites and others to attempt to date. I was planning on doing both mineral separations and whole-rock [dates]. Everyone regarded the dolerites as Mesozoic and related to the breakup of Gondwanaland. The Karroo had a spread [of ages] from 200 million years to the Tasmanian [dolerite] age; the South American dates were early Cretaceous."[126]

In the fall of 1960, McDougall attended the meeting of the Geological Society of America in Denver and heard Stanley Hart deliver his important paper on the suitability of hornblendes and pyroxenes for K-Ar dating (detailed in Chapter 2).[127]

I was quite impressed and encouraged to attempt to date the Tasmanian dolerites and separate the pyroxene from those rocks. It was evident that there was too much alteration except for the chilled marginal phase. I did plagioclase and pyroxene mineral separations. Those data were obtained in the second half of my time in Berkeley [spring semester, 1961]. In retrospect, I think Curtis's support was such that he should have been a coauthor on that paper on the dating of a basic intrusion that I published in 1961 [in *Nature*]. It developed while I was in Berkeley that I decided to date the Hawaiian Islands, but I can't remember the origin of the idea. I read the literature on Hawaii. I couldn't get funds to do it at first and finally convinced Jaeger. He was away on study leave, and Mervyn Patterson had taken over as acting head of the Geophysics Department at the A.N.U.

McDougall would not have been able to start collecting the Hawaiian basalts without Jaeger's intercession. Jaeger was very supportive of McDougall's efforts to date basalts. "If there were limited funds, Jaeger would go for the project that was speculative if it had promise of an important discovery—he'd back the risky ones."[128] Edward Irving, John Richards, and John Sass were all of the opinion that Jaeger did not fund staff members on the simple basis of seniority.

McDougall rode from Berkeley to Menlo Park with Sherman Grommé to meet Cox and Doell: "It was well known by that time that I intended to date the Hawaiian chain." Cox and Doell had sampled in Hawaii from the point of view of secular variations; McDougall noted, "I do not recall great discussions about polarity reversals. They were very interested in hearing about dates on Hawaiian rocks. The interest that they expressed to me was not in polarity studies but in secular variation. They were interested in getting dates."* McDougall met with Cox and Doell a number

*McDougall, t. 3, s. 2, viii-14-78. Recall that Doell and Cox were concerned with the possible competition implied by Arman Frederickson's intent to do polarity-reversal studies (Document No. 23) and regarded Martin Rutten as an "interloper" (unattributed on request). They were

of times, including a lunch with Jaeger, who was then visiting Berkeley, but in all of those meetings he could not recall "anything about talk on polarity reversals in relation to what I was going to do."

It is clear in my mind that I had decided to attempt to do this work in Hawaii not on the basis of polarity studies at all;* it was because of this question of whether one could obtain ages on volcanic rocks, and Hawaii was a classic place to attempt to do this. Excellent exposures, very good understanding of the geology, and, to boot, this geomorphologically recognized migration of volcanism along the chain—it was just an admirable situation, where you had geological control of a kind and good exposures, which I'd become very conscious of on my way to the States in 1960. I can remember so clearly flying down the island chain and seeing those fantastic exposures, particularly on Molokai. Not that I had any notion at that stage that I might try to do some dating, but once you've seen an area you recall such things.† My whole thrust at that time [in going to Hawaii to collect] was really a question of whether you could date these types of materials and, if so, could we demonstrate what the ages of volcanism in the Hawaiian Islands were and whether they stacked up against the geomorphologic evidence that had first been produced by [Charles Dwight] Dana in 1840 or thereabouts and subsequently supported by a large number of papers by diverse individuals including [Harold T.] Stearns [1946].‡

---

apparently eager for McDougall's dates on young basalts from the Pacific basin, but at the same time they did not want to alert potential competition by emphasizing polarity studies over secular variations. On the other hand, Doell noted (in Document No. 29A) that McDougall visited the magnetics laboratory in the summer of 1961 "to discuss radiometric dating of reversal [sic] horizon in Hawaii." If Doell and Cox did, in fact, fully inform McDougall of their intentions in polarity-reversal studies, it was most likely in the hope of securing data from one of the two centers (Berkeley and Canberra) they knew were potentially capable of dating very young rocks. (They were ignorant of the potential of Folinsbee's group at Edmonton.) There was then, as yet, no hint that anyone at the A.N.U., including Edward Irving, saw any virtue in polarity-reversal studies, and Doell and Cox may not have viewed the Australians as imminent competitors. David Hopkins recalled that "the Hawaiian rocks were regarded by Doell and Cox as difficult and perhaps impossible to date, based on low potash content of the whole rock and general lack of feldspar phenocrysts in those tholeiitic lavas (I think that Brent [Dalrymple] was more or less unwilling to tackle them)" (written comm., xii-10-78).

*Of McDougall's first two publications treating volcanic rocks from the Pacific (McDougall, 1963a and b, on Hawaii and Fiji, respectively), the first contained only a brief remark on the possibility "of dating polarity-reversals in Hawaii . . . by using the potassium-argon method," and no mention of rock magnetism appears in the Fiji paper.

†In Appendix C we learn of a similar important influence of the Hawaiian Islands on J. Tuzo Wilson with regard to his formulation of the hot-spot theory.

‡McDougall, t. 3, s. 2, viii-14-78. Dana and others were correct in surmising that the chain of volcanoes were sequentially graded in age, with the youngest still active. The Hawaiian chain appears to be a thread ridge or nematath that was produced by volcanic eruptions as the Pacific tectonic plate passed over a nearly fixed hot spot or thermal plume in the mantle; this idea was first advanced by J. Tuzo Wilson in 1963 (as is discussed in Appendix C). Isotopically dated rocks of different island chains indicate that the Pacific plate moved northward a few inches per year from about 100 to 60 million years ago, but changed direction to the northwest and slowed to about 2 inches per year about 40 million years ago.

During the spring of 1961, John Jaeger visited Berkeley, and in response to an application made earlier by McDougall for a staff position at the A.N.U., Jaeger made it clear that McDougall would shortly receive an appointment and return to operate the potassium-argon laboratory that Jack Evernden had assembled. McDougall was at that time "a poor impoverished student" and requested funds of Jaeger to collect basalts in Hawaii for one month during August before returning to Canberra.

At about that time, through correspondence with people at the A.N.U., McDougall learned that Don Tarling would be collecting basalts in Hawaii for his doctoral thesis, under Edward Irving, on paleomagnetic secular variations in the Pacific basin.*

On departure to Hawaii, McDougall agreed to keep Cox and Doell informed of his dating efforts in Hawaii.† Before he left, McDougall learned where Cox and Doell had collected in Hawaii;‡ he remembered "on the island of Kauai, seeing drilling holes from the collections produced by Doell on his visit there." None of the samples McDougall collected were oriented for magnetic determination; all were collected "specifically for dating." He spent about one month in the field; two or three days on Kauai (the oldest of the big islands), three or four days in western Oahu, two or three days on Molokai, a few days on western Maui, and then on to the Island of Hawaii, where the Hawaiian Volcano Observatory supplied a vehicle and Donald Richter, a geologist who had earlier accompanied Doell, took him to Kohala, the oldest volcano on the island.[129]

McDougall met Don Tarling in Honolulu during the Pacific Science Congress in August 1961 and "had conversations with him about where he was collecting and what he planned to do." They

*In a letter to Cox, on ii-19-60, Balsley touched on a "forthcoming trip to Hawaii . . . to do some paleomagnetic scouting for us" by members of the Rock Magnetics Project. Balsley explained that "our plans are a bit clouded at the moment because of the possible paleomagnetic expedition by the Australian National University worker to Hawaii. . . . I suppose they may regard our interest as claim-jumping. Jaeger, a month or two ago, mentioned the possibility of their doing some work in Hawaii. On the other hand, more than a year ago I asked Jerry Eaton [of the Hawaiian Volcano Observatory], when he was visiting Verhoogen in Berkeley, to send me some samples if he had the time (—he didn't), so I suspect we've all had our eye on Hawaii for some time." (The letter, which touches on other questions relating to Hawaiian rock magnetic studies, is Document No. 29.)

†McDougall, t. 3, s. 2, viii-14-78. In fact, he did not inform them of his dating efforts in Hawaii until ii-4-63 (letter from McDougall to Doell, Document No. 29-1), one week after the publication of his first Hawaiian dates in *Nature* (McDougall, 1963a).

‡The monthly report of the Rock Magnetics Project for the month ending vii-21-61 notes "Visitor: Ian McDougall (Australian National University) to discuss radiometric dating of reversal horizon in Hawaii." Doell also noted, "Measurements on the Hawaiian lava flows are continuing. Preliminary reports . . . on Hawaii are being prepared" (Document No. 29A).

did not attempt to coordinate their work. Tarling was "looking at secular variations," and McDougall was "interested in dating volcanism along the island chain." "Our contacts, which were no more than a few hours, were only about describing our respective projects. We were each trying to do our separate things." Evernden and Curtis had been "reasonably supportive" of McDougall's plan: "Curtis probably thought it was worth having a go but felt that there was small chance of success."[130] McDougall recollected that

> the attitude then was that it was important to try to get mineral separates. I had made quite extensive collections of basalts that had large plagioclase crystals, but I also collected a large amount of material that was fine-grained, typical basalts. The criterion then, as it is now, was that I just collected fresh material. Anything that was obviously weathered or oxidized or strongly iron-stained, I would never collect. I usually collected dense, fresh rocks. The nicest rocks were the things like the mugearites and hawaiites, which usually occurred higher up in the sequences. I was excited about the freshness of the samples, especially the mugearites—the likes of which I've never seen since.

McDougall spoke at length about his apprehensions regarding the dating of the Hawaiian rocks; those fears, present when he departed Berkeley in August 1961, were heightened upon discussing his plans at the Pacific Science Congress. He recalled that "people thought that what I was trying to do, in some senses, was crazy. Everyone knew you couldn't date [basaltic] rocks as young as the Hawaiian ones. I got so disillusioned by the general response from people who were knowledgeable about Hawaii that I just quit the conference after two or three days and went out into the field."

McDougall's fears concerning the dating of the Hawaiian rocks were a major factor in keeping him from an attempt to date them for almost a year after his return to Canberra in September 1961. He emphasized more than once that he did not have the confidence to attempt to date whole-rock samples of young basic rocks upon his return from Hawaii. During that year he dated older rocks—dolerites from Tasmania, and others from Antarctica and South Africa.[131] From these studies he grew increasingly confident in the laboratory, especially as a result of finding "good agreement" between the ages from whole rocks and mineral separates.

When he returned to the A.N.U., he found that John Richards and Allan Webb were dating eastern Australian rocks; thus he would need to build an additional extraction line. He was "staggered to find the oven covers had [just] fallen and smashed the flight tube of the mass spectrometer. The magnet was out of plumb, with an offset between the top and bottom pole piece." He organized a workshop to make the mass spectrometer functional and began the dating of the Karroo and Antarctic dolerites as his first piece of work.[132] John Richards noted that McDougall also dated dolerites from South America and talked about possibly dating the Deccan traps of India. "It was clear that the Karroo was dated for Gondwanaland studies."[133] Don Tarling confirmed Richards's remark.

> When McDougall came back to the A.N.U. from Hawaii, he was distinctly not interested in looking at the Hawaiian rocks. The point was that I was already looking at these Hawaiian rocks in terms of polarity changes and trying to get out relative dating between the Pacific islands on the basis that the polarity event (whatever was causing it) was at least a time marker; again, irrespective of age, it was a fairly clear marker which you could in fact use for dating and correlation between the different islands, even if you didn't know the precise age.* So basically, although I was originally looking for secular variations, in fact my interest quite quickly switched over to the use of reversals as a relative dating tool, even if at that stage we couldn't do absolute dating. Ted Irving was definitely pressuring Ian McDougall towards the end of this period [prior to McDougall's attempt at dating the Hawaiian rocks], which must have been about 1962, to try and do radioactive dating and to try and get at least the last event [the Matuyama–Brunhes boundary] as an absolute minimum.† I know that Ian was very much opposed to doing any of this at all. I'm not quite clear to what extent the actual reasons were political or personal. I tend to think possibly there was quite a lot of personal element in it, but also, of course, I think he wondered whether his techniques would be able to date these things. If he tried and failed he was probably frightened of looking a bit of a fool, even if he wouldn't have. [Tarling's surmise was largely borne out by McDougall himself.[134]] Ted Irving was trying to pressure McDougall into attempting to date my Hawaiian rocks for more than a year, and in this sense, I think it's quite clear that there was no real race [for the production of a polarity-reversal time scale].[135]

*Tarling's remarks concerning his interest in reversals at this early date, which was in marked contrast to the interests of Irving and other members of the English directionalist group (by whom he had been trained), are borne out by his infrequently cited publication of 1962, "Tentative Correlation of Samoan and Hawaiian Islands Using 'Reversals' of Magnetization."

†This appears to mark the time when Irving began to speak encouragingly of polarity-reversal studies, but it appears also that his deep commitment to the directional program precluded his possible collaboration with McDougall, thus that opportunity was left to Tarling. Irving noted that his interest in reversals was aroused in the course of his studies in the Paleozoic rocks of Australia early in 1962, which led to his discovery of the Kiaman Magnetic Interval (Irving and Parry, 1963; Irving, t. 2, s. 1, xi-28-78). However, Irving's major concern was, and still is, studies in the location of ancient magnetic pole positions.

The race was, in fact, about to begin. Tarling reiterated the important fact that they tried long and hard without success to get McDougall to date the Hawaiian rocks, and that their interest in them was aroused and maintained in ignorance, as far as Tarling was concerned, of the work of Cox and Doell. Tarling believes that Irving was also not fully aware of the work of Cox and Doell at that time (before the June 1963 publication of the first polarity-reversal scale by Cox, Doell, and Dalrymple).

McDougall recalled: "I began to get considerable confidence by the middle of 1962 that I could date whole basic rocks accurately if they were fresh. The greater concern was whether you could detect radiogenic argon in the Hawaiian rocks, because there were virtually no real age data on them. One really didn't know even the ballpark on age." In mid-1962 he ran the first Hawaiian samples, which were mugearites from western Oahu.* He crushed the rocks into large pieces (about 1 centimeter) for fear of atmospheric argon contamination.

If you crushed them too fine you'd increase the surface-area-adhering argon. . . . The first argon run staggered me. I knew immediately that I had detected radiogenic argon. From then on I started to work through the samples. I recognized that we were dating these quite young rocks because, first, the rocks were very fresh—the amount of air [argon] contamination in fresh stuff is an order of magnitude [ten times] less than in such things as biotite. The implication was that the air argon was much, much lower than in a biotite and you could therefore detect smaller amounts of radiogenic argon. This is what excited me and seemed to be the major breakthrough. What was important in fresh [whole] volcanic

*The time he started dating these rocks is important, given his promise to keep Doell and Cox informed about them. The first publication treating the Hawaiian rocks appeared on i-26-63 in *Nature*; on ii-4-63, McDougall wrote to Doell (Document No. 29-1) to report that he now had "a goodly number of results from all the main islands and Hawaii itself." He added that "Hawaii must be very young [less than 1 million years], as your results suggested." It is clear from the letter that McDougall's collaboration with Tarling had begun. In closing, McDougall asked Doell if he could let him know "what the polarity of the Makaweli volcanics [of Kauai] is," since "from the age of the Koloa speciment [also of Kauai], which was collected from the Lawaii Bay section," he expected it to be reversed, by comparison with results from the other islands. "Tarling reports the Koloa as having mixed polarities, but he did not collect from Lawaii Bay." Also included was the telling note that "the Kauai volcanics have been rather difficult to date," but McDougall revealed some success in deriving three Kauai dates, one from the Koloa series and two each from the Makaweli formation and the Napali formation. Two of the three dated horizons from Kauai were apparently discarded (they never were published), and only the Napali date appeared in modified form in the first scale of McDougall and Tarling on x-5-63. McDougall did not reveal any other of the "goodly number of results" from the other islands (Maui and Molokai) in that significant letter, which clearly indicated to the Rock Magnetics Project group that McDougall and Tarling were by then principally concerned with dating polarity reversals and had become serious competitors.

rock is that the amount of air argon 40 is much lower than anything else. That was the main reason why the whole mafic rocks became datable.[136]

McDougall reiterated that he had never seen fresher rocks than those he collected in Hawaii and further noted: "It's still a very subjective business. If you show me a series of thin sections in volcanic rocks I can tell you those that are datable and those that are hopeless—the real difficulty is where you draw the line on acceptability. I don't know of any objective way of doing this. I do it by feel."

McDougall regarded his undergraduate training, especially in petrology, as a crucial precondition to his successful dating efforts and remarked,

I would dispute the idea that just any geologist could select proper samples of whole basic rock for dating. [This view is in sharp contrast to that held by Jack Evernden.] The cardinal ability is to be able to date sufficient samples in known or related stratigraphic relationships so that you can make an assessment of the various assumptions that go into the calculation of an age. A single potassium-argon age can be more misleading than informative, simply because there are several assumptions going into the calculation of that age. The other point in relation to dating these young rocks was that I had expected small amounts of radiogenic argon relative to the air [argon 40], and to minimize the blank—the contamination from the handling in extraction—I used samples of 10 to 15 grams. This meant that the blank from the argon-extraction line was much smaller per gram of sample than if you had used one gram of material, which was the approach that a lot of people were using. . . . It took some little time to realize why we were being so successful in dating these young rocks, but in retrospect it's very clear. My confidence increased as time went on, as I got more and more data, particularly with the work that I did on Réunion in 1964.[137] The ages were beautifully consistent with the stratigraphy; those rocks too were remarkably fresh and on a par with those from Hawaii.[138]

Don Tarling's doctoral studies, under the direction of Edward Irving, were concerned with the collection of data from the low latitudes (Hawaii, Tahiti, Samoa, Fiji, etc.) in an attempt to model ancient secular variations of the earth's magnetic field. The dissertation lent itself indirectly to the directionality studies of Irving, who was primarily concerned with resolution of the continental drift question. (Recall also that Irving attempted to dissuade Cox and Doell from polarity-reversal research, which he regarded as not worthy of their efforts.) McDougall had sampled in Hawaii in 1961 because of "the prospect, however remote, of being able to date rocks from the Hawaiian volcanoes"; that attempt was conditioned largely by "the good understanding of Hawaiian geology" and the idea that the volcanoes grew older with distance from the active ones. McDougall remarked: "Reversals didn't play any role at all in what I was doing in Hawaii." Thus at the time of the first contact between McDougall and Tarling in August 1961 in Honolulu, McDougall was not interested in polarity-reversal studies, and Tarling was fully occupied with secular variations. After a half-day contact, in which they discussed Hawaiian geology, "no program was set up for collaborative work."[139]

After McDougall and Tarling returned to Canberra in September 1961, there was little contact between them. The potassium-argon dating laboratory at the A.N.U. was located in a temporary building that housed the Radiochemistry Department (before it was subsumed by Geophysics under Jaeger), in a different part of the campus from the paleomagnetism laboratory. The two men saw each other only at Thursday afternoon departmental seminars. McDougall's lack of interest in polarity-reversal studies, his preoccupation with dating Tasmanian dolerites, and his total neglect of the Hawaiian rocks even in the face of urging from Edward Irving, explain why McDougall did not seek an alliance with Tarling. However, once he entered polarity-reversal studies, McDougall collaborated briefly with a number of paleomagnetists in producing successively more complete versions of a polarity-reversal time scale in competition with Cox, Doell, and Dalrymple. McDougall's first two scales, the only ones treated in this book, were published with Tarling.

Donald H. Tarling came from a "working-class" home; his parents were separated when he was ten. He attended grammar school in Birmingham, England and "did arts at school at the ordinary level." He entered what is now the University of Keele and specialized in petrography and geomorphology. At graduation in 1957 he departed for the University of London to enroll in an M.Sc. program in geochemistry since he had come to feel, while doing petrology, that "all of the answers to petrological problems are geochemical." He was informed that there were no places available in geochemistry, but there was an opening in geophysics, which he entered, at Imperial College. During the first year, he did an applied geophysics course, followed by a second year of pure geophysics in which he went to Labrador as a field assistant to examine glacial valleys with a physical geographer. Upon return to complete the second year in 1959, which involved lectures for six months followed by a six-month dissertation, he looked for a project in which he could use much of his background. Tarling recalled: "I wanted originally to do a M.Sc. dissertation on continental drift, even in those early days, because I'd covered it to some extent in geography and geology. It was still not popular at that stage, but it made use of my background." The faculty thought his plan "not geophysical enough" and decided that he should write on continental drift and paleomagnetism. He did a thesis titled "Some Aspects of Paleomagnetism," which was "basically intended to be linked to continental drift." Upon completion of his M.Sc., the faculty suggested that Tarling join Edward Irving to work on a Ph.D. in paleomagnetism at the A.N.U. After writing to Irving, Tarling was offered a scholarship; he collected rocks in Devon and Cornwall to take to the A.N.U., but they turned out to be unusable. On the way to Australia, he stopped briefly at Aden and did the first paleomagnetics work on the opening of the Red Sea.

Upon arrival in Australia, Irving suggested that Tarling might be interested in working on secular-variation problems, more particularly, on low-latitude variations as evidenced in the islands of the Pacific basin. Tarling remarked: "I was fundamentally looking for secular variations, and reversals to a lesser extent." He was also still concerned with continental drift, and tried to determine if Fiji had been rotating about an axis. Most of the islands examined were too young to show evidence of continental drift, thus the derived data were of use mainly for study of secular variations, and reversals to some extent. "The basic objective was determined by Ted Irving—he in fact decided the project, although it was put to me as 'would I be interested in.'"[140] At one point during the three years Tarling worked on his Ph.D. degree, a fire in the department destroyed many of his maps and other materials, setting him back about six months.

Tarling was interested in polarity-reversal research before McDougall had returned from his Berkeley year; he placed the time of his initial interest in the subject in 1958 or 1959. He recalled,

> I was pretty sure that the polarity [reversal] scale was not every million years, which was the current feeling before that stage, based on the Icelandic data. I was fairly convinced it was an irregular scale, and therefore had quite a potential as a dating tool, from 1959 onwards.* Ted Irving was do-

*Tarling did not mention the evidence that "fairly convinced" him at that early date, when almost all others were in favor of regularity.

ing little actual collection of his own rocks and measuring them. A lot of the work he was doing was on previous collections. Essentially there was very little on which he could work with Ian McDougall. McDougall's interest in collaboration on polarity reversals arose partly from pressure from Irving and partly when he realized he could date Hawaiian rocks and that I could use his already dated material and put it into some sort of a scale on the polarity; then he became more interested.[141]

McDougall recounted:

I cannot recall how the collaborative effort with Tarling occurred. The collaboration may have been sparked by the first scale of Cox, Doell, and Dalrymple.* I may have realized that we had more data than they. It may be that after the successful dating of the Hawaiian rocks I concentrated on sequences that would be useful to polarity-reversal studies. It was almost fortuitous that we were each there doing our own work which lent itself to future collaboration. Perhaps in contrast to Cox and Doell, a lot of what happened in our case was not carefully planned; *it really was a result of this coincidental, fortuitous situation of Tarling collecting in Hawaii for secular variation and my collecting for dating. There are a remarkable number of fortuitous circumstances in this whole business.* Subsequent to the Hawaiian work a number of studies were initiated, initially to assist in distinguishing between the two hypotheses of reversals of the earth's magnetic field and self-reversal. The possibility of developing a time scale that would be used for other purposes just came about as a bonus. *Of course, I don't think any of us envisaged that it would become such a critically important part of the evolution of the whole revolution in earth sciences that took place in that period from 1964 to 1966.*[142]

In striking contrast to the "coincidental" and "fortuitous" collaboration of McDougall and Tarling, Cox's and Doell's attempt to decide the field versus self-reversal question was early conceived, planned, and executed in collaboration with Brent Dalrymple, though delayed for almost a year by circumstances over which the three had little or no control.

Earlier commentary has left little doubt as to the importance of John Jaeger as an administrative and intellectual influence in the A.N.U.; Jaeger was highly enthusiastic about the virtues and rich possibilities for the future application of potassium-argon dating in

*Document No. 19-1, noted earlier, clearly shows that McDougall was collaborating with Tarling some time before ii-4-63, which was four months prior to the publication of the first scale by Cox, Doell, and Dalrymple. But the December 1962 abstract of Doell, Cox, Kistler, and Dalrymple, which presented the first notice that a group was engaged in combining polarity data with radiometric ages, may have provided some of the impetus for the McDougall-Tarling collaboration (discussed in Chapter 6).

both old and young rocks. Jaeger offered McDougall "fatherly advice" to capitalize on his ability to measure the ages of young rocks and the evolution of volcanoes. McDougall recalled that

> Jaeger was very pleased about the outcome of the Hawaiian [dating] work, not the polarity-reversal side of it, as I demonstrated that we were able to measure the ages of these volcanic rocks and find a pattern down the island chain. He quite often said, "I don't think you should emphasize that [reversal] side at all." He tried to discourage us from putting too much effort in that direction. By that stage it had become obvious that reversal study was clearly important. I don't recall why Jaeger was dissuasive, whether again it related to his contacts with Berkeley people (he was very good friends with Turner and Verhoogen) or whether Irving's earlier view prevailed that people ought to be looking at directional problems [in relation to] continental drift.* Irving was never involved in any of the reversal studies in a direct way.[143]

In the United States, John Verhoogen had played a role similar to Jaeger's in attempting to dissuade Cox and Doell from pursuing polarity-reversal studies.†

By 1963, the stage was set, both in Menlo Park and Canberra, for an attempt to formulate an isotopically dated scale of reversals of the earth's magnetic field; success in that attempt would also remove the polarity of the field from the category of physical constants. The runners were on the blocks.

*In 1960, Jaeger published a "Report on Progress in Geophysics" of the solid earth, which was mainly a review of the A.N.U.'s work. In a section more than two pages long devoted to rock magnetism, he noted that "the major activity to date has been a stratigraphic survey to determine the direction of magnetization of . . . well-dated formations and . . . the position of the earth's dipole field relative to Australia at the corresponding time and to compare this with equivalent results from other continents." In a discussion touching on instruments, demagnetization, stress effects on directions of magnetization, and other points, he barely mentioned in passing that "repeated reversals have found to be common through a wide span of geological time. . . . However, at other periods the polarity is constant, for instance the whole of the Permian appears to be reversed." There was no mention of a plan or wish to do reversal studies of any sort.

†Verhoogen explicitly recalled that Cox and Doell had pursued reversal studies in spite of his opinion that it was not likely to be a fruitful area of research (Verhoogen, t. 2, s. 1, viii-11-77). Verhoogen was not convinced of a reversing earth field until the first scale of Cox, Doell, and Dalrymple appeared in June 1963 (oral comm., viii-11-77).

Chapter 6

The Scale Evolves

GEOCHRONOLOGY, or the assignment of dates to geologic events or features, lies at the center of many geologic studies and is an important component in most. The geologic time scale contains a number of intervals, each of which is taken to represent a specific fraction of the continuum of time (see Fig. 6.1). The segments of this time scale, however abstract and symbolically noted, are ultimately defined by events and their products, as preserved within rock bodies. Geochronologies differ from other time scales (which also may be rigorously defined both practically and philosophically) in that they are based in and grow from the phenomenology of the earth.*

The quest for tools with which to reckon earth history must surely have begun with the earliest thoughts about the earth that passed beyond simple description. But it was not until 1669 that Nicolaus Steno, through systematic observation of the processes in sedimentary rock formation, induced several fundamental ideas, among which the principle of superposition provided the first means by which a rock body could be dated, in any sense. Steno recognized that the oldest bed in a stratigraphic sequence is at the bottom and the youngest at the top.

Not until a century and a half later was knowledge of the succession of fossil faunas fitted to Steno's principle, almost simultaneously by William Smith in England in 1815 and Georges Cuvier and Alexandre Brongniart in France in 1811, to yield a principle for correlating rocks over great distances. There were, to be sure, severe

*Savage (1975) contains an extremely lucid account of geochronologies.

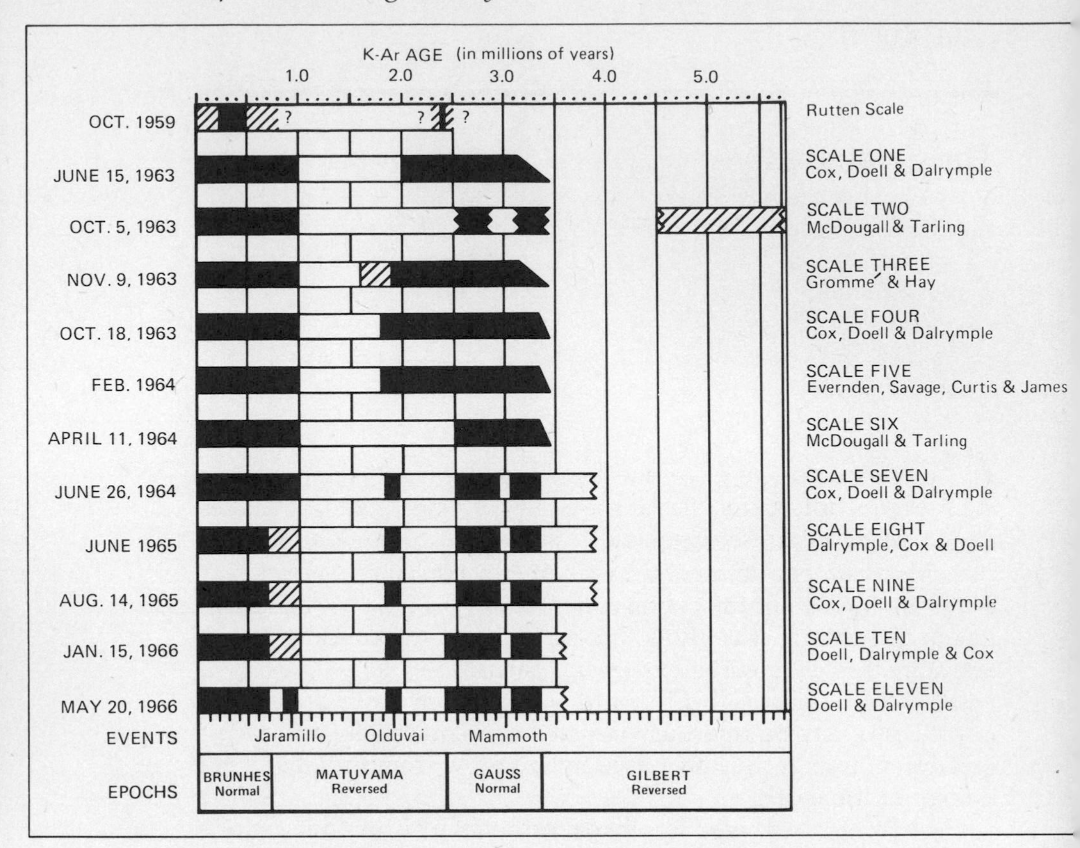

Fig. 6.1. The sequence of geomagnetic polarity-reversal time scales, 1959–1966. The scales are numbered as they are referred to in the text. The black intervals represent times of normal polarity; the white intervals show reversed polarity. Intervals of uncertain polarity are diagonally lined. Note that scale three, though published later than scale four, was written and submitted earlier. The polarity epoch names and the Olduvai and Mammoth event names were first introduced in the publication of scale seven by Cox, Doell, and Dalrymple. The Jaramillo polarity event was named in Doell and Dalrymple's presentation of scale eleven.

limitations to the usefulness of these early biostratigraphic methodologies; some were overcome, in the case of Cenozoic rocks, by Charles Lyell's contribution, in 1833, of a method in which the percentage of living species in a fossil fauna is a guide to age. Lyell's method facilitated the first subdivision of a period (the Tertiary) into epochs.

The method of paleontologic zonation conceived by Albert Oppel, in the course of his doctoral dissertation at Tübingen in 1856–58, demonstrated his remarkable understanding of almost all the variables assessed in modern biostratigraphic practice. It required Oppel's synthesis of an abstruse tool that in part overcame the dearth of time-diagnostic fossil species, largely through delineation of small chronostratigraphic intervals by overlapping the ranges of species. Unfortunately, it was little understood by most, still less put to practice, and not employed in the New World until 1938 by Robert Kleinpell in California. Oppelian zonation defines still the limits of biostratigraphic refinement.

The rise and growth of radiometric dating during the twentieth century provided the means to calibrate in years and verify the correctness of the relative geologic time scale that had grown from geologic studies during the previous two centuries. Nonetheless, for all that progress, much more often than not diagnostic fossils were absent, and rock units did not lend themselves to radiometric definition. New means of dating and correlation have always been and remain greatly sought after.

The evolution of life, the radiogenic alteration of certain isotopes, and the markings within minerals produced in the radioactive decay process are ongoing, progressive phenomena whose particular constituent events, when arranged on the continuum of time, are nonrepetitive. They would be, in fact, ubiquitous and reliable processes by which to measure time and date rocks, but their effects or remains are too seldom preserved. In contrast, the effects of the earth's magnetic field are recorded in many types of rock, all over the globe, in almost every environment. The imprint of the field, effected by the ordering of magnetic components within the rock, is unique at each location for each given instant, and remains stable, usually with little change in magnetic orientation and intensity, over long periods of time. Except for relatively brief intervals, the geomagnetic field is dipolar. Hence in general no two rock bodies of synchronous origin may have opposite magnetic polarities, except in the very rare cases of self-reversal, which may now be recognized (discussed in Chapter 4).

The sequence of magnetic reversals is ordered temporally by the superposition of rocks (lithostratigraphy), by biostratigraphy, or by isotopic dating, but the age and duration of each magnetic interval can be calibrated in years only by means of radiometry. A reversal of the magnetic field is not an intrinsically unique event, thus it is not distinguishable from reversals of the same sign at other times.

The time significance of the scale of polarity reversals grew from and is based upon two features: the exact dating of polarity reversals by the potassium-argon method, and the irregularity or randomness of the little-understood reversal mechanism. If the field reversed at precise, regular intervals, its usefulness as a tool in stratigraphic correlation would be reduced, because two rock sequences not datable by potassium-argon could not be correlated. The unique duration of each magnetic interval, and its place within a continuous sequence of magnetic intervals, makes each interval recognizable on a relative basis within that sequence. Thus polarity information provides only a place in the sequence, not a date per se, for the stratified rock from which it was derived, and the magnetic sign alone cannot date a stratum. Nevertheless, the scale of polarity reversals determined largely from basalt flow sequences on the continents became the key to worldwide correlation of a body of uninterpreted magnetic data derived from deep-sea sediment cores and magnetometer surveys of the basaltic ocean floor. The earth's magnetic field had imprinted remarkable patterns, containing sequences of positive and negative intervals, but the data remained baffling until the polarity-reversal scale, derived in the main from terrestrial igneous rocks, became refined enough to permit the correlations that confirmed seafloor spreading.

A Simple Model

SCALE ONE (COX, DOELL, AND DALRYMPLE): PERIODICITY?

Because of their training and experience in a field-oriented geology department at Berkeley, and their experiences elsewhere, Richard Doell and Allan Cox were early aware of the potentially broad applicability a scale of magnetic reversals would have in geologic science. The power of such a globally useful tool was apparent to almost anyone with geological perspective, but, as earlier explained, the rock-magnetics research community was small, and few even among them subscribed to a belief in reversals of the earth's magnetic field. Among those latter few who early perceived the important need to resolve the field reversal versus self-reversal question, only Cox and Doell demonstrated singularity of purpose and effort, by forming the Rock Magnetics Project largely toward that end.

In a little-known and still less-cited abstract in the *Transactions of the American Geophysical Union* of December 1962, Doell, Cox, Ronald Kistler, and Dalrymple presented the first formal notice that a group was engaged in combining polarity data with radiometric ages:

The magnetic properties of six igneous rock units from California with known K-Ar radiometric ages were investigated in order to determine time intervals between successive reversals of the geomagnetic field during the last several million years. Emphasis was given to mineralogical and thermal magnetic studies designed to detect self-reversals. Xenoliths [rock fragments of earlier origin incorporated into an igneous rock] with different mineralogies, which cooled with some of these rocks, were also studied to determine whether reversals of polarity were mineralogically controlled. All rock units from which both reliable magnetic data and radiometric ages are available can be fitted to a time scale with alternating magnetic polarity epochs of 0.45 to 0.50 million years. Although consistent with Khramov's discovery of magnetic polarity in sediments which varies periodically with depth, the limited data now available are also consistent with non-periodic time intervals. They do not constitute, therefore, an independent demonstration of periodicity of geomagnetic field polarity.*

Their new roles as pioneers led Cox, Doell, and Dalrymple to a somewhat proprietary attitude toward reversal studies in young rocks. Thus the eventful letter from McDougall to Doell of February 4, 1963,[1] which confirmed the collaboration of McDougall and Tarling in reversal studies, produced for Cox and his colleagues a heightened sense of urgency to accelerate their studies and publish their manuscript containing their polarity-reversal scale, a draft of which had been written by January 1, 1963. The final draft was mailed to *Nature* on March 19, 1963.† "Geomagnetic Polarity Epochs and Pleistocene Geochronometry" appeared in the June 15, 1963, issue of *Nature*; in it Cox, Doell, and Dalrymple presented their first polarity-reversal time scale (see Fig. 6.2). The geological commu-

*Khramov's (1958) influential surmise concerning periodic reversals was here opened to question for the first time.

†Cox's letter of submittal to *Nature*, dated iii-19-63, is Document No. 32A. On v-3-63 Cox wrote *Nature* (Document No. 32B) to inquire about the disposition of the manuscript, requesting a response by wire "at our expense." At this time, in a letter to R. M. Moxham, Doell wrote that "this is our first summary on the dating of reversals and we're rather anxious to get it out before our competitors in Canberra get something together." Their sense of urgency may also have been one reason they used Rutten's dates without confirming them at Berkeley (although relations with Jack Evernden were already strained by this time). In Document No. 32 (i-1-63), N. King Huber in a prepublication review of that paper noted that "the title ["Radiometric Dates of Geomagnetic Reversal Epochs"] is misleading. You are not giving radiometric dates of paleomagnetic reversal epochs at all. You are merely presenting radiometric dates of samples (most of which fall within epochs) from which you deduce the periodicity and lengths of epochs. Suggest you consider something like 'Geomagnetic reversal epochs as indicated by radiometric dates'."

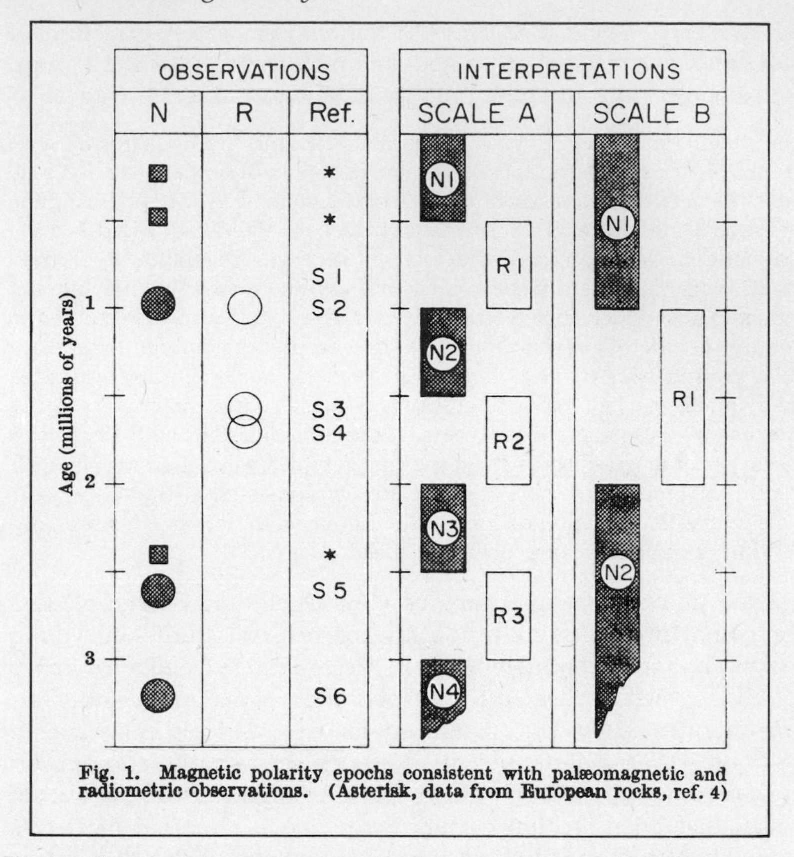

Fig. 1. Magnetic polarity epochs consistent with palæomagnetic and radiometric observations. (Asterisk, data from European rocks, ref. 4)

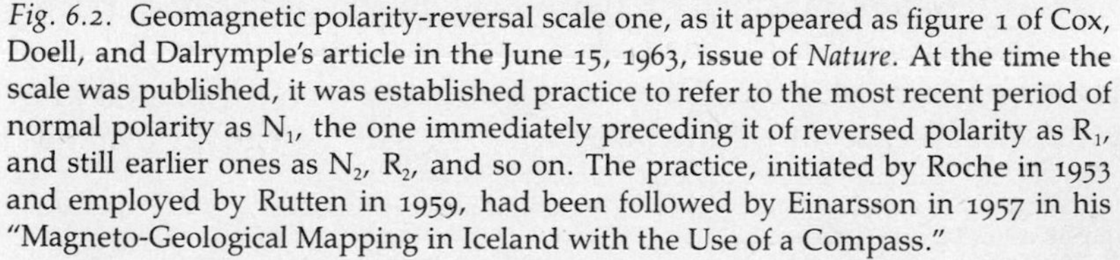

Fig. 6.2. Geomagnetic polarity-reversal scale one, as it appeared as figure 1 of Cox, Doell, and Dalrymple's article in the June 15, 1963, issue of *Nature.* At the time the scale was published, it was established practice to refer to the most recent period of normal polarity as N_1, the one immediately preceding it of reversed polarity as R_1, and still earlier ones as N_2, R_2, and so on. The practice, initiated by Roche in 1953 and employed by Rutten in 1959, had been followed by Einarsson in 1957 in his "Magneto-Geological Mapping in Iceland with the Use of a Compass."

nity was not at that time, in the main, convinced of the reality of a reversing magnetic field.* Sixty percent of that historic four-page paper was devoted to a discussion of the field-reversal versus self-reversal question; the authors were in effect prefacing the presentation of their scale with an argument against the possibility of self-

*John Verhoogen, for instance, said that he was not convinced of field reversals until Cox, Doell, and Dalrymple published the first version of the polarity-reversal scale.

reversal for the specimens from which the data had been derived. The paucity of reliable data points available at that time is apparent in that only nine served to construct the first scale: three Italian dates (discussed in Chapter 4) were taken from Rutten's 1959 paper (without verifying them with Evernden; see Chapter 5), and the remaining six were from California (see Fig. 6.3). Of the California dates, No. S1, from Owens Gorge, dated at 0.98 million years from sanidine, was derived from a preprint of Evernden, Savage, Curtis, and James's paper of 1964; No. S2, from Big Pine, dated at 0.99 ± 0.04 million years from obsidian,[2] was the first date that Dalrymple ever did specifically for the reversal scale; Nos. S3 and S4, from Sutter Buttes, dated from biotite at 1.57 ± 0.24 and 1.69 ± 0.10 million years respectively, had been published by Evernden, Curtis, and Lipson in 1957; No. S5, from McGee Mountain (2.6 ± 0.1 million years), and No. S6, from Owens Gorge (3.2 ± 0.1 million years), both done by Dalrymple, were two of the earliest dates from whole-rock basalt samples ever done;* they had been earlier published in the Geological Society of America *Bulletin* in April 1963 and were also part of Dalrymple's Ph.D. thesis.†

Aleksei Khramov's comprehensive, detailed study of 1958 (translated to English in 1960) greatly influenced Cox and Doell with respect to the question of reversal frequency and periodicity. Khramov had suggested, because his upper three contiguous magnetic zones in sedimentatry beds, including the present normal epoch, were of nearly equal thickness, that periodic reversals were implied. The few data points available to Cox, Doell, and Dalrymple permitted them to surmise only that "if the more recent polarity epochs are equal or nearly equal in length, the data of the present study indicate that these epochs must be either 1/2 or 1 million years long." Their nine dates, they noted, were "inconsistent with [regular] periods less than 1/2 million years in length, greater than 1 million years in length, or between 1/2 and 1 million years in length." They clearly expressed a preference for Scale B (1 million year-long epochs) over Scale A, and added a note in proof that additional, recently derived data also strongly favored Scale B. Mag-

*Dalrymple noted that Garniss Curtis encouraged him to date whole-rock basalt; in contrast, Evernden eschewed basalts as late as 1970 (Evernden and Evernden, 1970).

†Watkins (1972) said that the six Californian dates had been done by Dalrymple, who in fact was responsible for only three of them: S2, S5, and S6. The other three Californian dates and the Italian dates had been done by Evernden and Curtis; although the three from Italy were cited from Rutten's paper of 1959, they were neither credited to nor verified with their real source.

Table 1. PETROGRAPHIC DESCRIPTIONS, MAGNETIC PROPERTIES, AND RADIOMETRIC AGES

No.	Locality*	Age‡ (10^6 y.)	Reference for age	Polarity	Palæomagnetic sampling extent	Petrographic description of magnetic samples	Ferromagnetic minerals	Curie temp.
*S*1	Owens Gorge	0·98 (sanidine)	10	Normal	30 miles laterally, 600 ft. vertically, 90 oriented samples	Ignimbrite, marked increase in degree of welding with depth; abundant lithic fragments and inclusions	Magnetite or titanomagnetite, grain diameter 2–500μ. Minor alteration to hæmatite	570° C
*S*2	Big Pine†	0·99 ± 0·04 (obsidian)		Reversed	1 mile laterally, 400 ft. vertically, 17 oriented samples	Pumiceous to dense and banded rhyolite flow with varying degrees of perlitization	Magnetite or titanomagnetite, grain diameter 1–40μ. Opaque grains rare	530° C
*S*3	Sutter Buttes	1·57 ± 0·24 (biotite)	11	Reversed	35 ft. laterally, 10 ft. vertically, 4 oriented samples	Vesicular hornblende andesite flow with large poikilitic phenocrysts of plagioclase	Isotropic grains 2–45μ in diameter, pinkish brown to grey in colour, probably titanomagnetite. Minor deuteric hæmatite in biotite	575° C 675° C
*S*4	Sutter Buttes	1·69 ± 0·10 (biotite)	11	Reversed	110 ft. laterally, 25 ft. vertically, 7 samples	Slightly vesicular dacite intruded at shallow depth; phenocrysts of biotite, plagioclase, and quartz	Earthy red hæmatite dust, rare grains of magnetite 3–10μ diameter	560° C
*S*5	McGee Mtn.	2·6 ± 0·1 (whole rock)	12	Normal	430 ft. laterally, 90 ft. vertically, 20 samples	Fine-grained holocrystalline olivine and biotite-bearing basalt flow	Magnetite grains 1–60μ in diameter with occasional rims and veinlets of hæmatite; both minerals also appear as rims around olivine	520° C 675° C
*S*6	Owens Gorge	3·2 ± 0·1 (whole rock)	12	Normal	100 ft. laterally, 80 ft. vertically, 8 samples	Vesicular porphyritic holocrystalline olivine basalt flow	Magnetite or titanomagnetite, grain diameter 1–35μ; extensively replaced by hæmatite	575° C 675° C

* These localities (all in California) serve to identify the lava flows in the references cited, which give complete geographic and stratigraphic descriptions.

† This age, not previously published, is based on obsidian sample *KA*1069 from a rhyolite flow (ref. 13) in the N.W. ¼ Sec. 30, T. 10 S., R. 34 E. Big Pine 1 : 62,500 quadrangle. Sample weight 11·38 g, % K = 3·74, % Ar^{atm}_{40} = 67, $Ar^{r}_{40}/K_{40} = 0{\cdot}578 \times 10^{-4}$. Absence of perlitization is indicated by low weight loss on heating to 900° C (< 0·25%) and by refractive index (1·486 before heating, 1·483 after heating).

‡ (±) figures are standard deviations for the precision of the analyses.

Fig. 6.3. The localities, ages, magnetic properties, and petrography of the new rock specimens on which geomagnetic polarity-reversal scale one was based were listed in table 1 of Cox, Doell, and Dalrymple's June 15, 1963, article in *Nature*.

netic epochs of unequal length were also consistent with their results, they admitted, and "additional measurements are being made to clarify the important problem of periodicity and also aid in evaluating the accuracy of the techniques used for dating young rocks." Cox revealed the important fact that before and for a time after the development of the scale of June 1963, the group held the "working hypothesis that the field is periodic or nearly so."[3]

It was amazing that the early time scales [could be] so accurate [when] based on so few data points. It was really lucky that reversals didn't turn out to be the way we wanted them to be. The reason that we hoped they would be simple was that we thought they'd be more useful stratigraphically. In fact, at times, one might have even imagined a perfect periodic geomagnetic dynamo . . . with one-million-year markers—like a pulsar beating away. In a sense, that wouldn't have been as useful as the reversals turned out to be, because, although it takes the most work to determine a random pattern, if you want to use it for purposes of identification, a perfectly random pattern is the most useful pattern you could possibly have, sort of like fingerprints. . . . In 1964 [scale seven], we announced the short polarity "events," so we knew then it wasn't periodic; it was [then] a question of how many short events there would be. I wasn't happy to see new ones showing up, because it was making the time scale less and less regular. It seemed that if it got cut up with too many short events, interpretation would be very ambiguous.* (I think there is sort of an optimum level.) We see this now [1978], working on excursions of the field where the field will not really go reversed, but point in some odd direction for maybe a thousand years and come back. These are quite a controversial subject right now.[4]

In response to a question concerning the effect that early work in magnetostratigraphy (especially that of the Icelanders and Japanese noted earlier in Chapter 4) had on the thinking of his group at the time of their first scale, Cox recalled:

The main thing I got out of the earlier work was some feeling of the scale at which reversals occurred. There was reason to have a feeling that they may not be periodic, but they weren't broken up on too fine a scale. That's an important point, because once we began to see the polarity *events* coming in, our thinking changed. For a long while we tried to preserve a two-level hierarchical scheme of *epochs*, which were roughly a million years long, varying maybe plus or minus 50 percent, and *events*, which were 100,000 or 50,000 years. If at the beginning we had thought it was going to be twice or three times as random as it is, with much shorter spacings be

*Watkins (1972) discusses the problem of the definition of short polarity events from both the practical and philosophic viewpoints.

tween these random times, then it would have seemed hopeless, and we certainly would not have gotten out the early version of the time scale.

The important thing that I carried over from the work that had come before was a rough idea of when the field had reversed, and then a sense of the internal coherency of the data: what one should expect on the basis of what had been published. You wouldn't expect the field to reverse very often. There was a lot of give and take around the lab about how fine-scale these events were going to be and how many would occur. I, at least, always hoped there wouldn't be too many.*

For a long while I saw them as noise. For example, during the last 700,000 years if the field was normal all of that time and then reversed for a long time before then, that would be a very definite stratigraphic marker. If it had turned out that there were lots of little reversed events within the Brunhes Epoch (N_1) in 1964 or 1965, my approach toward this would have been that the more of these we found, the less useful the reversal time scale would be, because it would introduce much more ambiguity.[5]

Doell recalled that he did not think the first reversal scale very important at the time it was written, because reversals were not the only thing they were working on. He considered secular variations as equally important, since they contributed to confidence in paleomagnetic methods, which at that date were still open to some question.† Doell was then also engaged in sampling Permian and Triassic age rocks in Alaska in order to evaluate Warren Carey's Orocline concept.‡

SCALE TWO (McDOUGALL AND TARLING): SURPRISE

On October 5, 1963, Ian McDougall and Don Tarling published a four-page paper in *Nature* titled "Dating of Polarity Zones in the Hawaiian Islands." It contained a summary of magnetic data published by Tarling in 1962, but also a large number of new dates de-

*Cox's reaction to the data representing the crucially important Jaramillo event, discussed under polarity scale number eleven, was quite in keeping with this remark.

†"Doell's interests were strongly focused on secular variations in 1964 and 1965—much more so than were Cox's. They were his big interest in Iceland, almost an end in themselves" (D. M. Hopkins, written comm., xii-21-78).

‡Carey called an orogenic (mountain-building) belt with a sharp curve in it an orocline and believed that it marked horizontal bending of the crust after the folding of the orogenic belt. The orocline is a key element in Carey's expanding-earth theory (Carey, 1958, 1976). The rocks that Doell had sampled were thought to be of Permian age and were tested for natural remanence; he did no further work on them when new fossil evidence indicated their age was Triassic, since he had hoped to use them in fixing a Permian magnetic-pole position. At Grommé's suggestion, John Hillhouse studied Doell's cores at Menlo Park; in 1977 he published the strongest evidence to date suggesting that terranes in Alaska did not originate there (written comm., Grommé, iii-19-81; oral comm., Hillhouse, iv-1-81).

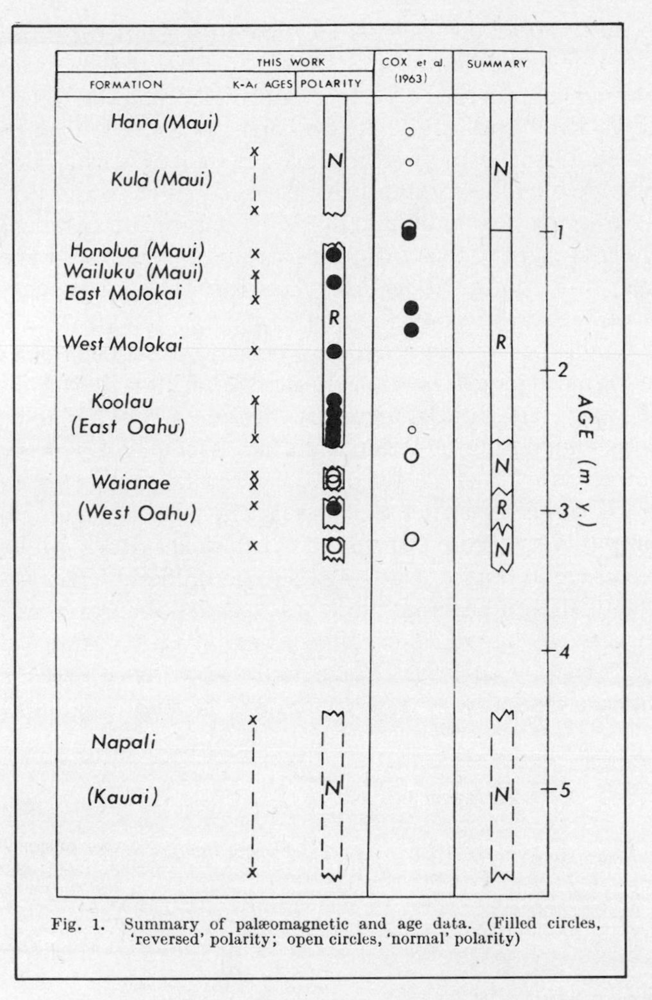

Fig. 1. Summary of palæomagnetic and age data. (Filled circles, 'reversed' polarity; open circles, 'normal' polarity)

Fig. 6.4. Geomagnetic polarity-reversal scale two, as it appeared as figure 1 of McDougall and Tarling's October 5, 1963, piece in *Nature*.

termined on whole-rock samples. The results were congruent with those of the scale of Cox, Doell and Dalrymple, but the first Australian scale indicated two possible additional polarity changes and also allowed "more precise dating of each level of polarity change." It was based on the previous nine points of Cox, Doell, and Dalrymple, plus twelve new data points from Hawaiian rocks (see Figs. 6.4 and 6.5).

The more complete Australian data suggested that the N_2–R_1 boundary be moved back from 2.0 to 2.5 ± 0.1 million years. A reversed interval was found between normally magnetized basalts at 2.95 m.y., thus leading the Australians to conclude that the magnetic field behaves more complexly than had earlier been supposed. McDougall, working from more dates than the Rock Magnetics Project had, was able to draw the significant conclusion that their results were consistent with "varying frequency" of reversals rather than a change about every million years, as suggested by Cox, Doell, and Dalrymple.

The publication of the Australian polarity scale was not expected by the Menlo Park group. Dalrymple had written to McDougall on July 30, 1963, with special regard to McDougall's earlier-mentioned difficulties in dating the Kauai rocks.[6] McDougall's response to Dalrymple on August 11, 1963, touched on the subject of reversals only once, but in that brief mention lay the portent, again, of growing competition: "From our results, periodicity of 1/2 million years in the change in polarity also seems to be unlikely." The next word of McDougall and Tarling's work was the October 5 polarity scale. Doell wrote to McDougall on November 13, 1963: "Your *Nature* paper caught us by surprise!! You should let your colleagues Irving and Richardson know what you are up to so they could spread the

Table 1. PALÆOMAGNETIC AND AGE DATA ON ROCKS FROM THE HAWAIIAN ISLANDS

Formation	Polarity (*N*, 'normal'; *R*, 'reversed')	No. of palæomagnetic sites	K–Ar age (m.y.)	No. of samples dated, or sample No.
Hana (East Maui)	*N*	3	< 0·4	—
Kula (East Maui)	*N*	6	0·86, 0·43	2
Honomanu (East Maui)	*R*	1	> 0·86	—
Honolua (West Maui)	*R*	1	1·15 ± 0·02	4
Wailuku (West Maui)	*R*	5	1·29 ± 0·03	3
Lanai	*R*	3	—	—
East Molokai	*R*	6	1·3 – 1·5	5
West Molokai	*R*	4	1·85 ± 0·01	1
Koolau (East Oahu)	*R*	8	2·2 – 2·5	5
	N	1	2·76 ± 0·02	*GA* 556
	N	1	2·84 ± 0·02	*GA* 809
Waianae (West Oahu)	*R*	8	2·95 ± 0·06	*GA* 557
	N	2	3·27 ± 0·04	*GA* 810
Koloa (Kauai)	*N* and *R*	2	—	—
Napali (Kauai)	*N*	3	4·5 – 5·6	3

Fig. 6.5. In table 1 of their October 5, 1963, paper, McDougall and Tarling presented the new data on which scale two was based.

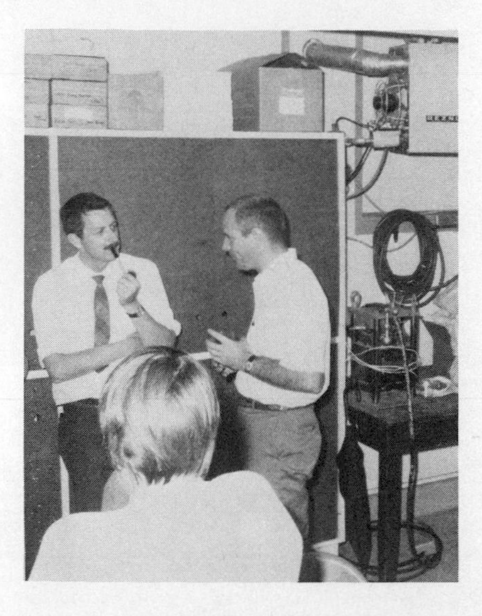

Ian McDougall, at left, visited Brent Dalrymple in the Rock Magnetics Laboratory at Menlo Park in 1968.

word of your good work at international meetings. As you note, there is disagreement between about 2.0 and 2.5 [million years]."

SCALE THREE (GROMMÉ AND HAY): MESSAGE FROM OLDUVAI

McDougall and Tarling's first scale had appeared in print on October 5, 1963; 13 days later Cox, Doell, and Dalrymple's second scale appeared in *Science* (see Fig. 6.6). This scale incorporated ten new data points from California and a new data point supplied to them, before publication, by Sherman Grommé and Richard L. Hay, both of U.C. Berkeley. Although Grommé and Hay's paper appeared in *Nature* on November 9, 1963, shortly after Cox's second scale, it had been submitted several months earlier* and contained important paleomagnetic data for a Pleistocene basalt flow from Tanganyika (now Tanzania) *that had then recently been redated at 1.8 million years by Evernden and Curtis* and reported earlier that year by Hay in *Science*.† The Grommé and Hay paper is thus here referred to as the third polarity-reversal scale.

*Grommé and Hay's paper was submitted to the editors of *Nature* on vi-20-63 (Document No. 30A); its receipt was acknowledged six days later by a postcard, and no further word was received by Grommé and Hay for three months. They wrote *Nature* on ix-26-63 to inquire about the delay.

†In 1947, Kenneth Oakley had collected lavas from Olduvai that were dated at 1.3 million years by G. H. R. Koenigswald, W. Gentner, and H. J. Lippolt in 1961; however, the stratigraphic position of the samples was not accurately determined, and the basalt had been altered and made questionable for dating (written comm., Gentner to Evernden in Evernden and Cur-

Richard Hay on a hillside above Olduvai Gorge in Tanzania in 1973. The basalt specimens on which the Olduvai event is based were collected out of sight, upstream at the bottom of the gorge. Hay recalled that he bet Jack Evernden, who redated the rocks with Garniss Curtis, that they would be less than 2 million years old—they were about 1.8 million.

In 1962, Richard Hay was planning volcanological research in Olduvai Gorge, prompted by the studies of Louis Leakey, Evernden, and Curtis. At that date, no adequate stratigraphic and mapping study had been done there. Grommé recalled: "John Verhoogen suggested to Richard Hay that he get samples and ship them back."* In January 1963, Grommé measured the polarity of

---

tis, 1965). Based on basalt collected by Curtis, he and Evernden published in *Nature* the following year ("Age of Basalt Underlying Bed I, Olduvai"), reporting some dates in excess of 4 million years. (All of Evernden and Curtis's Olduvai dates were published in 1962 in a major monograph in *Current Anthropology* crammed with an amazing volume of data produced in the Berkeley lab.) When Hay examined the volcanic sequence near the floor of the gorge in 1962, he collected a tuff from beneath the overlying basalts that Curtis had collected and dated with Evernden; the tuff was only 1.9 million years old. Since the field evidence indicated that the tuff and overlying lava-flow rocks had been extruded in no more than "a few tens of years," Hay believed that redating would show that they were all about the same age. Evernden and Curtis redated the basalts at 1.8 million years. In 1965, Evernden and Curtis explained in detail the difference between the procedures they had used in obtaining the original and the later dates (p. 352).

*Sherman Grommé was completing his doctoral dissertation in paleomagnetism under John Verhoogen; Verhoogen simultaneously suggested the Olduvai project to both Hay and

the rock samples sent by Hay and decided with Hay to send a note to *Nature* as a "very matter-of-fact thing." Grommé continued: "The project was undertaken *before* the publication of the first Cox, Doell, and Dalrymple time scale. Our idea at that time was just to contribute another brick in the structure. We were motivated chiefly by the scarcity of K-Ar dates. We were aware in January 1963 that Cox, Doell, and Dalrymple were working on reversals; their manuscript was available in preprint to me before we wrote our Olduvai paper."[7]

John Verhoogen recalled:

> Dick Hay brought back lava from Olduvai which had been dated at 1.8 million years, and this fell in the middle of the Matuyama [Reversed Epoch] that Cox, Doell, and Dalrymple had just worked out. Grommé measured the lava and found it to be normal, whereas everybody would have expected from the data that it would tend to be reversed; and I remember our grilling Hay—did he remember that in Olduvai the sun is in the north and not in the south? was he sure that his compass was working right? and so on and so forth. That was the first instance of one of those short episodes of reversal . . . that was the Olduvai thing [event].[8]

Grommé and Hay concluded their brief paper with the essential surmise that gave form to their version of the third polarity-reversal scale: "The two tentative geomagnetic polarity chronologies proposed by Cox *et al.* represent apparent periodicities of either 0.5 or 1 million years. The present result is consistent with both. In terms of the latter scheme, which is the simplest and which Cox *et al.* favor, the data from Olduvai Gorge combined with their data conservatively restrict the time of the next-to-most recent polarity reversal (the N_2–R_1 boundary) to between 1.9 and 1.6 million years."

SCALE FOUR (COX, DOELL, AND DALRYMPLE): GROWING COMPLEXITY

The Olduvai date was important in moving Cox, Doell, and Dalrymple to reconsider their earlier subscription to the idea (largely conditioned by the work of Khramov of 1957 and 1958) that polarity epochs were periodic. They had in fact constructed their alternative scales A and B with nearly equal intervals of time be-

---

Grommé. Grommé had first met Hay in his senior year at Berkeley, when enrolled in Hay's sedimentology course. Verhoogen remarked of Grommé: "He is as clever a man with instruments as you can imagine. . . . He really doesn't believe anything until he's seen it. And in that sense he was a very useful adjunct to our group, . . . because with Grommé around, you can keep your feet on the ground" (Verhoogen, t. 3, s. 1, viii-16-77).

tween reversals. That acceptance of uniform periodicity in drawing the first scale was not derived from nor greatly supported by the then available nine data points. Neither Cox, Doell, and Dalrymple (October 18, 1963) nor Grommé and Hay (November 9, 1963), when confronted with the normally magnetized Olduvai basalt, dated at 1.8 million years, were ready at that stage of development in their thinking about the behavior of the field to consider short periods of reversal that might lie within much longer intervals of a dominant polarity. The data base for such a conceptual leap was still lacking. However, Cox, Doell, and Dalrymple still were explicit in noting that "additional polarity epochs with lengths of the order of [100,000] years or less might be inserted at several points where gaps exist in the present radiometric age data, although the investigation of sediments by Khramov (1958) suggests that such extremely short polarity epochs either do not exist or else are confined to times when no sedimentation occurred, which is unlikely."* They also concluded that results to that date "confirm the hypothesis of geomagnetic field reversal, and the three most recent polarity epochs *are probably not all of equal length*" (emphasis added).

Both the scale of Cox, Doell, and Dalrymple of October 18, 1963 (scale four) and McDougall and Tarling's scale of October 5 (scale two) were in press at the same time. Upon publication it was immediately apparent that the two time scales were not in agreement in one important respect. McDougall and Tarling placed the boundary between epochs N_2 and R_1 at 2.5 million years on the basis of five reversed-polarity flows with ages ranging from 2.2 to 2.5 million years. In contrast, Cox, Doell, and Dalrymple had suggested that the boundary lay at approximately 1.85 million years.†

*The capture of short polarity intervals within terrestrially derived rock sequences is less likely than in deep-sea sediments, which often contain unbroken sequences of layers. The role of the remarkably useful deep-sea core data is treated in Chapter 8.

†Within days of seeing McDougall and Tarling's scale, Jack Evernden wrote to McDougall (letter of xi-1-63, Document No. 30C) obviously in demonstration of his sustained mistrust of whole-rock basalt dates: "I must say that I do not believe your data published in *Nature* to be accurate. In fact, I am virtually certain that they are invalid. I have now done 28 samples between 200,000 and 3.4×10^6 years from several parts of the world and have obtained beautiful agreement of results between K/A age and paleomagnetic direction. The schedule of reversals is that of Cox *et al*'s last paper in *Science*, October 18, 1963." He also requested that McDougall send "small pieces of each sample which you dated" and closed with the note that "if your data prove to be valid for Hawaii, the whole picture becomes chaotic." Evernden's strong defense of scale four of Cox, Doell, and Dalrymple likely reflected the fact that by that date he had already written his own scale (scale five), which is virtually identical to scale four of x-18-63. This important letter also indicates that Evernden was not in communication with Cox, Doell, and Dalrymple inasmuch as he was not aware that they had already found that the discrepancy between the Australian and American scales was due to sampling errors on the part of the Menlo Park group.

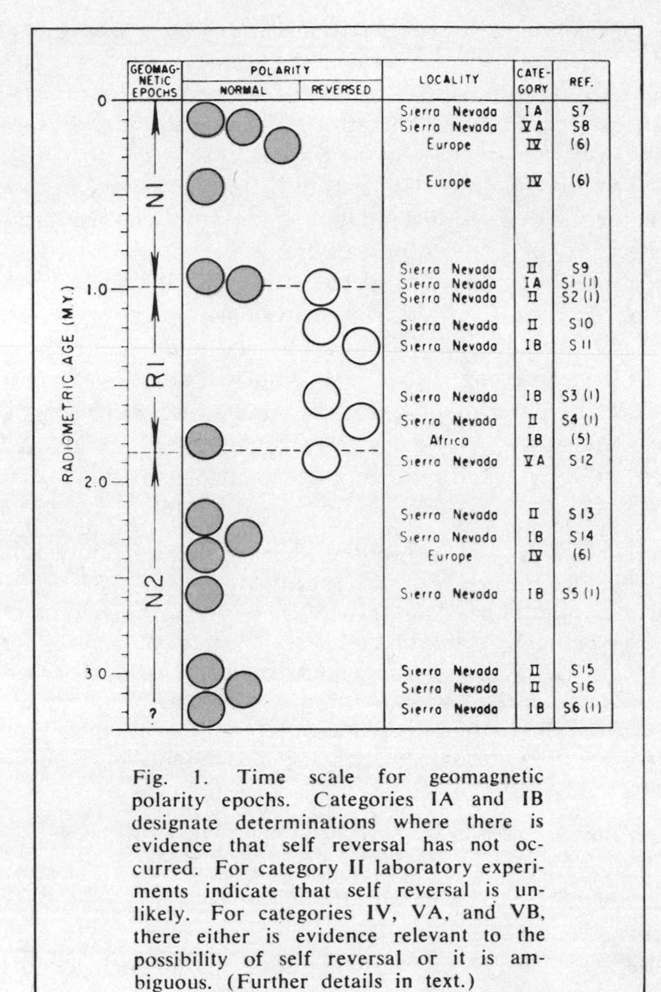

Fig. 1. Time scale for geomagnetic polarity epochs. Categories IA and IB designate determinations where there is evidence that self reversal has not occurred. For category II laboratory experiments indicate that self reversal is unlikely. For categories IV, VA, and VB, there either is evidence relevant to the possibility of self reversal or it is ambiguous. (Further details in text.)

Fig. 6.6. Geomagnetic polarity-reversal scale four appeared as figure 1 of Cox, Doell, and Dalrymple's October 18, 1963, paper in *Science*. Although the third polarity-reversal scale, by Grommé and Hay, was actually published shortly after scale four, it had been prepared and submitted for publication several months earlier.

The discrepancy between the two scales arose from sampling errors on the part of Cox, Doell, and Dalrymple, who found that samples S12, S13, and S14, all from the Lake Tahoe area, were open to question. Doell and Dalrymple returned together to check the field localities against the field notes they recorded earlier. They reported in a letter to *Science* that at site S13 they had mistakenly "sampled different flows only a few meters apart," one of which was normally magnetized and the other reversed. "For S14, because of dislocations by frost riving at the radiometric sampling site, they collected samples at an undisturbed outcrop a kilometer away."[9] The two outcrops had earlier been placed within the same geologic formation,[10] and they were seemingly from the same flow unit. Scrutiny of the isotopically dated outcrop revealed that the directions of multiple samples were all reversed, not normal, as earlier reported. Sample S12 showed a tendency toward self-reversal and was set aside. The samples that were magnetically determined from S13 and S14 were thus shown to be normal, but the radiometrically dated ones with the same sample numbers earlier reported had, in fact, come from nearby flow units, which were reversed in polarity. Doell's recall of the event was vivid:

> When we got McDougall and Tarling's paper, it was the first time that there were discrepancies between the data we were preparing for the time scale . . . two of ours were really discrepant. We got out all the information from the files, and it turned out those were two of a relatively small number in which Brent Dalrymple had not been able to accompany either Allan Cox or me or both of us back to the outcrop, to show exactly where he'd taken his samples. Allan and I had sampled these localities without Brent. We [Doell and Dalrymple] went tearing up to the Sierra with a field magnetometer, and sure enough, we'd sampled different lava flows in both locations.[11]

Dalrymple's memory was equally clear: "The samples were taken literally only inches apart but from different flows. That terrible mistake resulted in our developing rigid criteria for time-scale acceptability. We made a resolution shortly afterward to buy a Polaroid camera and to photograph all outcrops at each locality and draw right on the photo where the sample came from."[12] At the time of this reconciliation of the two scales, the polarity scale in its most complete form comprised only 35 data points extending to 3.5 million years before the present. Cox, Doell, and Dalrymple closed their explanatory and conciliatory note to *Science* (January 24, 1964) as follows: "The fact that independent investigations in California and Hawaii have enabled us to promptly identify an error of the type here described may serve as a convincing, though embarrassing, demonstration of the usefulness of paleomagnetism for precise worldwide stratigraphic correlation." This was a remarkable demonstration of grace in adversity!

SCALE FIVE (EVERNDEN, SAVAGE, CURTIS, AND JAMES): NEW DATA—NO CHANGE

In February 1964, Jack Evernden, Donald Savage, Garniss Curtis, and Gideon T. James published "Potassium-Argon Dates and the Cenozoic Mammalian Chronology of North America" in the *American Journal of Science*. The 53-page paper was laden with data

Magnetic Polarization and K/A Ages

Run no.	Sample	Locality	Age (m.y.)	Magnetic Direction	K/A Ref.	Mag. Ref.
KA 853	leucite	Italy	0.095	Normal	(3)	(4)
KA 971	basalt w/r	California	0.15	Normal	(1)	(1)
KA 1215, etc.	basalt w/r	Hawaii	0.25	Normal	*	(2)
KA 855	leucite	Italy	0.27	Normal	(3)	(4)
KA 1185	sanidine	Italy	0.43	Normal	(3)	(4)
KA 348	leucite	Italy	0.71	Normal	(3)	(4)
KA 305, etc.	sanidine	California	0.98	Normal	*	(5)
KA 1069	obsidian w/r	California	0.99	Reversed	(5)	(5)
KA 1162	sanidine	Italy	1.1	Reversed	(3)	(4)
KA 1181	sanidine	Italy	1.2	Reversed	(3)	(4)
KA 1078	basalt w/r	California	1.2	Reversed	(1)	(1)
KA 1211	basalt w/r	Hawaii	1.2	Reversed	*	(2)
KA 1094	basalt w/r	California	1.3	Reversed	(1)	(1)
KA 1188	basalt w/r	Idaho	1.4	Reversed	*	(6)
KA 101	biotite	California	1.5	Reversed	*	(5)
KA 65	biotite	California	1.6	Reversed	*	(5)
KA 1184	basalt w/r	France	1.6	Reversed	*	(7)
KA 490	biotite	California	1.9	Reversed	*	(2)
KA 1097	basalt w/r	California	1.9	Reversed	(1)	(1)
KA 1100	basalt w/r	Africa	1.9	Normal	(3)	(8)
KA 1102	basalt w/r	California	2.3	Normal	(1)	(1)
KA 302	sanidine	Italy	2.3	Normal	(3)	(4)
KA 973	basalt w/r	California	2.6	Normal	(5)	(5)
KA 1135	plagioclase	California	3.0	Normal	(1)	(1)
KA 1103	basalt w/r	California	3.1	Normal	(1)	(1)
KA 1013	basalt w/r	California	3.2	Normal	(5)	(5)
KA 1213	basalt w/r	Hawaii	3.3	Normal	*	(2)
KA 1172	basalt w/r	Idaho	3.4	Reversed	*	(6)

(*) this paper.
(1) Cox, Doell, and Dalrymple (1963b).
(2) Doell and Cox (personal communication).
(3) Evernden and Curtis (1963).
(4) Rutten (1959).
(5) Cox, Doell, and Dalrymple (1963a).
(6) H. Powers (personal communication).
(7) Kloosterman (1960).
(8) Grommé and Hay (1963).

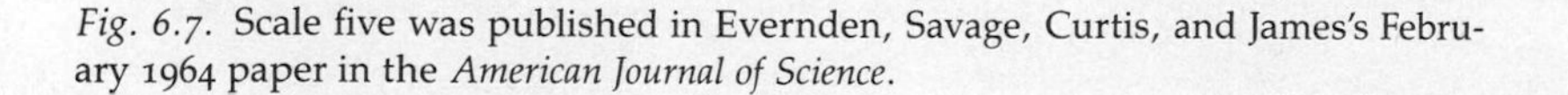

(9) Cox, Doell, and Dalrymple (in press).

Fig. 6.7. Scale five was published in Evernden, Savage, Curtis, and James's February 1964 paper in the *American Journal of Science*.

important to the paleontologic and stratigraphic communities, as implied by the title, and became a very important and frequently cited reference.* A one-page section, added after the paper had been submitted,† was devoted to paleomagnetic reversals and included dates and polarities on 28 samples, 13 of which had earlier been published by Cox, Doell, and Dalrymple (see Fig. 6.7). In keeping with earlier commentary regarding friction between Evernden and the Rock Magnetics group, the authors pointedly mentioned that "though accidentally obscured by Cox, Doell, and Dalrymple, the fact is that every age determination used by them in [scale four] was done at Berkeley in our laboratory, either by Dalrymple under our supervision or by ourselves." The polarity boundaries were not different from those of Cox, Doell, and Dalrymple.

SCALE SIX (McDOUGALL AND TARLING): A CAUTIOUS APPROACH

Shortly after Cox, Doell, and Dalrymple published corrections to their October 1963 time scale (scale 4), McDougall and Tarling published again in *Nature*, on April 11, 1964. Their paper contained the sixth polarity-reversal scale and a note that "the two sets of data are now found to be in good agreement." There was no essential difference between the new scale and their earlier one of October 5, 1963 (scale two). Their new scale had a simple pattern, as did their earlier one, showing normal polarity from the present to approximately 1 million years ago, a reversed interval from about 1 to 2.5 million years ago, and a normal interval from 2.5 to 3.2 million years ago (see Fig. 6.8). They did, however, emphasize that among the 35 age and polarity measurements on which their scale was based, two results did not agree with the simple pattern of the scale: Grommé and Hay's normally polarized basalt from Olduvai Gorge, dated at about 1.85 million years, in scale three, and the reversely magnetized Hawaiian lava, dated at 2.95 million years, from their earlier scale (scale two).

McDougall and Tarling's uneasiness in presenting their simple-patterned scale in the face of the Olduvai data was apparent: "Owing to the remarkable consistency in the age range of 1.0 to 2.5 m.y. of reversed polarity in rocks from Hawaii and the Sierra Nevada, it is of considerable importance that the Olduvai basalt should be re-

*James and Evernden did most of the dates; Savage and Curtis were in Italy during much of the time in which it was compiled (C. A. Repenning, written comm., vii-5-78).

†A preprint of the paper *lacking* the polarity information was made available to several researchers, including Dalrymple.

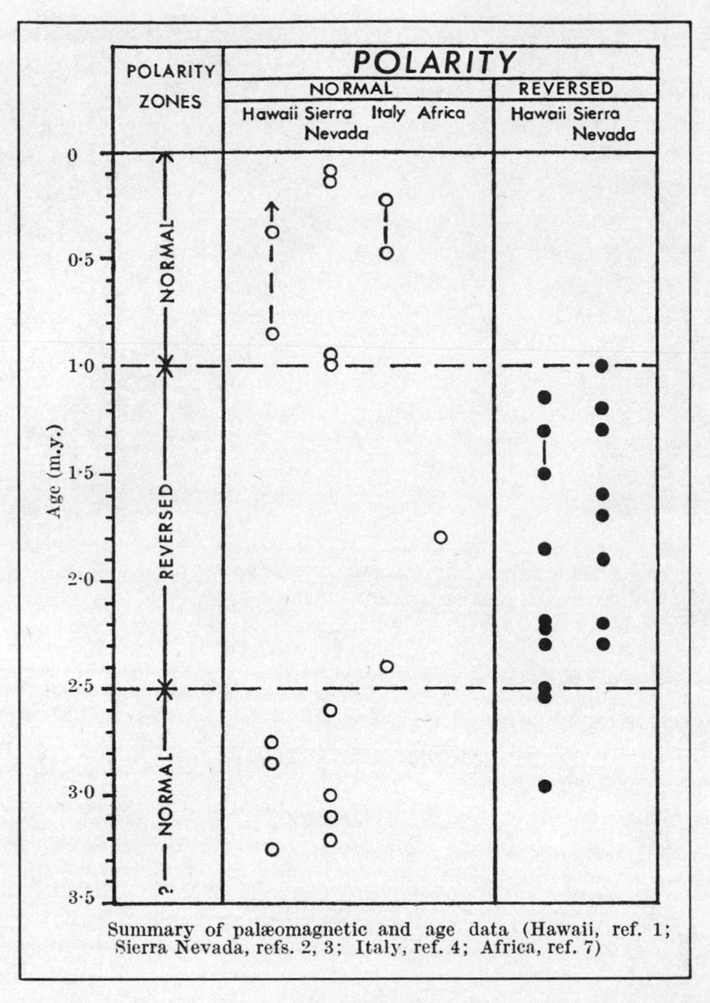

Fig. 6.8. Scale six, from McDougall and Tarling, appeared in *Nature* on April 11, 1964.

examined in detail." They also noted the exceptional character of the Hawaiian lava, the reversed polarity of which was not in question, but they did think the age uncertain (it fell in the middle of N_2) because of possible argon loss by diffusion, "although the measured age is reproducible." McDougall's dating work was commented upon by a number of professionals, who thought him an extremely careful worker; and that is apparent in his approach to these two disparate data points. It is, however, noteworthy that

"We're always looking for some kind of regularity in nature, and it turns out that reversals are as disorderly as they can possibly be. So the order we found, a very funny kind of order, was almost perfect disorder." Allan Cox in his office in 1980, newly appointed dean of the School of Earth Sciences at Stanford University.

they closed the paper with no mention of a future search for possible short polarity events that might have been represented by the two enigmatic dates. Instead, they implied that the scale as presented would eventually prevail: "Further work undoubtedly will resolve the few remaining discrepancies."

The way those disparate data were treated in terms of the available data base provides insight into McDougall and Tarling's approach. Two things may be inferred: first, that the simple pattern of long, unbroken polarity intervals, however new and data-poor, had become a strong influence in their thinking, as it indeed also had for Cox, Doell, and Dalrymple; second, that they were very cautious in not inferring more from the two points, which were grossly anomalous in terms of the simple scale.

There was a third disparate data point in terms of the scale drawn—the 2.4-million-year-old Italian sample dated by Evernden and reported by Rutten in 1959 (Berkeley sample KA 302, detailed in Chapter 4)—but McDougall and Tarling did not weight it heavily because "it was determined in the field by compass." That historically important datum, which was cited in all early versions of the scale, remains anomalous and unexplained still, but the adoption of more stringent criteria of quality led to its being discarded in 1966 (scale ten).

The simple scale concept warrants scrutiny. Allan Cox commented directly:

> One thing was happening, I believe, that often happens in science. As you begin gathering data, you first try to fit it with a very simple model, and then if a more complicated pattern emerges, you have to make a more complicated model. Of course, if you make a model that has more parameters than are justified by your data, it's very creative but not very scientific. Looking back at the successive time scales to see how they got more and more complicated, I think the two groups, Menlo Park and Canberra, working on it started out with very simple models, and then as we discovered events, anomalies in terms of the simple model, we were forced to make a more complicated kind of model. The Olduvai event . . . was first found in the data of Dick Hay and Sherman Grommé from Olduvai Gorge. . . . They didn't think it was an event. They thought it was helping to tie down the Matuyama-Gauss [N_2–R_1] boundary. That's how uncertain things were at that time, because the data were very sparse. . . . It turned out that the pattern that we were looking for was as random as it could possibly be. I think in any kind of experimental project it's very rare to find a completely random pattern in nature.
>
> We're always looking for some kind of regularity in nature, and it turns out that reversals are as disorderly as they can possibly be. So the order we found, a very funny kind of order, was almost perfect disorder.[13] . . .
>
> It's an order of magnitude easier to match up shaky data with a pattern you're sure of, than to interpret shaky data in terms of a model that hasn't been worked out.[14]

Order in Disorder

SCALE SEVEN (COX, DOELL, AND DALRYMPLE): INNOVATIONS

On June 26, 1964, Cox, Doell, and Dalrymple published a scale in *Science* that was a noteworthy departure from previous ones, in both concept and terminology (see Fig. 6.9). On the basis of a summary of all results then available plus many new and previously unpublished data (the result of Dalrymple having his own laboratory), they emphasized their belief that the earth's magnetic field had reversed at irregular intervals. They had recognized shorter intervals of opposite polarity within the previously established longer ones; they named the longer intervals "polarity epochs" and the shorter ones "events." The Olduvai data (of Grommé and Hay) were confirmed by additional results from North America

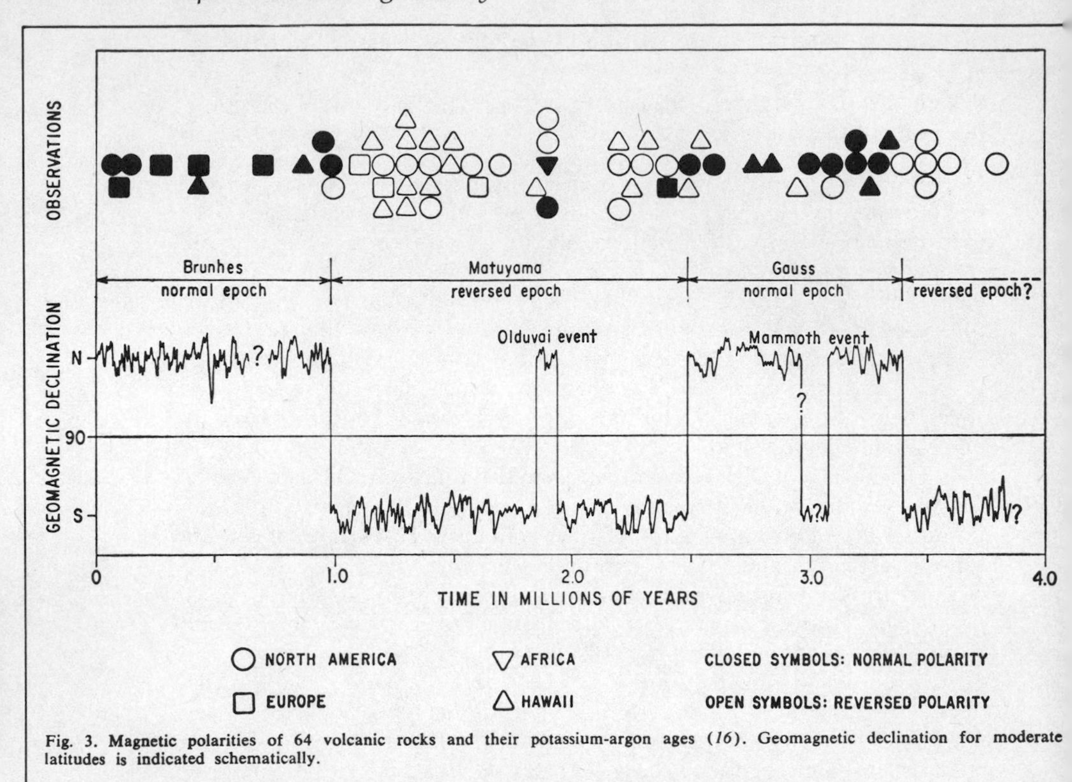

Fig. 3. Magnetic polarities of 64 volcanic rocks and their potassium-argon ages (*16*). Geomagnetic declination for moderate latitudes is indicated schematically.

Fig. 6.9. Scale seven, published by Cox, Doell, and Dalrymple as figure 3 of their June 26, 1964, paper in *Science*, differed from its predecessors in both concept and terminology.

(the new data came from the Pribilof Islands of Alaska but were not reported fully until 1965 in support of scale nine).* They called the short period of normal polarity at 1.9 million years the "Olduvai event," thereby establishing the convention of naming polarity events for the geographic site from which they were recognized. "Less securely documented" was the Mammoth event, "based on two points of reversal near 3.0 million years ago." Because the time interval between these points is about equal to the precision of the potassium-argon dating method, additional data were required "to

*Scale number seven was used in an attempt at correlation with magnetic patterns on the seafloor by Fred Vine and J. Tuzo Wilson in 1965 and in a correlation with reversal ratios in deep-sea sediment cores by Christopher Harrison in 1966. The failure to demonstrate close correlation, in both cases, was due mainly to the inadequacies of the scale (discussed in Chapter 7).

determine whether these two points represent one or two events." The discovery of 3.0-million-year-old lava flows at Mammoth Mountain in California seemed to confirm the reliability of the reversed Hawaiian flows of that approximate age reported but questioned by McDougall and Tarling the previous October 5. Both the Mammoth and Olduvai events are treated further in the discussion of polarity-reversal scale number eight. The new detail in scale seven was due mainly to Dalrymple's having his own gas-extraction line and mass spectrometer. The matter of how to refer to the magnetic intervals was a concern. Doell recalled:

> Between the end of 1963 and the middle of 1964, we just about doubled the amount of data that was published. . . . By the time we wrote [scale 7] we were certainly fairly convinced that we knew what we were doing. One of the reasons for abandoning the N_1–R_1 [designation] was that as new polarity data came in, we were well aware of the irregularity of the length of those things; we thought a numbering scale became somewhat difficult. I think we were probably influenced by traditional geologic thinking about naming things, too. I remember having long discussions with our colleagues, like Dave Hopkins and Joe Hoare and others, about whether we should name them or not. We got a lot of argument about it after we did; people like Keith Runcorn and others thought it was a very poor thing to have done. I don't know whether it was or not, but we did it.[15]

Dalrymple, too, noted that "Allan, Dick, and I spent many hours trying all sorts of complicated numbering schemes that would allow insertion of new events. In order for them to work, they had to be too complicated. Names were just simpler, and that's why we started naming them."[16]

The decision to adopt a new magnetostratigraphic nomenclature was made only after long deliberation both within the Rock Magnetics Project and with a number of specialists from outside the Survey representing related disciplines. As early as 1963, at the Berkeley meeting of the International Union of Geodesy and Geophysics, an informal meeting of those interested in the naming conventions of polarity intervals was held in the Earth Sciences Building around the old, round Andrew Lawson table. Present were Martin Rutten, Cox, Doell, Ted Irving, John Verhoogen, Grommé, Dalrymple, and others. The discussion centered on the problems of using the N_1–R_1 system. Dalrymple noted, "At that meeting the strongest opponent to names was Ted Irving. Allan, Dick, and I were arguing that names were the only workable way. The meeting ended by us simply disagreeing. We also kept point-

Sherman Grommé, at left, and Edward Irving at the Doells' in August 1963.

ing out to Ted [Irving] that he had named the Kiaman [magnetic interval in the Permian], but somehow that seemed different to him."[17] Grommé recalled, "I don't remember what was decided, but I do know that Cox and Doell did not abide by any such decision in their later publications."[18]

Cox, Doell, and Dalrymple were employed by the Survey and were thus required to follow Survey rules in nomenclatural problems of any sort. In January 1964 they discussed their magnetic studies and taxonomic problems with Rudolph W. Kopf, then the Menlo Park representative of the Geologic Names Committee. Kopf liked and approved the scheme of Cox, Doell, and Dalrymple, but consulted George V. Cohee, chairman of the committee, in a letter of February 20, 1964:

At first it was decided to use N_1 and N_2 vs. R_1 and R_2 to designate the 1st and 2nd times of normal and reverse magnetic polarities, but as new "normal" and "reversals" [sic] are found, these subscripts require revision, or retention with redefinition, thus compounding the problem of conveniently referring to any given time interval. It was suggested that perhaps the time of reversal could be designated quantitatively by decimals, referring to millions of years, such as 0.9, 1.1, 2.0, etc., but since these are generally based on K/Ar dates which are subsequently revised, a new numerical designation may imply, to some, a revised dating, or, to others, a new dating not previously recognized. . . . The concept is still in its infancy, so that suggestions now would be most appropriate. They [Cox, Doell, and Dalrymple] wondered if they could use some time terms, such as epoch and interval. . . . Allan [Cox] wondered about honoring deceased geophysicists by applying their names to these broad time intervals. . . . It will be several years before the issue will come up,* but terms, either formal or informal, will have to be coined to apply to these time periods of magnetic intervals. Whatever system is used, it must be plastic enough to allow subdivision, preferably as intervals as well as revision by redefinition, so as to remain useful as the basic data become continuous rather than from the present 25 [or so] isolated samples.[19]

Kopf also noted in postscript that he had just reviewed the paper in which scale number seven was presented.† Cohee replied to Kopf on February 25, 1964:

I have discussed this matter of recognizing periods of reversal of magnetic polarity within the geologic past with several of the Names Committee members. They do not feel that the terms epoch and interval should be used for this purpose but suggest that the reversals be named, such as the John Doe reversal, etc. and the time or period between reversals be referred to as inter-reversals. For example, they could be designated inter-reversal, A, B, C, etc. and it might develop they would want to use A–1, A–2, A–3, etc. and inter-reversal B–1, B–2, B–3, etc.[20]

In their scale of June 26, 1964, Cox, Doell, and Dalrymple presented the following terse rationale:

We have given names rather than numerical or sequential designations to the polarity epochs for the following reasons. The previously used numerical systems count back sequentially from the present at each change in polarity (first normal, first reversed, second normal, and so on). However, if even a short-lived polarity interval is missed when a numerical system is set up, all older designations must be changed when the new polarity interval is identified. This would undoubtedly introduce considerable confusion into attempts to use the epochs for purposes of stratigraphic correlation. For example, in the first four articles linking paleomagnetic and radiometric results, most of the polarity data in the period 1.0 to 2.5 million years ago were reported; yet the significance of the short normal-polarity event at 1.9 million years ago [the Olduvai event] was missed. The numerical systems also preclude designation of any given interval between polarity changes until all later intervals have been recognized.

*Scale number seven, which established the use of the terms "epoch" and "event" in magnetostratigraphic subdivision, was published four months later.

†Kopf, as representative of the Geologic Names Committee, had the authority to approve this paper, which he did, in the absence of a formal Committee policy ruling.

The Rock Magnetics group had given much thought to the taxonomic and nomenclatural questions likely to arise as further data were integrated into increasingly complex polarity-reversal scales. They were at the center of the problem, most intensely involved, long counseled in applying the scale to stratigraphic correlation, and unencumbered by a formal community decision on the question; thus they felt free to use subdivisional designations of their own choice, contrary to the inclinations of many.*

SCALE EIGHT (DALRYMPLE, COX, AND DOELL): REDATING THE BISHOP

It was mentioned earlier that the Bishop Tuff was important in fixing a minimum age for the earliest North American glacial episode, and that this rock body figured significantly in the formulation of the polarity-reversal time scale, because it lent itself to isotopic dating. Evernden and his coworkers had reported ages of 0.87 to 1.2 million years from the Bishop Tuff (see Fig. 6.10).[21]

Charles M. Gilbert had reported in 1938 that the Bishop Tuff contained xenoliths of both granitic and metamorphic rocks and also xenocrysts.† When Curtis and Evernden first dated the tuff at Berkeley in 1957, it was thought that the extremely high temperatures present during the emplacement of welded tuffs and lava flows would degas older, foreign materials such as xenoliths and xenocrysts and thus preclude the inheritance of argon 40 from them.‡ In June 1965, Dalrymple, Cox, and Doell reported in the *Bulletin* of the Geological Society of America: "Because of the importance of the age of the Bishop Tuff and because contamination by xenoliths always poses a problem in radiometric dating of ash flows, we reinvestigated the potassium-argon age of the Bishop

*As a great number of data have mounted, nomenclatural problems have become heightened; especially is this true for the Mesozoic (Creer, 1971). Pecherski addressed the problem in 1970, and McElhinny and Burek formally proposed alternative solutions in 1971. The I.U.G.S. International Subcommission on Stratigraphic Classification has compiled a number of proposals regarding magnetostratigraphic nomenclature (Hedberg, 1977), including the carefully conceived one of Oriel et al. (1976). The Subcommission proposed that the name "polarity zone" be applied to magnetostratigraphic units, that the term "chron" be used for the geochronological equivalent, and that "chronozone" be reserved for the chronostratigraphic equivalent. If this proposal is adopted, the terms "chron" and "subchron" will probably replace the now widely used terms "epoch" and "event," first employed by Cox, Doell, and Dalrymple.

†Respectively, rock fragments and crystals foreign to the igneous rock in which they occur.

‡This according to Curtis (1966). Dalrymple remarked: "I showed in January 1964 that even basalt temperatures were insufficient to degas an old xenolith [Dalrymple, 1964a]. This led eventually to the redating of the Bishop Tuff. Note that the work for this paper was done while I was at Berkeley (probably at Garniss Curtis's suggestion). The sample was collected in 1961 and run as #KA 980 . . . probably in 1962" (written comm., xii-21-78).

Tuff using only primary pumice fragments free of xenolithic inclusions" and found that it was "about 0.7 million years old."*

Dates from the Bishop Tuff had received wide attention because of its field relations with the Sherwin Till glacial deposits.[22] The Bishop Tuff was a natural choice for dating in the early efforts of Curtis and Evernden at Berkeley. Cox, Doell, and Dalrymple used it from their first time scale on, and so did McDougall and Tarling. As each additional publication employs a widely cited datum (such as that of the Bishop Tuff), a curious phenomenon takes place in a research community. Norman D. Watkins, in 1971, called it the "reinforcement syndrome." He noted that a first published description has great leverage in that it enables "workers pondering their own 'curious' data to realize its real (?) meaning. But what if the behavior described in the initial publication is in fact erroneous? A substantial trap will have been laid." Watkins cited as a significant example of such insidious reinforcement the classically defined and deeply established concept that the Pleistocene epoch was punctuated by four glacial periods, which were confirmed by many different studies. About seven are now recognized!

Dalrymple, Cox, and Doell's redating of the Bishop Tuff, based on duplicate analyses of three samples of sanidine,† yielded an average value of 0.71 million years from the six determinations (see Fig. 6.11). The dating was a concomitant of a paleomagnetic study of 103 oriented samples collected from five outcrops along much of the lateral extent of the Bishop Tuff. The entire stratigraphic thickness was sampled at two outcrops along the Owens River. All of the samples were normally magnetized, and the small degree of scatter in magnetic direction among the samples clearly indicated that all the sampled outcrop rocks had been emplaced "within several centuries or less." The results of the redating signaled the dan-

*That date of 0.7 million years was reported as a "note added in proof" in another paper of 1965 by Cox, Doell, and Dalrymple, discussed here as scale number 9, which was completed before vi-2-65. The volume it appeared in is dated only "1965"; thus one is likely to surmise incorrectly from the "added in proof" note that the redating of the Bishop Tuff was first reported there. A letter of vi-2-65 from Cox to Prof. H. E. Wright, Jr., confirms the manuscript completion date and reveals that copies of the manuscript were being sent to Donald Savage, Charles Repenning, Clyde Wahrhaftig, N. King Huber, and Robert L. Smith. Cox's note—"We will appreciate comments from any other Pleistocene stratigraphers who might be interested in glancing at the manuscript"—attests, as earlier noted, to the group's close association with and dependence on stratigraphers, especially those practiced in the later Cenozoic.

†The sanidine (for two of the samples) came from air-fall pumice immediately beneath the welded tuff. It was assumed to be, and proved to be free of detrital and soil materials, which had been picked up and incorporated into the moving flow from which the overlying welded tuff was formed (D. M. Hopkins, written comm., xii-21-78). The third sample came from a large pumice block in the welded tuff at the top of the unit (G. B. Dalrymple, written comm. iv-18-81).

TABLE 1. PUBLISHED POTASSIUM-ARGON AGE DETERMINATIONS ON THE BISHOP TUFF, CALIFORNIA

Reported K-Ar age (10^6 years)	Identification number	References*	Remarks
0.681; 0.783 } 0.87	KA 210R†	(1) p. 2	The higher value was obtained by correcting the lower value for 15 per cent loss of radiogenic argon during preheating
0.830; 0.955 }	KA 210 R1†	(1) p. 2	
0.9	KA 277	(2)	
1.2	KA 278	(2)	KA 305–KA 328 represent determinations on different size fractions of the same sample. Different values for the same determination have been reported in different publications.
0.96; 0.98	KA 305	(2); (3) p. 175	
0.96; 0.98	KA 320	(2); (3) p. 175	
0.91; 0.93; 0.91	KA 321	(2); (3) p. 161; p. 175	
0.91; 0.92	KA 328	(2); (3) p. 175	

* References: (1) Evernden and others (1957); (2) Evernden and Curtis (in press); (3) Evernden and others (1964).

† These ages were calculated using a different decay constant for electron capture ($\lambda_\epsilon = 0.557 \times 10^{-10}$ yr^{-1}) than has been used since 1958. The effect of the more recent decay constant ($\lambda_\epsilon = 0.585 \times 10^{-10}$ yr^{-1}) would be to decrease the calculated age by about 4½ per cent.

Fig. 6.10. The original dates on the Bishop Tuff reported by Evernden and his coworkers from 1957 to 1964 were summarized in table 1 of Dalrymple, Cox, and Doell's June 1965 article in the *Geological Society of America Bulletin.*

TABLE 2. NEW ANALYTICAL DATA FOR POTASSIUM-ARGON AGE DETERMINATIONS ON THE BISHOP TUFF, CALIFORNIA

Sample no.	Material analyzed	Per cent K_2O (1)	(2)	Average	Wgt (gms)	Argon analyses Ar^{40}_{rad} (moles)†	$\frac{Ar^{40}_{rad}}{Ar^{40}_{total}} \times 100$	Calculated age* (10^6 years)
4G001	Sanidine	10.63	10.56	10.60	6.734	7.758×10^{-11}	61.6	0.736 ± 0.07
					7.837	9.250×10^{-11}	45.8	0.754 ± 0.08
4G002	Sanidine	10.69	10.72	10.70	5.671	6.544×10^{-11}	56.0	0.730 ± 0.07
					6.799	7.441×10^{-11}	78.6	0.692 ± 0.06
4G003	Sanidine	10.97	10.94	10.96	7.533	7.799×10^{-11}	42.2	0.639 ± 0.07
					8.535	9.904×10^{-11}	60.9	0.717 ± 0.07

* Using $\lambda_\epsilon = 0.585 \times 10^{-10}$ yr^{-1}, $\lambda_\beta = 4.72 \times 10^{-10}$ yr^{-1}, and atomic abundance $K^{40}/K = 1.19 \times 10^{-4}$. Each age represents a calculation using the argon analysis with the average K_2O value for that sample. The ± figure is our estimate of the precision of the determination at the 95 per cent confidence interval.

† Corrected for the extraction line blank of 1.12×10^{-12} moles Ar^{40}_{rad}.

Fig. 6.11. New dates for the Bishop Tuff—the basis for scale eight—as they appeared in table 2 of Dalrymple, Cox, and Doell's June 1965 article.

gers in selecting samples for dating from ash flows. Most important, it led the Rock Magnetics group to conclude that "the boundary between the Brunhes normal and Matuyama reversed polarity epochs is now uncertain within the limits of 0.68–1.0 million years."

SCALE NINE (COX, DOELL, AND DALRYMPLE)—AND SEVEN REVISITED

"Quaternary Paleomagnetic Stratigraphy," which was published in time to be presented to the participants in the Seventh Congress of the International Association for Quaternary Research on August 14, 1965, contained scale nine (see Fig. 6.12). Scale nine was very similar to scale seven, with the following exceptions. First, in a note added in proof, Cox, Doell, and Dalrymple, in light of the redating of the Bishop Tuff at about 0.7 million years, regarded the age of the Matuyama–Brunhes boundary as "uncertain within the interval 0.85 ± 0.15 million years."* Second, the paper presented an expanded rationale for the naming of the Brunhes, Matuyama, and Gauss epochs and the Olduvai and Mammoth events. Third, the Gilbert epoch was formally named.† Fourth, a refined estimate of the time required for a transition was facilitated by the set of 62 volcanic units for which both potassium-argon dates and paleomagnetic measurements (none of which were intermediate) were known. Fifth, the scale nine paper was ambitious in scope: in the course of 14 lengthy pages, which included an introductory section tantamount to a primer on paleomagnetism, they documented many pertinent details concerning the new data that had been announced only briefly in the paper of scale seven.

They explained their choice of nomenclature as follows:

> To the two latest epochs we have assigned the names Brunhes and Matuyama, to honor two geophysicists who were among the first to recognize the significance of reversed magnetization in rock (Cox *et al.*, 1964b [scale seven]). To the second youngest normal epoch we have previously assigned the name Gauss (Cox *et al.*, 1964b) and to the second youngest reversed polarity epoch we here assign the name Gilbert, in honor of the geophysicists K. F. Gauss (1777–1855) and W. Gilbert (1544–1603), who pioneered the study of the earth's magnetic field.

*Watkins (1972) incorrectly cited scale nine, and Dalrymple (1972) mistakenly referred to scale ten, as the first published source for the redating of the Bishop Tuff and repositioning of the Matuyama-Brunhes boundary.

†Dalrymple (1972) erroneously cited scale ten for the naming of the Gilbert epoch.

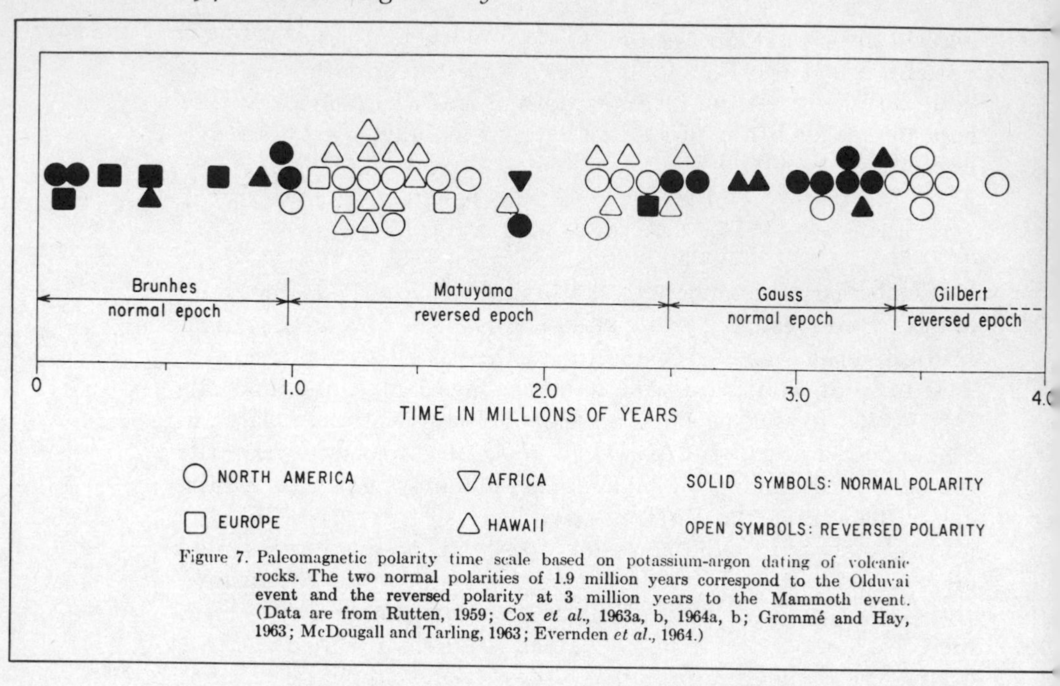

Figure 7. Paleomagnetic polarity time scale based on potassium-argon dating of volcanic rocks. The two normal polarities of 1.9 million years correspond to the Olduvai event and the reversed polarity at 3 million years to the Mammoth event. (Data are from Rutten, 1959; Cox *et al.*, 1963a, b, 1964a, b; Grommé and Hay, 1963; McDougall and Tarling, 1963; Evernden *et al.*, 1964.)

Fig. 6.12. Scale nine, as Cox, Doell, and Dalrymple presented it in their August 1965 paper for the Seventh Congress of the International Association for Quaternary Research, published in *The Quaternary of the United States* by Princeton University Press.

Their rationale for using names repeated in large part their explanation in the paper of scale seven. They explained more fully that polarity-epoch names are preferable to numbers for the reasons that arise from polarity intervals of unequal length.

The exact maximum length for a polarity epoch is not yet precisely known, but at least one epoch several tens of millions of years long occurred in the Permian Kiaman Magnetic Interval of Irving and Parry, 1963.* Such long epochs are separated by times when the earth's field switched polarity rapidly, but it is doubtful whether radiometric dating techniques will be able to resolve these shorter polarity intervals in rocks older than Pliocene. The longer epochs, however, are stratigraphically useful, and it seems reasonable to identify them with names as Irving and Parry (1963) have already done for the long polarity epoch within the Permian.

*Note Irving and Parry's choice of the term "Interval," which they formally designated before Cox, Doell, and Dalrymple introduced their terms in scale seven. The Rock Magnetics group had evidently already decided on the terms "epoch" and "event" in 1963, "at about the time of their second time scale" (scale four; unattributed on request).

As in the paper of June 1963, setting forth scale seven, there was a very brief discussion of the two events, and little of that was new. The Olduvai normal event within the Matuyama reversed epoch, reckoned at about 1.9 million years, first identified in a lava from Tanzania by Grommé and Hay in 1963, was corroborated by study of several basalt flows from the Pribilof Islands of Alaska (see

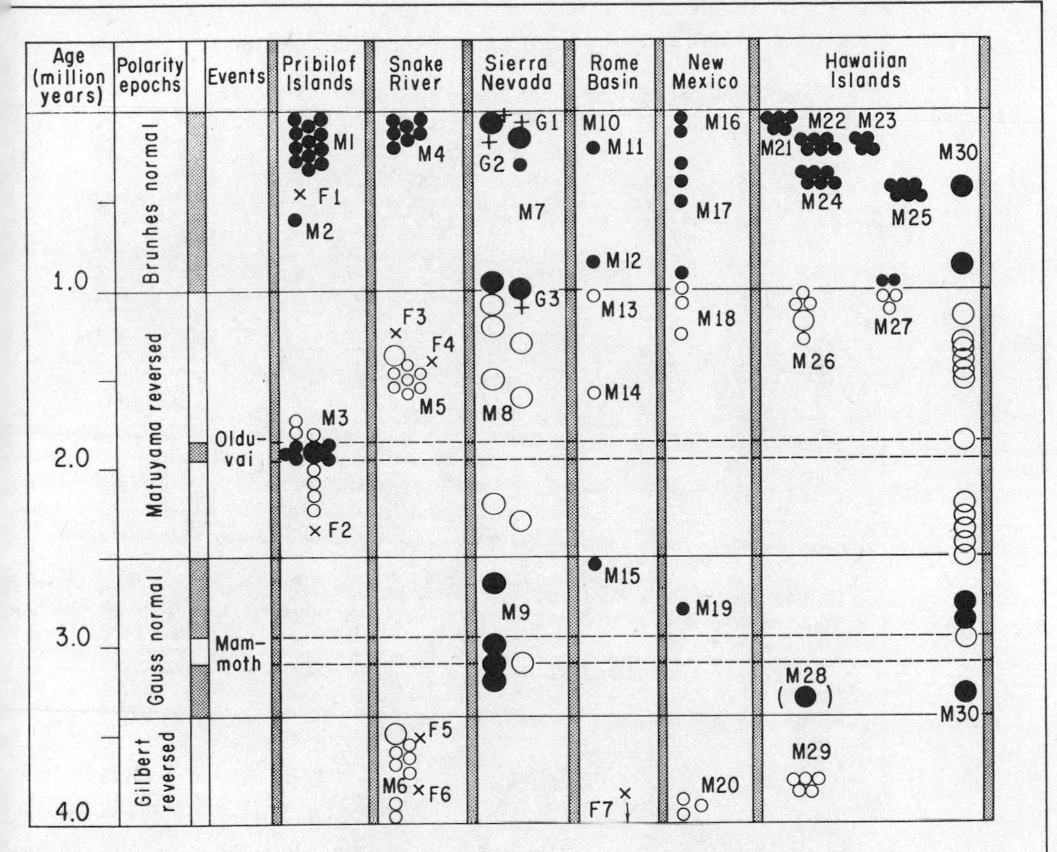

Figure 8.

Paleomagnetic correlation of the Pribilof Islands, Alaska; Snake River, Idaho; Sierra Nevada, California; Rome Basin, Oregon; New Mexico; and Hawaiian Islands.

Small circles, no radiometric dates on paleomagnetic samples; large circles, radiometric dates on paleomagnetic samples; solid circles, normal magnetization; open circles, reversed magnetization; +, glacial deposits; ×, fossils.

Fig. 6.13. When they presented scale nine in August 1965, Cox, Doell, and Dalrymple incorporated important data from the Pribilof Islands of Alaska to corroborate the existence of short events and confirm the view of a highly irregular field-reversal pattern. The expedition to the Pribilofs was conceived by David Hopkins.

Fig. 6.13). The Mammoth event of reversed polarity within the Gauss normal epoch was also identified, but with less assurance than the Olduvai event. The event, based on a basalt flow from Mammoth Lake, California, lacked full confirmation because of the stratigraphically (not radiometrically) questionable character of McDougall and Tarling's (scale two) reversed-polarity lava from Oahu, dated at 2.95 ± 0.06 million years. The authors also reiterated that "events are distinguishable from epochs solely on the basis of their duration; the lengths of the last three epochs range

David Hopkins in 1972 in an archaeological excavation at the Chaluka mound, western Umnak Island, Aleutian Islands, Alaska. Hopkins, who is the foremost student of the geology of the Bering Land Bridge, was the in-house stratigrapher for the Rock Magnetics Project at Menlo Park; he was an important influence in the early career of Allan Cox.

from 0.9 to 1.4 million years, whereas events appear to be about one-tenth as long."

The data derived from the Pribilof Islands were extremely important; they provided solid stratigraphic and potassium-argon evidence for the reality of short events* and at the same time confirmed the new view of a highly irregular field-reversal pattern adopted by the Rock Magnetics group at the time they drew up scale seven (June 26, 1964).

The expedition to the Pribilof Islands was conceived by David Hopkins, described earlier (in Chapter 5) as the Rock Magnetic group's "in-house stratigrapher" and an early influence on Cox. T. F. W. Barth in 1952 described marine mollusks interbedded with lava flows, which Hopkins thought potentially valuable in his attempt to develop a history of the Bering Sea region. Hopkins also thought that a paleomagnetic study in the Pribilofs by Cox would lead to the first radiometric dating of marine sediments of the Bering Sea. Cox, Hopkins, and Edward Roth therefore spent 23 days on St. Paul Island and 8 days on St. George during the summer of 1962. Hopkins recalled:

> On St. Paul everything was disappointingly simple; all of the rocks were normal and evidently quite young. Then we went to St. George, and we were elated to find that the very first rocks we looked at were reversed. Almost the second rocks we looked at near the village were normal. This was wonderful; we had reversed on [top of] normal, and we figured this was what we then called R_1 and N_2. Then when we got out on the east end of the island on Tolstoy Point, we found more reversed rocks under the normal rocks. It turned out that we had a thin normal sequence that we could trace extensively around the coast of the island, with a thicker reversed sequence above and reversed rocks below;† we didn't know whether [the rocks below] were thick or not. At Tolstoy Point we had the sequence reversed, normal, reversed.[23]

*Recall that in 1964, when the Olduvai and Mammoth events were first defined, most of the available paleomagnetic evidence was in favor of long polarity intervals without short polarity events. A notable exception, though not radiometrically dated, which must have provided Cox, Doell, and Dalrymple with evidence from a stratigraphically well-defined section, was provided by Wensink (1964a), who found within the N_2 and R_1 of central Iceland several basalt flows that were opposite in polarity to the series in which they were interbedded: "Apart from the main geomagnetic periods indicated by the notations introduced by Roche [1953], extra short periodical reversals of the geomagnetic field have been found in Eastern Iceland both in the geomagnetic N_2 and in the geomagnetic R_1 series. Up to now these extra reversals have not been met with in areas outside Eastern Iceland. As these period are short lived in relation to the duration of the main geomagnetic periods we have not introduced other notations" (p. 383).

†The magnetostratigraphy was established in the field using a small fluxgate magnetometer that had been developed by the Rock Magnetics Project.

TABLE 2. Summary of Paleomagnetic-Radiometric Data

Polarity Epoch	K-Ar Age, 10^6 years	Magnetic Polarity*	Magnetic** Category	Material Dated	Locality	Identification Number†	References	
							Age	Magnetics
	0.00	*Normal*	III	Basalt	Hawaii	KA1218[6]	6	9
	0.07	*Normal*	III	Basalt	Hawaii	KA1215[6]	6	9
	0.075	Normal	IA	Basalt	Sierra Nevada§	S7[2]	2, 15	2
	0.095	*Normal*	IV	Leucite	Italy	KA853[6]	6	10
	0.150	Normal	IB, VA	Basalt	Sierra Nevada	S8[2]	2, 15	2
	0.27‡	*Normal*	IV	Leucite	Italy	None[10]	6, 10	10
Brunhes (normal)	0.28	**Normal**	IA	Obsidian	Sierra Nevada	S17		
	0.43	*Normal*	III	Basalt	Hawaii	KA1219[6]	6	9
	0.43‡	*Normal*	IV	Sanidine	Italy	None[10]	6, 10	10
	0.45	Normal	III	Basalt	Maui, Hawaii	None[4]	4	4
	0.53	*Normal*	III	Basalt	Hawaii	KA1220[6]	6	9
	0.68	**Normal**	IA	Sanidine	Sierra Nevada	S1[1]	16	1
	0.71	*Normal*	IV	Leucite	Italy	KA348[6]	6	10
0.85 ± 0.15								
	0.94	*Normal*	II	Plagioclase	Sierra Nevada	S9[2]	2, 15	2
	0.99	Reversed	II	Obsidian	Sierra Nevada	S2[1]	1	1
	1.1	*Reversed*	IV	Sanidine	Italy	KA1162[6]	6	10
	1.14	**Reversed**	VB	Basalt	Sierra Nevada	S18		
	1.16	Reversed	III	Basalt	Maui, Hawaii	None[4]	4	4
	1.2	Reversed	II, VB	Latite	Sierra Nevada	S10[2]	2, 7	2
	1.2	Reversed	III	Basalt	Kauai, Hawaii	KA1211[6]	6	
	1.2	*Reversed*	IV	Sanidine	Italy	KA1181[6]	6	10
	1.29	Reversed	III	Basalt	Maui, Hawaii	None[4]	4	4
Matuyama (reversed)	1.3	Reversed	IB	Latite	Sierra Nevada	S11[2]	2, 7	2
	1.3–1.5	Reversed	III	Basalt	Molokai, Hawaii	None[4]	4	4
	1.4	Reversed	III	Basalt	Idaho	KA1188[6]	6	11
	1.5‡	Reversed	IB	Biotite	Sierra Nevada	S3[1]	6, 12	1
	1.6	*Reversed*	IV	Basalt	France	KA1184[6]	6	13
	1.6‡	Reversed	II	Biotite	Sierra Nevada	S4[1]	6, 12	1
	1.64	**Reversed**	II	Olivine latite	Sierra Nevada	S19		
	1.85	Reversed	III	Basalt	Molokai, Hawaii	None[4]	4	4
	1.9	*Reversed*	(See text)	Biotite	Sierra Nevada	KA490[6]	6	
	1.9	Reversed	IB, VA, VB	Latite	Sierra Nevada	S12[2]	2, 7	2
Olduvai event (normal)	1.9‡	Normal	IB	Basalt	Africa	None[5]	5, 6	5
	2.2	**Reversed**	IB, II	Andesite	Sierra Nevada	S13[3]	2, 7	3
	2.2–2.5	Reversed	III	Basalt	Oahu, Hawaii	None[4]	4	4
	2.3	*Reversed*	IV	Andesite	Sierra Nevada	S14[3]	2, 7	3
	2.3‡	*Normal*	IV	Sanidine	Italy	None[10]	6, 10	10
2.4 ± 0.1								
	2.46	**Normal**	IB, II, VA(?)	Andesite	Sierra Nevada	S20		
	2.6	Normal	IB	Basalt	Sierra Nevada	S5[1]	8	1
Gauss (normal)	2.76	Normal	III	Basalt	Oahu, Hawaii	GA556[4]	4, 14	4
	2.84	Normal	III	Basalt	Oahu, Hawaii	GA809[4]	4, 14	4
	2.95	*Reversed*	III	Basalt	Oahu, Hawaii	GA557[4]	4, 14	4
	3.0	Normal	IB, II	Plagioclase	Sierra Nevada	S15[2]	2, 7	2
Mammoth event (reversed)	3.06	**Reversed**	IA	Basalt	Sierra Nevada	S21		
	3.1	Normal	II, VB	Basalt	Sierra Nevada	S16[2]	2, 7	2
	3.2	Normal	IB	Basalt	Sierra Nevada	S6[1]	8	1
	3.27	Normal	III	Basalt	Oahu, Hawaii	GA810[4]	4, 14	4
	3.3	**Normal**	II	Basalt	Sierra Nevada	S23	7	
	3.3	*Normal*	III	Basalt	Kauai, Hawaii	KA1213[6]	6	
	3.32	**Normal**	II, VB(?)	Basalt	Sierra Nevada	S22		
3.35 ± 0.1								

TABLE 2. (Continued)

Polarity Epoch	K-Ar Age, 10^6 years	Magnetic Polarity*	Magnetic** Category	Material Dated	Locality	Identification Number†	References: Age	References: Magnetics
	3.4	Reversed	III	Basalt	Idaho	KA1172[6]	6	11
	3.48	**Reversed**	II	Basalt	Sierra Nevada	S28		
Gilbert (reversed)	3.5	**Reversed**	IB	Basalt	Sierra Nevada	S24	8	
	3.5	**Reversed**	II	Basalt	Sierra Nevada	S25	7	
	3.5	**Reversed**	IB	Basalt	Sierra Nevada	S26	8	
	3.6	**Reversed**	IA	Basalt	Sierra Nevada	S27	8	

* Bold face type identifies data presented in this paper; italics identify less reliable data (see text).

** Indicates type of magnetic data available. See text for details.

† References are those where magnetic and radiometric data first appear together; the identification number is that given in these references.

‡ Indicates revised(?) dates from reference 6, which differ from those given in earlier publications.

§ Indicates Sierra Nevada region. Includes regions as far east of the Sierra Nevada as the Carson and Virginia Ranges.

REFERENCES

1. *Cox et al.* [1963*a*].
2. *Cox et al.* [1963*b*].
3. *Cox et al.* [1964*a*].
4. *McDougall and Tarling* [1963].
5. *Grommé and Hay* [1963].
6. *Evernden et al.* [1964].
7. *Dalrymple* [1964*a*].
8. *Dalrymple* [1963].
9. *Doell and Cox* [1961].
10. *Rutten* [1959].
11. *Cox* [1959].
12. *Evernden et al.* [1957].
13. *Kloosterman* [1960].
14. *McDougall* [1964].
15. *Dalrymple* [1964*b*].
16. *Dalrymple et al.* [1965].

Fig. 6.14. Scale ten, as it appeared in table 2 of Doell, Dalrymple, and Cox's January 15, 1966, paper in the *Journal of Geophysical Research.*

Although the crucially important samples from the Pribilof Islands were brought back to the Rock Magnetics Project at Menlo Park in 1962, they went undated until late 1964.* The recognition and confirmation of the Olduvai event on St. George was important because the striking radiometric results could be corroborated by clear physical stratigraphy.

SCALE TEN (DOELL, DALRYMPLE, AND COX): NEW CRITERIA

The recalibrated scale ten was presented in the January 15, 1966, issue of the *Journal of Geophysical Research* (see Fig. 6.14). It barely differed from scale nine, but it was attended by a greater certainty arising from new data for 12 rock units from eastern California and

*An account of the difficulties in securing dating of samples by the Rock Magnetics Project was given in Chapter 5. A great backlog of magnetically determined samples had accumulated from several localities, and Dalrymple was delayed in dating the samples that confirmed the Olduvai event. "Age Calculation Sheets" in the Rock Magnetics Project records indicate that the Pribilof sample from site P–20 was run on x-2-64, and P–3 and P–19 on xii-15-64 (site numbers and other data on the Pribilof rocks are given in Cox, Hopkins, and Dalrymple, 1966).

Nevada that sharpened the definition of the Gilbert–Gauss boundary at 3.35 ± 0.1 million years. The Mammoth event, which up to that time had been based on a single age analysis of 3.1 ± 0.1 million years,[24] was confirmed by a second date from the same volcanic unit, made on one of nine paleomagnetic samples collected near the Mammoth Mine in Mono County, California; the average of the two dates was 3.06 ± 0.1 million years. Most important in what might be termed a review paper was a table of paleomagnetic and radiometric data, which were included only after meeting criteria designed to guard against future confusion resulting from inadequate sampling and analysis. The newly applied criteria included the following:

Samples for dating and magnetic analysis must come from the same cooling unit.*

Extensive laboratory testing should validate magnetic stability.

Radiometric ages of 2 million years or less require standard deviations of 0.1 million years or less; standard deviations of older ages must not exceed 5 percent.

K-Ar dates must be superpositionally comformable within the cooling unit and within the local rock sequences.

Careful scrutiny of rock samples must ensure against inheritance or loss of argon 40.

The adoption of these criteria resulted in the rejection of eight dates that had been used in scale seven; scale ten, presented as a table of 56 data points with the epoch boundaries and events clearly drawn and dated, was infused with a new measure of certainty, painstakingly acquired.

In view of what would follow, one important conclusion of the paper should be set off: "For parts of the last 4 [million years], the available radiometric data are inadequate to preclude the possibility of additional events. However, because there is no evidence for additional events in other undated stratigraphic sections of this general age range, *their existence is unlikely*."[25] Recall that Cox began his time-scale work with a simple model in which simplicity implied regularity, and that he had difficulty in accepting a randomly reversing magnetic field. At the time scale ten was made, there was a goodly number of dates, distributed through the last 4 million years in such a way that no significant gaps appeared, in

*Recall the chagrin suffered by the Rock Magnetics Group because of their failure to meet this criterion (discussed under scales two and four).

Allan Cox collecting basalt samples in the Galapagos Islands in 1966 using a portable diamond drill.

which an event might be hidden (according to Cox's idea of the duration of an event). The fact that almost a dozen events are presently recognized forces the conclusion that Cox and his coworkers could not have assessed, at that date, the variables and the possibilities they implied. But there was likely more to the conclusion quoted above than simple objective reckoning; it likely grew from Cox's intellectual fix against further disorder, which he expressed both before and after this time.

SCALE ELEVEN (DOELL AND DALRYMPLE): THE JARAMILLO EVENT

In 1965, while Allan Cox was abroad in the Galápagos Islands and later working with Joseph Hoare on Nunivak Island and near

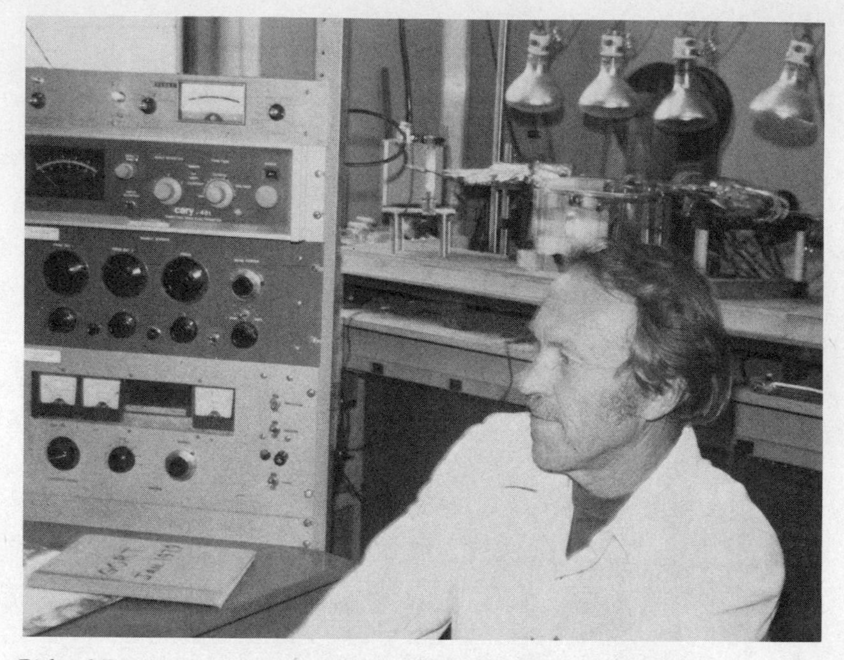

Richard Doell, retired from the U.S. Geological Survey, was a visiting professor at Berkeley when this photo was taken in Garniss Curtis's potassium-argon laboratory in 1981.

the Yukon River in Alaska, Richard Doell and Brent Dalrymple continued work on a series of rock samples from the Valles Caldera, part of a volcanic complex comprising the Jemez Mountains, 35 miles northwest of Santa Fe, New Mexico. Their attention to this seemingly fruitful region had been drawn by the work of Robert L. Smith, Roy A. Bailey, and Clarence S. Ross published in 1961.* Dalrymple, Doell, and Cox reported preliminary results from that area in November 1965, at the meeting of the Geological Society of America in Kansas City, Missouri. Dalrymple read the paper, entitled "Recent Developments in the Geomagnetic Polarity Epoch Time-Scale"; that short paper (for which only an abstract was published) contained the brief note: "There may be another event at about 0.9 million years, although it is not yet confirmed. Because of the short duration of these events, the chances of finding another

*In 1968, Doell and Dalrymple published (with Smith and Bailey) a full, 36-page account of the Valles Caldera work.

one are rather small; thus subsequent work may turn up additional events that are as yet unrecognized."* On May 20, 1966, a two-page-long paper was published in *Science* that was destined to become a historical marker in the geological sciences; it was entitled "Geomagnetic Polarity Epochs: A New Polarity Event and the Age of the Brunhes–Matuyama Boundary."†

Doell and Dalrymple determined the age and paleomagnetism of 19 Pleistocene volcanic rock units, six of whose ages fell between 0.7 and 1.0 million years. Recall that until scale eleven more than 15 reversely magnetized, well-dated volcanic rocks were known from the interval between 1.5 and 1.0 million years, but only the Bishop Tuff provided an age datum between 0.7 and 1.0 million years. The new dates were therefore of great value in defining the still questionable Matuyama–Brunhes boundary. The samples from the six rhyolite domes in the Valles Caldera were not questionable with respect either to potassium-argon or to paleomagnetic determination. Of these six, three reversed specimens were dated between 0.71 and 0.73 million years, an intermediate-polarity rock was dated at 0.88 million years, a normal specimen fell at 0.89 million years, and a reversed rock was dated at 1.04 million years. Doell and Dalrymple did not find these results surprising, since the rocks "were formed near the time of the last polarity transition." They were, however, pressed to explain "the normal and intermediate directions bracketed between reversed directions," which they thought suggested three possibilities: "(1) The precision of the potassium-argon age measurements is not sufficiently high to distinguish between the ages of the units, that is 4D057 and 3X178 [the intermediate- and normal-polarity samples] are really younger than the other domes; (2) one or more of the domes may have self-reversed magnetization which was carefully precluded by testing; or (3) there may be a short polarity event near the Brunhes–Matuyama boundary."

Since the first five domes listed in their Table 1 were established

*Dalrymple, Doell, and Cox's unpublished paper, read on xi-4-65 at 9:00 A.M. (Document No. 38), is further discussed in Chapter 7.

†The six specimens on which the Jaramillo event is based were collected in the summer of 1964. All were dated using the mineral sanidine, save one—sample 3X194. Their natural remanent magnetism was determined from December 1964 through March 1965; demagnetization was done during that March, and heating experiments to guard against self-reversal were not done until August 9 and September 3, 1965 (Rock Magnetics Group work record sheets, New Mexico I and II). The manuscript that was submitted to *Science* on January 26, 1966, had been completed by December 23, 1965, and began the review process within the Survey at that time;

Table 1. Potassium-argon ages and polarities of six volcanic units from the Valles Caldera, New Mexico.

Unit No.	K-Ar age (millions of years)	Polarity
4D049	0.71	Reversed
3X122	.72	Reversed
4D074	.73	Reversed
4D057	.88	Intermediate
3X187	.89	Normal
3X194	1.04	Reversed

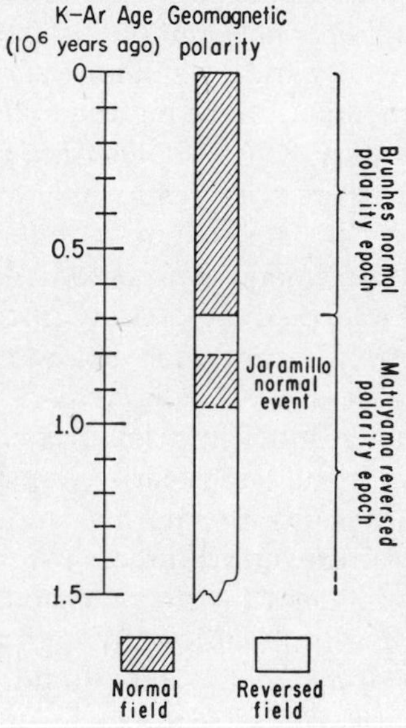

Fig. 1. Suggested sequence of the most recent changes in polarity of the earth's magnetic field.

Fig. 6.15. Table 1 and figure 1 from Doell and Dalrymple's May 1966 paper in *Science*, which presented scale eleven and the Jaramillo event.

as younger than 3X194 (the oldest, reversed unit), and four other normally magnetized domes in the Valles Caldera that lie stratigraphically above the six domes under discussion were dated between 0.43 and 0.54 million years, the ages Doell and Dalrymple presented are compatible with the stratigraphy (see Fig. 6.15). In view of that fit and the requirement that "at least two of the calcu-

final approval was given on January 18, 1966. Doell and Dalrymple had first thought of the name Abrigo (after one of the volcanic domes of the Valles Caldera) for the Jaramillo event, but the name was preempted.

lated ages would have to be in error by more than four times their standard deviations," they were led to reject the first hypothesis of imprecision in dating. After due note of the well-established Olduvai normal and Mammoth reversed events, they concluded that the new data, when considered with previously published information, suggest "the sequence for the more recent polarity changes shown in Fig. 1" (see Fig. 6.15). They "preferred" to place the Matuyama–Brunhes boundary at 0.7 million years; then they stated: "*We here name the normal event near 0.9 million years the 'Jaramillo normal event'* after Jaramillo Creek, which is approximately 3 km south of the locality of unit 3X187. From the present data it is not possible to tell whether the intermediate direction represents the transition to or from the Jaramillo normal event, nor therefore whether the event occurred just before or just after 0.9 million years ago."[26]

Upon discovery of the Jaramillo event, and before submitting the paper of scale eleven to *Science*, Doell and Dalrymple sent a telegram to Allan Cox, then in an Eskimo village on the Yukon. The telegram read, "You owe me a martini." Cox recalled, "That meant that Dick [Doell] had done the heating experiments and there was no indication of self-reversal." Cox had believed that the rock representing the Jaramillo event was a self-reversed one. Cox continued:

> It was so close to the boundary; I didn't want it to be an event. I didn't want to have events that close to the boundary, because I knew they were going to be really mean to work out. In fact, once you get back beyond the Brunhes, if you get events close to a boundary it's really hard to identify them, because it looks like they're in the dating [imprecision in dating within the limits of standard deviations might reverse the order]. So the only way you can pick up a quick flip of the field that occurs close to a boundary is if you happen to be very lucky and get a stratigraphic succession. So you prove it from stratigraphy. Then you get the date roughly from potassium-argon.[27]

Almost two decades later Doell recollected that "Allan was a little doubtful about the Jaramillo event at that time. I can remember that night, having the bet for a martini about that. He thought we were going too far out, to base this just upon a few rocks. 'But, hell,' we argued with him, 'we had no more rocks than that when we hypothesized the whole thing [the polarity-reversal scale] in the beginning!'"[28]

Doell and Dalrymple closed the paper of May 20, 1966, with a note of modest hope that others might find their recent data useful for "geologic correlation or other purposes." The "other purposes" were to include the serendipitous use of that scale, containing the Jaramillo event, in a most amazing way. No one dreamed of what was about to unfold!

PART III

Turning the Key: Applying the Scale

Chapter 7

The Vine-Matthews-Morley Hypothesis

A HIGHLY speculative and poorly received hypothesis was advanced in 1963 by Fred J. Vine and Drummond H. Matthews, and independently by Lawrence W. Morley. It held that the solid rock of the ocean floor is imprinted with the record of field reversals in the form of a sequence of alternately magnetized stripes; the stripes, with widths proportional to the alternating intervals of the polarity-reversal scale, formed as newly created ocean floor spread from mid-ocean ridges. In February 1966, geomagnetic polarity-reversal time scale eleven, containing the Jaramillo event, was successfully correlated with magnetic-anomaly profiles across mid-ocean ridges. Overnight the Vine-Matthews-Morley hypothesis was confirmed; the acceptance of seafloor spreading thus became inescapable. The revolution had been triggered. Almost simultaneously, from a third independent source, the polarity intervals of the reversal time scale were demonstrated in deep-sea sediment cores.

A look at certain research areas outside of geochronology and magnetic-reversal studies will clarify the significance of those astonishing correlations. It is important to know that when the confirming polarity-reversal evidence for seafloor spreading was advanced, there were already large and growing bodies of other kinds of favorable evidence available. Among them were structures indicative of tension and high values of heat flow at the axes of the mid-ocean ridges, the youthfulness of seafloor rocks, Benioff-Wadati seismic zones in conjunction with large gravity anomalies and reduced heat flow at the deep-sea trenches, and the growing plausibility of convection in the underlying mantle as an engine.* Singularly influential was the body of paleomagnetic data accumulated largely from 1954 onward by the English directionalists,[1] which increasingly suggested that continental movements were required to make sense of ancient pole positions. Many arguments in favor of drift, including the older classical ones advanced by Wegener[2] and others before him, and newer kinds of studies, largely grown from oceanographic surveys in the 1940's and 1950's, attracted increasing attention. Slowly and gradually the evidence in favor of mobilism† grew, but as late as 1965 and early 1966, most of those in the North American geological community were still skeptical "anti-drifters," and although communities abroad, notably the English, were more sympathetic to it, the question of drift was far from decided even there. The diverse mass of evidence in favor of mobilism was considered by most to be lacking in the numerical precision required for acceptance. Those other research areas, contemporaneous with polarity-reversal studies, provided the necessary complementary data that were correlated with the polarity-reversal time scale.

The recognition and correlation of the polarity-reversal intervals from three independent sources turned the tide of opinion in favor of continental drift within a few short years. That precipitous reversal of belief is especially poignant against the background of a series of important papers that were presented in 1964 at the "Symposium on Continental Drift."‡ The list of renowned contributors and panelists included several paleomagnetists, but none of them spoke directly about polarity reversals; they were all directional-

*A Benioff-Wadati (seismic) zone is a plane of earthquake activity that slopes at about 45°, extending from a deep-sea trench under the adjacent continent. The theory of plate tectonics holds that lithospheric plates (composed of the crust and the uppermost, rigid part of the mantle) plunge into the asthenosphere (upper mantle) along this zone and cause earthquakes. A gravity anomaly is a measurement value that is greater (positive anomaly) or smaller (negative anomaly) than the predicted or expected value at a particular place. The strong negative anomalies beneath the deep-sea trenches led Felix Andries Vening Meinesz in 1930 to conclude that some force was holding up the earth's surface in its irregular shape against the tendency of gravity to flatten it. He thus lent support to the convection-current theory of Arthur Holmes.

†"Mobilism," a loose term that came into use during the 1960's, is an approximate synonym for drifting; mobilists hold that great lateral displacement of crustal (or lithospheric) segments (or plates) has occurred. The opposing view was held by the fixists (anti-drifters).

‡Organized for the Royal Society by Patrick Blackett, Edward Bullard, and Keith Runcorn (held iii-19–20-64) and published in the *Philosophical Transactions of the Royal Society* in 1965. The English were earlier inclined to be sympathetic to drift, due largely to the influence of the English magnetic directionalists, among whom were the prestigious group that convened the symposium. Sir Edward Bullard was accused by some of "loading" the symposium panel with "drifters" (Bullard, t. 1, s. 2, vii-2-79); in contrast, the conveners of the American Association of Petroleum Geologists' symposium on drift, held in New York in 1926, were said to have inadequately represented the viewpoint of the drifters.

ists. No workers in polarity reversals were present except Martin Rutten, who mentioned Icelandic reversal studies only peripherally. In that symposium volume, emanating from a community decidedly sympathetic to continental drift, there was not a hint of the crucial role that reversal studies were to play in effecting a revolution only a year after its publication.

The Hypothesis Elucidated

The Vine-Matthews-Morley hypothesis is equal in importance to any formulated in the geological sciences in this century. It was conceived during a time of great intellectual ferment and entailed the synthesis of several almost discrete and then questionable components. The idea was independently and almost simultaneously conceived by Fred Vine, working under Drummond Matthews at the University of Cambridge, and by Lawrence Morley of the Geological Survey of Canada. What is commonly called the Vine-Matthews hypothesis I will thus refer to as Vine-Matthews-Morley.

Vine and Matthews had priority in publication (through curious circumstance, treated below). On September 7, 1963, in *Nature*, Vine and Matthews published "Magnetic Anomalies over Ocean Ridges," much of which was based on the recent British Survey of the Carlsberg Ridge in the Indian Ocean.

> Work on this survey led us to suggest that some 50 percent of the oceanic crust might be reversely magnetized, and this in turn has suggested a new model to account for the pattern of magnetic anomalies over the ridges.
>
> The theory is consistent with, in fact virtually a corollary of, current ideas on ocean floor spreading* and periodic reversals in the Earth's magnetic field. If the main crustal layer (seismic layer 3) of the oceanic crust is formed over a convective up current in the mantle at the centre of an oceanic ridge, it will be magnetized in the current direction of the Earth's field. Assuming impermanence of the ocean floor, the whole of the oceanic crust is comparatively young, probably not older than 150 million years, and the thermoremanent component of its magnetization is therefore either essentially normal, or reversed with respect to the present field of the Earth. Thus, if spreading of the ocean floor occurs, blocks of alternately normal and reversely magnetized material would drift away from the centre of the ridge and parallel to the crest of it.
>
> This configuration of magnetic material could explain the lineation or

*Although greatly influenced by personal contact with Hess by the date of writing, Vine cited Dietz (1961a) in the matter of seafloor spreading, because Hess had not yet formally published (Vine, t. 1, s. 1). Morley was not informed of Hess's ideas but had read, and was greatly influenced by, Dietz's paper (Morley, t. 1, s. 1, iv-24-79).

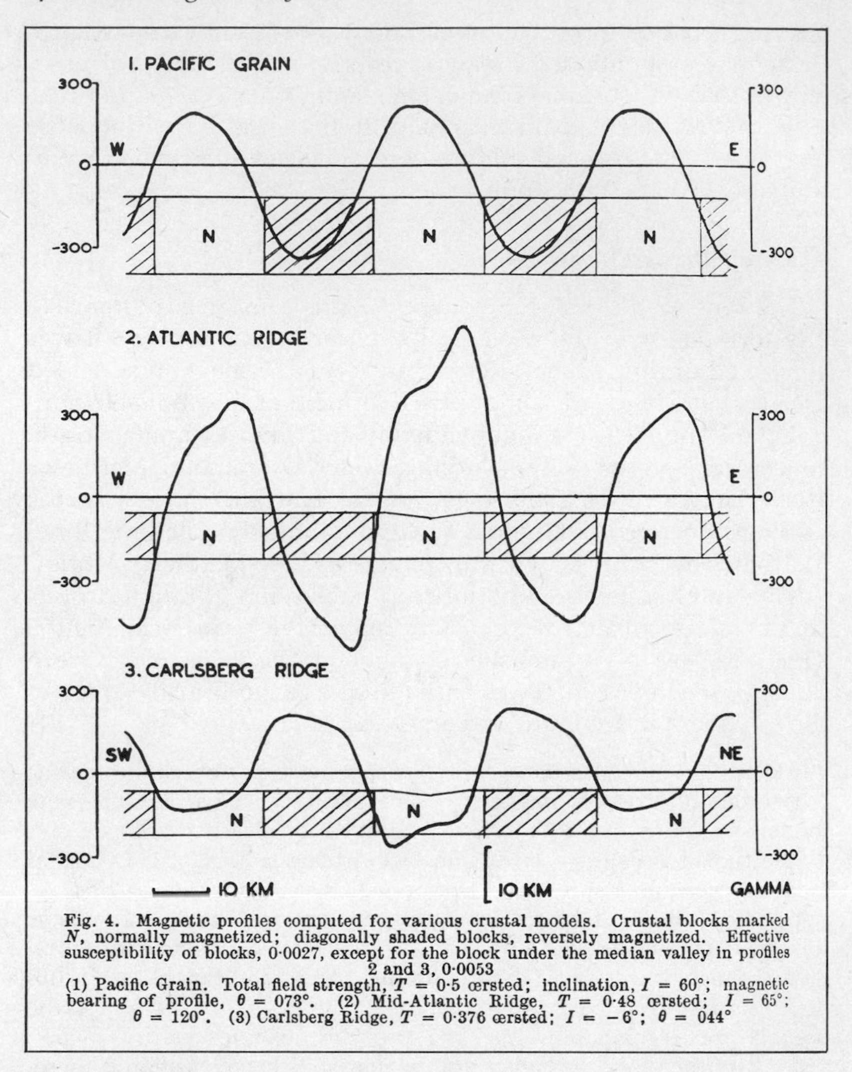

Fig. 7.1. Vine and Matthews's figure 4, from their September 1963 article in *Nature*.

"grain" of magnetic anomalies observed over the Eastern Pacific to the west of North America. . . . Here north-south highs and lows of varying width, usually of the order of 20 km, are bounded by steep gradients. The amplitude and form of these anomalies have been reproduced by Mason, but the most plausible of the models used involved very severe restrictions on the distribution of lava flows in crustal layer 2. They are readily explained in terms of reversals, assuming the model shown in Fig. 4 [see

Fig. 7.1]. It can be shown that this type of anomaly pattern will be produced for virtually all orientations and magnetic latitudes, the amplitude decreasing as the trend of the ridge approaches north-south or the profile approaches the magnetic equator. The pronounced central anomaly over the ridges is also readily explained in terms of reversals. The central block, being most recent, is the only one which has a uniformly directed magnetic vector. This is comparable to the area of normally magnetized late Quaternary basics in Central Iceland on the line of the Mid-Atlantic Ridge. Adjacent and all other blocks have doubtless been subjected to subsequent vulcanism in the form of volcanoes, fissure eruptions, and lava flows, often oppositely magnetized and hence reducing the effective susceptibility of the block, whether initially normal or reversed. The effect of assuming a reduced effective susceptibility for the adjacent blocks is illustrated for the North Atlantic and Carlsberg Ridges in Fig. 4.

In Fig. 4, no attempt has been made to reproduce observed profiles in detail; the computations simply show that the essential form of the anomalies is readily achieved. The whole of the magnetic material of the oceanic crust is probably of basic igneous composition; however, variations in its intensity of magnetization and in the topography and direction of magnetization of surface extrusives could account for the complexity of the observed profiles.

Vine and Matthews

In 1961, at the University of Cambridge, Drummond Matthews completed his Ph.D. thesis, *Rocks from the Eastern North Atlantic*. He was shortly thereafter appointed to a teaching post at Cambridge and told "to run the British part of the International Indian Ocean Expedition, which got under way before the Americans did largely because Maurice Hill, Head of Geophysics at Cambridge, had persuaded the Navy to lend a ship, H.M.S. *Owen*."[3] Matthews spent about six months at sea in 1961 recording bathymetry and gravity profiles across the Carlsberg Ridge in the northwest Indian Ocean. In keeping with Cambridge tradition at that time—doing detailed work because they knew that they "couldn't compete with the American labs in miles run"—he "planned to do a detailed square on the crest of the Carlsberg Ridge,"[4] the position of which he extrapolated from some Soviet and British soundings. Among other things, his measurements included seismic velocities and magnetization intensities. He was by that date "absolutely persuaded that the mid-ocean ridge was to be understood in terms of fissure eruptions, and [he] knew basalts of the seafloor were remanently magnetized."[5] In November 1962, he returned to Cambridge with what

Fred Vine, at left, and Drummond Matthews in the University of East Anglia Village in September 1970. Prior to publication of his hypothesis, Vine related it to Maurice Hill, head of geophysics at Cambridge: "I think [Hill] thought I was totally mad. He was polite enough not to say anything; he just looked at me and went on to talk about something else. . . . At the time I was quite keen to publish the idea with Teddy Bullard. [He] quite rightly said 'no way.' He didn't want his name on the paper."

was then the most detailed magnetic survey of a portion of the ocean floor; it ranks still among the best ever done. Matthews recalled that at the time of the Carlsberg Ridge survey in 1962, "we had no idea of connecting seafloor spreading and reversals—at least I didn't."

I came back to find that Fred Vine had been handed to me as my research student. (I'd been away, so Maurice Hill had been looking after him.) He had been working on a magnetic survey in the English Channel. The question that I asked Fred to solve was: okay, here's a magnetic survey, a good bathymetric survey, and gravity data too [from the Carlsberg Ridge]. We would like to produce the equivalent of a Bouguer anomaly map.* I

*A Bouguer anomaly is a gravity anomaly figured by considering the attraction effect of topography but not isostatic (flotational) compensation. Lateral variations in the Bouguer anomaly reflect lateral differences in the density of the underlying rock.

wanted Fred to find out what direction the structure was magnetized in and how strongly. The urge to do this came from our knowledge that the median valley was negatively magnetized. Fred got going a two-dimensional magnetic-anomaly program that Teddy Bullard had written and quickly showed that the negative anomaly was what you'd expect from a normally magnetized rock under the median valley. We were trying to strip off the surface topography in order to look deeper, to see the roots of the fissure volcanoes that were responsible for the ridge. Fred went to Imperial College and borrowed from [K.] Kunaratnam, a Ceylonese, a program which he had devised for determining the direction of magnetization of seamounts, if they had a reasonably isolated magnetic anomaly associated with them; if you knew the shape of the seamount you could fit the best vector to its magnetization. Fred was a very competent mathematician or computer handler.[6]

Fred Vine was of the opinion that

Matthews was ahead of his time in three respects at least. [1] He had worked on ocean-floor rocks—on dredged rocks—mainly on their mineralogy and alteration. [2] He had come to realize that remanent magnetization was important. At Cambridge there was an important rock and paleomagnetics group: Runcorn had been there and moved away to Newcastle; [Ronald] Girdler and [Jan] Hospers had also been there. Although the paleomagnetics people had left by the time I got there, Drum Matthews had been there with those people and [3] had become imbued with the ideas of the importance of remanence and the possibility of reversals of the magnetic field.[7]

Vine had been attracted to geology as early as the sixth form, and thus had done the Natural Sciences Tripos (math, physics, and chemistry) in his first and second years; by the third year he concentrated in igneous petrology and was "keen to do research in marine geophysics." When Vine entered the graduate program at Cambridge, it was Matthews who suggested that he first work on the relationship between oceanic magnetic anomalies and topography. Such correlations were a matter of course among gravimetrists fitting gravimetry data to topography. Vine noted that "about 15 years ago, Marine Geophysics was based mainly in a large house on Madingley Rise itself; in back of the house, in what were the stables, we used to have the marine sediment cores; they also housed Drum Matthews and myself in what would have been the coachman's apartment just above the stables. So the idea was literally hatched in the stables at Madingley Rise."[8] Although theirs was a research student–advisor relationship, Matthews had only

Harry H. Hess of Princeton University joined J. Tuzo Wilson (from Toronto) and Edward Bullard and Maurice Hill at Cambridge in 1965. Fred Vine was also there and recalled: "Wilson projected [the Juan de Fuca] ridge off the Oregon and Washington area in the 1965 paper. There was one morning when we were all together and Tuzo said, 'Look, there should be a ridge here,' and Harry Hess said, 'Well, if you're going to put a ridge there then there ought to be some magnetic expression of it on the Raff and Mason map.' He pointed out that 'one [should] apply the Vine-Matthews hypothesis if you've got a ridge and you've got magnetics.'"

just received his Ph.D. and was close to Vine in age; thus the two enjoyed a relationship of "enlightened equality."[9]

Vine's early undergraduate interest in marine geophysics had been quickened in January 1962, when Harry Hess of Princeton was invited to Cambridge as guest speaker at the Inter-University Geological Congress, organized annually by undergraduates. The 1962 Congress was entitled "The Evolution of the North Atlantic." (Hess's landmark paper of 1962 treating seafloor spreading was not yet published.) His talk had a profound effect on Vine, and the notion of seafloor spreading and oceanic rift zones came to occupy much of his thinking. In keeping with his new consuming interest, as its president, he delivered a special address to the geology club at Cambridge (The Sedgwick Club) in the spring of 1962 entitled "HypotHESSes"—a review of Harry Hess's life and work.*

An early draft of Hess's epochal "History of the Ocean Basins" had been written in 1959 and submitted for publication in Volume 3 of *The Sea*, which was being edited by Maurice Hill of Cambridge. Hill, who was Head of Marine Geophysics at Cambridge, had made parts of the manuscript available to students. Vine thus had an opportunity to read Hess's chapter as a graduate student in late

*Vine's address is noted in the minutes of the 870th meeting of the Sedgwick Club for v-15-62 (Document 34). Vine kindly provided a copy of the handwritten draft (Document 35), which reveals an inordinate familiarity with the literature in general for an undergraduate. Among the 31 cited references in that ten-page paper are 22 publications by Hess.

1962 before it was withdrawn from *The Sea*; it was eventually published in the Buddington Volume of the Geological Society of America instead.* Vine was also familiar with the work of Raff and Mason and Vacquier, who treated the unprecedented and enigmatic magnetic-anomaly patterns, offset by what appeared to be great lateral faults, recently found off the coast of California. He had also read Tuzo Wilson's and Warren Carey's ideas on drift. Brian Harland was openly sympathetic to drift in teaching him structural geology at Cambridge, and he had heard Keith Runcorn lecture there on several visits.

Vine was in fact convinced of drift while still an undergraduate. But it is curious that both he and Matthews were barely read in or informed of the magnetics work of the Icelanders, and ignorant of that of Roche and Khramov, who had demonstrated much of the strongest evidence in favor of polarity reversals.† The Cambridge faculty, however, was fairly well convinced of reversals, owing to the doctoral work there in the early 1950's of Jan Hospers and Edward Irving. The English, by the late 1950's and early 1960's, were not only generally more sympathetic to drift than most other national communities, but were also distinctly more receptive to the notion of polarity reversals.

Vine recalled that the major impetus that "led to the Vine-Matthews hypothesis was the detailed study of part of the Carlsberg Ridge."

> I first started research in October 1962 in order to devise interpretation methods for marine magnetic anomalies. Drum Matthews was eager to devise analog methods for direct modeling, whereas this was a time when digital computers were just coming in. We realized that the obvious way to go was using digital computers. I did two-dimensional and three-dimensional programs for interpreting magnetic anomalies. The mathematics was not my own; some was due to Teddy Bullard, and Victor Vacquier had devised one of them; he had not yet published at the time,‡ but I wrote to him and got a preprint.[10]

*The late Bruce Heezen (oral comm., viii-23-76) noted that Hess's classic paper was dated December 1960; it appeared in mimeographed form in early 1961, but bore a research contract date in order to meet his contractual obligation. Edward Bullard confirmed that Hess distributed an unpublished version of his postulate in 1960 (oral comm., vii-2-79). Henry Frankel (1979, p. 352) examined the 1960 version and noted, "There are no substantive changes between the 1960 and the published version of 1962."

†Matthews was in the Marine Geophysics group while Hospers worked in the Paleomagnetics group; the two were physically separated, treating different problems, and not in sustained, close contact.

‡Published by Vacquier in 1962 (1962a).

Vine, in search of useful techniques and programs, traveled about in 1963 speaking to still others. The three-dimensional programs he eventually used in the hypothesis paper had been largely devised by K. Kunaratnam of Imperial College in the course of his Ph.D. work;* they involved a Ferranti Mercury computer, and were supplied to Vine before Kunaratnam had published. Vine continued:

If you had a pronounced topographic feature such as a seamount with pronounced magnetic anomalies, then you could fit the magnetic anomalies to the topographic irregularities by the least-squares method and model the resulting magnetization. To interpret the magnetic anomalies as due to remanent or paleomagnetism was new at that time; up to about then magnetic anomalies had been interpreted as due to [differences in magnetic] susceptibility. With the exception of a paper by [Ronald] Girdler and [George] Peter in 1960, oceanic magnetic anomalies had been interpreted in terms of induced magnetization.†

I applied this [new] method to two seamounts on the Carlsberg Ridge, which Matthews had mapped in some detail. In applying it to the area discussed in the 1963 paper, I deduced that one of the features was reversely magnetized and the other one essentially normally magnetized. It was that which led to the idea—having got very specific evidence for reverse magnetization in the ocean crust. Although in a way it was a big leap, for me it was a fairly small leap to put that together with spreading and reversals because as you have gathered, I believed in spreading, or wanted to believe in spreading. If you worked very closely with the data and the essential problem that was assigned to me as a graduate student, of interpreting these anomalies, then I think you'd realize that almost out of desperation there was no other way of interpreting them.[11]

It was a very gross thing—to generate strong vertical magnetization contrasts which would explain the anomaly gradients without having to have improbable petrologic contrasts. You must have lateral contrasts to generate such anomalies. The only models that had been suggested were very improbable structures or very improbable petrology contrasts. With reversals, you could throw all of that out of the window and have a perfectly uniform crust laterally and [still] produce the magnetization contrast. And that was the whole essence of the idea. . . . We were in no position then, because of the scruffy profiles available, to say whether [the pattern] was clearly written or symmetrical or whether it correlated with any time scale.

*Kunaratnam's Ph.D. thesis at the University of London was titled "Application of Digital Electronic Computers to Gravity and Magnetic Interpretation."

†Vine elsewhere cited Laughton et al. (1960) as another exception. Both papers are discussed further below.

The data we had from the Indian and Atlantic Oceans at that stage are very noisy data. The record of anomalies [was] not clearly developed with details such as the Jaramillo event, so as a result of working in the Indian Ocean as Matthews and I had done, we didn't think of this process as being regular and simple, as it's now known to be. I'm not sure that we even expected that there would be a very clear symmetry, or, again, as it became crucial in 1965, if Cox was right [about the time scale of reversals].[12]

In keeping with the belief of many others, Vine clearly expressed his sustained doubt of the validity of the polarity-reversal scale. Like most of those outside of polarity-reversal studies, he was not convinced until the report of the Jaramillo event and its recognition in the *Eltanin* 19 magnetic profile (treated in the next chapter).

Although Vine could not recall the exact circumstances in which he first conceived of the hypothesis, he noted that

Drum Matthews was on his honeymoon—he was away. In fact there were very few people, for some reason, to talk to about it. When he returned from his honeymoon, I showed him the paper. I don't think he was too wild about it. [In corroboration, Matthews noted that "it was undoubtedly Fred who put all those ideas into one piece and came along with a scrap of paper, which was a draft of the *Nature* paper."[13]] I remember talking to Maurice Hill; I think he thought I was totally mad. He was polite enough not to say anything; he just looked at me and went on to talk about something else.[14]

Vine regarded Hill's lack of response to his hypothesis as typical of physicists who are not greatly interested in the geologic aspects and results; he saw Hill as

disinterested rather than dissuasive. . . . I remember discussing it with Teddy Bullard, who was much more encouraging, although he realized it was a bit of a long shot. At the time I was quite keen to publish the idea with Teddy Bullard. As a graduate student I thought it would look great: "Bullard and Vine." Teddy Bullard quite rightly said "no way." He didn't want his name on the paper. [Vine thought that Bullard probably regarded the hypothesis as quite far-fetched, but was polite enough not to say it to his face, and "could see the possibilities."*]

I wrote the hypothesis up and Hill got Drum Matthews to change it—to make it more respectable, basically: to give it more substance in terms of

*John Verhoogen remarked, "I thought Vine and Matthews's hypothesis rather ridiculous" (oral comm., viii-16-77). Cox, Doell, and Dalrymple, on the other hand, were "very excited about the Vine and Matthews paper" when it appeared (Dalrymple, written comm., viii-6-79).

Sir Edward C. Bullard, the pioneering and innovative Cambridge geophysicist, in 1970. Among his many contributions was a model of a self-exciting dynamo, based on Walter Elsasser's theory that the geomagnetic field is produced by electrical currents in the outer core, which is both liquid and electrically conducting. Bullard, who died in 1980, was an immediately enthusiastic supporter of the Vine-Matthews-Morley theory and did much to further studies of continental drift in the early 1960's.

quoting data. I think it was because of that that we had no difficulty in getting it published. Hill made the point that one had to include enough hard data beyond wild speculations. Hill and Matthews essentially agreed on that point—they were quite right.

The manuscript was sent to *Nature* no later than late July, because from the beginning of August to the beginning of December 1963, I was at sea in the Indian Ocean as part of the 22nd International Indian Ocean Expedition with Maurice Hill and Drummond Matthews.[15]

It is of interest that neither Vine, at the time he conceived of the hypothesis, nor Matthews, when shown the first draft of it by Vine, were apprised of the work of the Rock Magnetics Project at Menlo Park. Vine had noted his intellectual debt to Cox, Doell, and Dalrymple[16] in the published account of the hypothesis in 1963, however, he remarked in a taped interview that "their paper was very useful, although I had the idea before that." (Such was also the case in its conceptualization by Lawrence Morley.) Vine added, that "When the Cox, Doell, and Dalrymple paper was published, it wasn't a clincher by any means, but it was a great help in the sense that it was all one had to go on."[17]

The response to Vine and Matthews's paper was modest, even in the friendliest quarters. Vine reminisced:

It must have been a year after we had written the paper, that Drum Matthews was writing a review article in the summer of 1964 and got involved in the problem again. He came in and said, "That was a very good idea you had over a year ago." It was as though the penny had just dropped. I think it would be fair to say that at about that time I was getting pretty discouraged and beginning to lose faith myself. It went over like a lead balloon; in some ways there was no response. People just sort of turned away.[18]

Interest would rise again as research, much of it at Cambridge, caught up with his idea.

The Cambridge Geophysics Influence

Cambridge had enjoyed a notable history as a center of geophysical research with a statistically oriented paleomagnetics group that included Jan Hospers, Keith Runcorn, Ken Creer, and Edward Irving as research students during the 1950's. Edward Bullard's interest in marine geophysics prompted him in the early 1950's to build, with Arthur Maxwell, one of the first probes* for the measurement of heat flow from the ocean floor.[19]

The geophysics group had long had an interest in the region of the Carlsberg Ridge; it was an outgrowth of studies in that part of the world that began with the pioneering efforts of Bullard in the 1930's in the East African rift valleys.† His data, derived by use of the Cambridge Pendulum, along with those of Köhlschutter in 1899 and 1900, demonstrated major Bouguer gravity anomalies, which Bullard in 1936 believed supported a compressional origin for the rift valleys (in keeping with the theory of Edward Wayland and Bailey Willis[20]), but the origin of the rift valleys, whether tensional or compressional, remained an open question until seismological studies in the 1950's showed that the dominant type of faulting is normal, resulting mainly from tension.‡

In 1937, John Wiseman and R. Seymour Sewell, in their report, "The Floor of the Arabian Sea," surmised that the Gulf of Aden connected the valley along the northern part of the Carlsberg Ridge

*The first heat-flow measurements in the ocean bottom were made in 1947 and 1948 by the Swedish Deep Sea Expedition, using an unreliable, 36-foot-long instrument. They got only two acceptable measurements, both higher than they expected at that date since the granitic continents, with greater radioactivity, were believed to generate more heat than the basalts of the seafloor (Pettersson, 1949, p. 469).

†See Girdler (1964) for a concise but detailed history of studies of rift valleys.

‡Studies of continental rift valleys led to two markedly different theories of their origin; Suess (1904), Gregory (1921), and others believed that the floor of the rift was a graben, dropped between parallel gravity faults due to horizontal tension. In contrast, Wayland (1930) and Willis (1936) ventured that the floor of the rift was depressed and held down between reverse faults by horizontal compression. De Bremaeker's (1956) and Sutton and Berg's (1959) studies indicated that the major movement along the faults of the rift valleys was a combination of dip-slip and strike-slip. Most now consider the rifts of East Africa to represent an earlier stage in the separation of continental crust than the Red Sea and Gulf of Aden exemplify.

with the rift valleys of Africa. They also remarked on the similarities among the Carlsberg, Mid-Indian Ocean, and Mid-Atlantic Ridges. In 1954, Jean Pierre Rothé suggested that the East African rift system was a continuation of the Mid-Atlantic rift by way of the Mid-Indian Ocean Ridge and the Gulf of Aden. Rothé's surmise grew from his discovery that the rift valleys were virtually congruent with a slim, continuous zone of shallow earthquakes. By 1956 it was suggested by Maurice Ewing and Bruce Heezen that the mid-ocean rift system was a continuous world-encircling structure.

Growing concern with rift structures arose partly from their connection with revived interest in continental drift and expanding-earth theory, and partly from the fact that certain available gravitational and seismological data raised more questions than they answered. The origin and character of the global mid-ocean rift system were quite open to question; the lack of conclusive data of almost any sort, therefore, elicited speculation in terms of analogies drawn from continental geology. Matthews said, "I remember being taken aside by Maurice Ewing at the First International Oceanographic Congress in 1958 and patted firmly on the shoulder and told to forget the idea that the mid-ocean ridge was to be understood in terms of fissure eruptions. Ewing believed it was due to the folding of limestones, and I was a bit shattered by it.* In Cambridge, we accepted it as a volcanic structure."[21]

Heezen, Ewing, and Edward Miller first described the pattern of magnetic anomalies associated with the Mid-Atlantic Ridge in 1953; the anomalies were discovered during their pioneering tow of a magnetometer from a ship bound from Dakar to Barbados. Not until 1957 was the next report of large magnetic anomalies over rift valleys made by Ewing, Heezen, and Julius Hirshman. They found a positive anomaly over the axial valley of the Mid-Atlantic Ridge along the entire length of the surveyed area.

On October 9, 1957, Ronald W. Girdler of Cambridge read a provocative paper to the Geological Society of London. In it the Red Sea was shown to have a positive gravity anomaly along its center, in contrast to the East African rift valleys, which have negative anomalies.[22] The two features are nevertheless apparently structurally continuous. Sir Edward Bullard thought that no one "could have suspected that the gravity field over the Red Sea would be so different from that over the rift valleys to the north and south."[23]

*By 1959 Ewing had altered his thinking; he and his brother John demonstrated geophysical evidence from the axis of the Mid-Atlantic Ridge that basalt must be several miles deep there.

Ronald W. Girdler, at Menlo Park in 1961, was photographed by Richard Doell, who used a Polaroid camera to collect a "rogues' gallery" of visitors to the Rock Magnetics Project.

Girdler advanced a tensional theory to account for the formation of the rift and believed that the sharp positive anomalies over the center of the Red Sea "are due to an intrusion of basaltic material through the down-faulted basement rocks."[24] The surmise that basalt had been emplaced was supported by "frequent reference to basalts on numerous volcanic islands" in the deepest parts of the Red Sea and also by the fact that the Red Sea trough "breaks through the Pre-Cambrian shield rocks of the Arabo-Nubian massif."[25] As an appendix to that important paper, Girdler included "An Aero-Magnetic Section from Berbera to Aden and Its Interpretation," displaying a profile of total magnetic intensity derived by use of a fluxgate magnetometer flown at a height of 1,350 meters (see Fig. 7.2). It showed a large residual anomaly of 1,800-gamma amplitude bilaterally symmetrical over the center of the Gulf of Aden, which was interpreted by use of the Press and Ewing formulae of 1952.* Girdler noted that

> the same assumption is made for the Gulf of Aden as for the Red Sea gravity profile, i.e. that the anomaly is due to an intrusion of rectangular cross-section, infinitely long and reaching the surface. The formulae given by Press and Ewing are simplified and the computations are a little laborious. . . . Further, it is noted that the assumption of the longitudinal intrusion is supported by the fact that positive gravity anomalies have

*Frank Press and Maurice Ewing's paper, "Magnetic Anomalies Over Oceanic Structures," not only provided computations of theoretical magnetic anomalies for typical ocean-floor landforms and structures, but also stressed the importance of the airborne magnetometer as a geophysical tool.

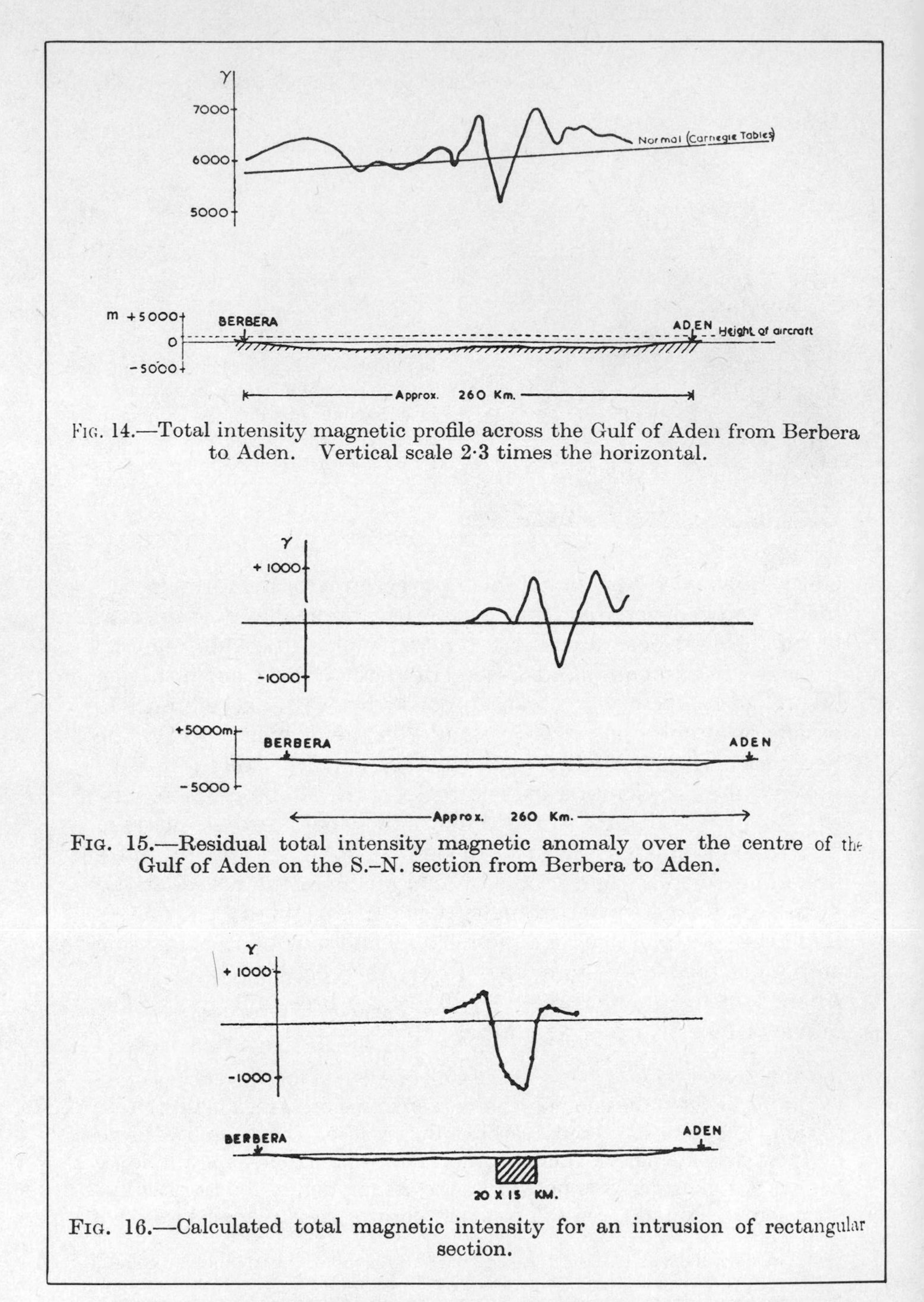

Fig. 14.—Total intensity magnetic profile across the Gulf of Aden from Berbera to Aden. Vertical scale 2·3 times the horizontal.

Fig. 15.—Residual total intensity magnetic anomaly over the centre of the Gulf of Aden on the S.–N. section from Berbera to Aden.

Fig. 16.—Calculated total magnetic intensity for an intrusion of rectangular section.

Fig. 7.2. Girdler's total-intensity magnetic-anomaly profile and residual total-intensity profile across the Gulf of Aden were presented in figures 14 and 15 of the paper he read at a meeting of the Geological Society of London on October 9, 1957, which was published on November 21, 1958, in the Society's *Quarterly Journal*. Girdler believed that the anomaly was caused by an intrusion of rectangular cross-section, infinitely long and reaching the surface (fig. 16, bottom).

been observed along the centre of the Gulf of Aden. The most notable feature of the magnetic interpretation is that the intrusion giving rise to the anomaly is narrower than that giving rise to the gravity anomaly in the Red Sea.[26]

In 1959, researchers from Cambridge presented new magnetic and seismic data that confirmed Girdler's earlier postulate, clarified the nature of the intrusions, and delineated large magnetic anomalies over the deep axial trough running the length of the Red Sea.[27] The data, along with similar data published that year by John and Maurice Ewing for the Mid-Atlantic Rift Valley, suggested an intimate connection between global fracture zones under tension and the character of the intruded rocks. In the same year, Ewing, Hirshman, and Heezen reported that "on all Atlantic profiles north or south of the 25th parallel [north], a high-intensity positive [magnetic] anomaly is observed. This anomaly is broader and more intense than other anomalies in the mid-oceanic ridge profiles, and is easily recognized. It is shown that the anomaly is not caused by topographic relief, or by simple fault structures, *but is due to a subsurface body of high magnetic susceptibility*."[28]

By 1962, Thomas D. Allan had preliminary results (known to the Cambridge group) of a magnetic survey in the Red Sea and Gulf of Aden. And in 1963, Edward Bullard and Ronald Mason published two striking profiles of Allan's across the Red Sea showing anomalies of about 800-gamma amplitude over the central trough. The seismic profiles and earlier evidence indicated that "the structure beneath the axial deep trough is completely different from that beneath the marginal shelves." Early in 1963, Girdler unequivocally stated:

It is clear from the nature of the magnetic anomalies over the Red Sea that the deep axial trough is underlain by a series of parallel dykes or fissures. These partly fill a wide tensional crack in the continental crust. The evidence . . . suggests that the zone is of tensional origin, and the systematic widening of the zone towards the south of the Red Sea and to the east of the Gulf of Aden suggests that this is associated with a rotational movement of Arabia relative to Africa.[29]

The detailed seismic-refraction results permitted the interpretation that the "depth of the main intrusions is approximately 2 km beneath the sea floor." The inferred rock density, magnetic intensity, and speed of seismic waves showed that olivine basalt, an undifferentiated igneous rock generally of deep origin, was the most likely rock type. Girdler emphasized that the region is one where the global oceanic rift system intersects a continental area: "The Gulf of Aden is continued by the Carlsberg Ridge, Mid-Indian Ocean Ridge, Atlantic-Antarctic Ridge, and the Mid-Atlantic Ridge."[30] John Ewing and Maurice Ewing in 1959 had clearly demonstrated a positive gravity anomaly and a magnetic anomaly, and their seismic data suggested that basalt extends several kilometers deep along the axis of the Mid-Atlantic Ridge. Rock samples had been dredged from the Carlsberg Ridge by John Wiseman and Seymour Sewell in 1937, and both Samuel James Shand (in 1949) and Harry Hess (in 1954) had brought up specimens from the Mid-Atlantic Ridge. From all these studies, it appeared that the igneous rocks gabbro, basalt, and serpentine were common to the worldwide oceanic rift zone.

All of the foregoing suggests that Matthews's interest in the region of the Carlsberg Ridge and the plan to survey it as part of the International Indian Ocean Expedition in 1962 were largely a natural outgrowth of an almost decade-long assault on the geophysical and structural problems of that part of the world by the Cambridge group and their associates.

The first magnetic measurements made in the oceans with a towed magnetometer were mainly reconnaissance profiles, and were widely separated. It was thought impossible, at that stage, to correlate magnetic features from one profile to another. In 1954, Maurice Hill,[31] who had originally reported the existence of the Mid-Atlantic Ridge's axial or central valley, failed to find a consistent magnetic anomaly there.* Until 1960, sharp magnetic anomalies over the seafloor were mainly interpreted in terms of differences in rock types and magnetic susceptibility in adjacent blocks of oceanic crust; the Vine-Matthews-Morley hypothesis held to an alternative cause.

In the course of interviews, both Fred Vine and Drummond Matthews mentioned the studies of Laughton, Hill, and Allan and Girdler and Peter, both in 1960, as important elements precursory to their hypothesis.[32] Both studies—the former treating a magnetic anomaly on a seamount 150 miles north of Madeira, the latter interpreting a great anomaly in the Gulf of Aden—suggested that the rocks giving rise to the anomalies were reversely rather than normally magnetized. These conclusions, however open to question at

*Lynn Sykes (1967) showed many offsets along the ridge, which satisfactorily explained the seeming absence of the central anomaly in some of Hill's localities.

the time, represented the first departures from the earlier petrologic or susceptibility interpretations.* There are no unique solutions in the interpretation of magnetic anomalies, because many different geologic models can give rise to them; field data may be interpreted in more than one way.

Those interpretations of oceanic magnetic anomalies as owing to remanence or paleomagnetism, rather than induced magnetism, were a new twist. Although polarity reversals were by then strongly suggested by continental rock studies and subscribed to at Cambridge (because of the earlier influence of Hospers, Irving, and others), the inductive leap—ascribing marine anomalies to reversed magnetization—had not been taken. Thus, Vine and Matthews were intellectually indebted to Girdler and Peter and to Laughton, Hill, and Allan.

Edward Bullard and Ronald Mason, in a 1963 article entitled "The Magnetic Field over the Oceans," noted that "little information has been published on the magnetic properties of submarine rocks: this is partly because of the lack of representative material to study."[33] They reported that Mason had found basalts dredged from the Mendocino fault escarpment, off California, to have extremely strong remanent magnetization† and correspondingly high ratios of remanent to induced magnetization. The Mendocino rocks were "more than three times as magnetic as those from Mull," which Mason had also measured.[34] Most of the material taken in dredging elsewhere was variously weathered and had lost some of its magnetization. Drummond Matthews had found in 1961 that magnetic susceptibilities varied by as much as ten times‡ among weathered basalts from the North Atlantic, and the same had been found in the Pacific. The question of susceptibility as the cause for magnetic anomalies remained open, and the little available evidence did not preclude, and even hinted at, the possibility that great susceptibility contrasts (beyond those known from terrestrial rocks) could account for marine magnetic anomalies.

James M. Ade-Hall (now Hall) discussed "The Magnetic Proper-

*Ronald Girdler wrote to Allan Cox (viii-14-79): "In 1959, George Peter and I saw very clearly that there must be reversals in the Gulf of Aden. . . . We sent the paper to *Geophysics* and after a long time they wouldn't publish it. A much watered down version was eventually published in *Geophysical Prospecting* [Girdler and Peter, 1960]. As Cox and Doell published more and more work on reversals I became more and more convinced!!" (Document No. 36). It is, however, noteworthy that Girdler *did not* invoke field reversals to explain the axial anomalies of up to 1,200 gammas in the Red Sea in his paper of 1963 (1963a, pp. 39–42). He was apparently dissuaded from that by G. Peter (the surmise is supported by the oral comments of a former member of the Cambridge group).

†As much as 0.3 e.m.u./cm_3.

‡Susceptibilities ranged from 0.002 to 0.025.

ties of Some Submarine Oceanic Lavas" in a manuscript submitted to the press in July 1963 (it was published in February 1964). The study was prompted by Ade-Hall's discovery, in 1961, that basalts dredged from the eastern Pacific had natural remanent magnetizations one or two orders of magnitude greater than basalts from land. The 92 samples of the later study, although derived from only one Atlantic and two Pacific sites, were thought to be "typical of basic submarine lavas as a whole."[35] Compared with petrologically similar continental lavas, the submarine rocks showed a *lower* range of low field susceptibilities and a similar distribution of intensity of remanent magnetization, thus casting doubt on earlier assumptions of great susceptibility contrasts in oceanic rocks. Although not cited in Vine and Matthews's original presentation in September 1963 as a major element of their hypothesis, Vine did note in his paper of December 1966 that "of the three basic assumptions of the Vine and Matthews hypothesis, field reversals *and the importance of remanence* have recently become more firmly established and widely held."[36] (Emphasis added.) He cited Ade-Hall's work in support.

With the petrology-susceptibility argument in doubt, it remained for empirical evidence to be found that seafloor basalts might be reversely magnetized. The first such evidence was provided by Cox and Doell in 1962, who found that basalt samples from hole EM7, part of the experimental drilling phase of the Mohole project, showed reversed magnetization. Arthur Raff in 1963 found that fact "quite interesting in view of the total field anomaly over the area"[37] of the northeast Pacific, which had also been surveyed topographically. The magnetic anomalies in that region (treated shortly) are elongated and roughly parallel. As if to emphasize the great uncertainties at that date in the interpretation of marine magnetic data, Raff surmised that "the measured magnetic anomaly and the measured remanent magnetization of the EM7 samples allow for *only three* model solutions";[38] he did not, because he could not, order them by probability.

A Magnetically Lineated Seafloor

In late 1952, at a time when the question of reversely magnetized rocks was in debate, and marine magnetic studies were in their infancy, vessels of the Scripps Institution of Oceanography began routinely towing a magnetometer astern in the course of their oceanographic surveys. Those studies, however, were largely restricted to long, widely spaced profiles; because of the gaps in the coverage of the seafloor, the data when plotted on a map do not indicate directions of anomalous magnetic trends or their lateral distributions, and geological interpretation is thus precluded. Ronald G. Mason of Imperial College of the University of London, shortly after earning his doctorate in geophysics in 1951, was invited to work as a guest at Scripps by its director, Roger Revelle. Mason borrowed a rather primitive fluxgate magnetometer from Maurice Ewing at Columbia University for use on the Capricorn Expedition in 1952. From then until 1954, Mason acquired magnetometers from Westinghouse Corporation, but found them not entirely suited for work at sea.

Arthur D. Raff was first hired by Revelle as an electronic technician, to work with Russel Rait on bathymetry measurements. In 1952 Raff was aboard the *Horizon* on the Capricorn Expedition collecting seismic data; there he became acquainted with Mason, who was engaged in towing a magnetometer from Samoa to the United States. Mason later hired Raff to modify the Westinghouse fluxgate magnetometers and place them in pressure cases; the two began a collaborative effort that lasted several years and led to a most important discovery.

The proton magnetic-resonance method was first used by Martin Packard and Russell Varian in 1954 to measure the magnetic field over the land.* Varian Associates requested that Mason and Raff test the proton-precession magnetometer on board a ship in 1954 on a Scripps expedition to Acapulco. Packard brought the instrument to Acapulco and made magnetic measurements in the hills above the city. While Packard was recording intensities on land, the Scripps group was making magnetic measurements just offshore with a fluxgate magnetometer to compare the two machines. The records, Raff recalled, were "quite compatible with each other."[39] The proton-precession instrument, unlike the fluxgate magnetometer, had no drift problem and did not require calibration. The fluxgate magnetometer was therefore compared on shore with a proton-precession instrument at approximately

*The proton-precession magnetometer, which has largely replaced the fluxgate magnetometer in use at sea, has a sensing head consisting of copper wire coiled around a bottle holding proton-rich water. A strong magnetic field is applied to the bottle by electrifying the coil. Upon removal of the field, a small proportion of the protons precess, and their motion induces an electromagnetic force in the coil—which by a rapid switching device is able to both apply the polarizing field and sense the force due to precession (Montanari and Allan, 1963).

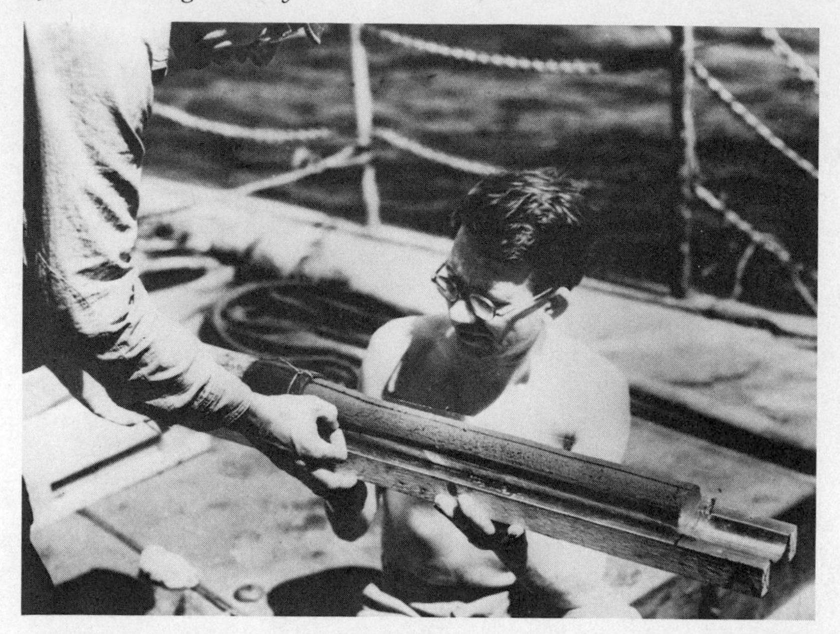

Ronald G. Mason, center, and Arthur D. Raff extruding a sediment core aboard the *Baird* in 1954. Although cores were taken for paleomagnetic measurement by Mason and Raff and others, no polarity reversals were detected in cores until the 1964 discoveries of Christopher Harrison and Brian Funnell, who both worked at Scripps while on leave from Cambridge.

monthly intervals during survey work at Scripps, and it provided the marine magnetists with a new measure of confidence in their data.

In 1955, Mason learned from members of the U.S. Coast and Geodetic Survey that they had been commissioned by the Navy to make a very large survey off the west coast of North America for purposes of submarine operation safety. The Navy wanted to map the bathymetry from the beach to a distance of 300 nautical miles (550 km) offshore, extending from Mexico's Guadalupe Island to the Queen Charlotte Islands of Canada. For the purpose, the Navy provided the U.S.S. *Pioneer*, a 310-foot submarine tender from World War II that could cruise at 17 knots and was suited to such a large area survey.

Mason was two months late in getting a magnetometer aboard the *Pioneer*, and was thus unable to study the area off Guadalupe. By the time the ship was surveying off San Diego, Mason and Raff had their equipment aboard and ran east-west survey lines, 8 kilometers apart. Raff recalled, "Mason worked up his data as they were collected, and almost immediately [by the late fall of 1955] saw that the magnetic anomalies beyond Catalina to San Clemente [Islands] in the San Diego coastal region had a north-south lineation. Nobody had ever seen anything like this in the history of the geophysical sciences. He realized he had something very important."[40] However, the results of that historic survey, representing the first attempt to produce a detailed magnetic map over a large area of the ocean floor, were not published by Mason until 1958. The data, derived with a fluxgate instrument that was frequently calibrated against a proton-precession magnetometer, provided a measure of the total magnetic field with small standard deviation and great stability over short periods.

Mason was at first cautious in reporting his results but was encouraged to pursue his studies further by Roger Revelle. His teaching commitments at Imperial College (where he had been a member of the geophysics faculty continuously since 1947) and a fear that Scripps might not continue the survey northward to the Queen Charlotte Islands brought him almost to the point of abandoning it. He was simply going to analyze what results he had at hand and publish them. At that point, Raff became intensely interested in the problem. Mason had returned to England and Raff had no immediate supervisor, so he simply continued the work that had been started on the *Pioneer*. The director of the Marine Geophysical Laboratory at Scripps, Alfred Focke, admonished Raff to spend less time on the *Pioneer* and the magnetic survey, and more time at his workbench in the Marine Geophysical Laboratory doing "some useful things." Raff recalled, "I ignored his admonition and simply went ahead and scheduled myself to work on the *Pioneer*, and he did not stop me in any way until we were able to accumulate all the data."

The heavy seas of the winter months precluded the derivation of useful bathymetric data, and work was halted until the spring of 1956, when Raff returned to the *Pioneer* to resume the survey. During the whole time Raff was collecting the data, they were sent back for reduction and analysis to Mason in England. Raff showed the resulting map to Alfred Focke, who responded, as Raff recalled, "I don't believe this; you have some sort of fluke or instrumental error. This couldn't be real; there's no such thing on the surface of the earth."[41] Raff, who had built a great part of the instruments used in gathering the data, "knew it was not instrumental error and that it was real."

In the summer of 1956, the Navy provided the *Rehoboth*, a sister ship of the *Pioneer*, for work in the survey; since two complete fluxgate magnetometers were not available at that time, Raff "borrowed a nuclear-precession magnetometer from Varian Associates, put the precession bottle in the extra 'fish,' used the fluxgate towing cable, and instrumented aboard the *Rehoboth* with nuclear-precession electronics."[42] The nuclear-precession system failed during the one-month cruise because Raff had used insulating material to hold and position the precession bottle in the fish (see the photo of Raff and the magnetometer on p. 293); the material held the generated heat in the bottle, causing it to explode. He consequently had to redo the unsurveyed oceanic area using a Scripps ship on which they had their own homemade precession magnetometer. Raff noted that Mason, with "amazing eyes and hands," worked assiduously plotting intensity values as close as 1/32nd of an inch on the anomaly maps extending from San Diego to Cape Mendocino. By 1961, Mason and Raff had surveyed and published total-magnetic-anomaly maps of an area of the Pacific Ocean floor off the coast of North America, approximately 5 degrees of longitude wide and extending from 32° to 50° N. latitude.[43] The anomalies formed a giant pattern, oriented north-south and composed of stripes that varied greatly in width (from a few up to 80 kilometers) and were thousands of kilometers in length (see Fig. 7.3). Equally surprising, the magnetic stripes were fragmented and offset by great, east-west-trending fault zones, which had earlier been discovered from bathymetric data published by H. William Menard in 1955. Because the lineations of the magnetic pattern are approximately perpendicular to the faults, Victor Vacquier and others were able, by extending the map in the area of the faults and by matching the magnetic pattern across the faults, to demonstrate that there is great horizontal offset of what appear to have been once-continuous topographic ridges.[44] That demonstration spoke strongly for the idea of profound lateral displacements of parts of the ocean floor, and thus played an important role in attracting further interest to drift, for objections to drift had been based in large measure on the absence of evidence of great lateral crustal movements.

The accumulating data from the oceanographic research expeditions of the 1950's had led to a series of remarkable discoveries. The early 1960's were thus a time of widespread speculation on the sig-

Arthur Raff, at far left holding cable, during the launching of the magnetometer "fish" over the fantail of the U.S. Coast and Geodetic Survey ship *Explorer* in the area of the Florida Keys, 1960.

nificance of the new data from the seafloor. The new facts were integrated around 1960 (though not published until 1962) by Harry Hess, with components of earlier works,* into what at that time was considered a highly speculative synthesis; Hess referred to it as "geopoetry." He hypothesized that ocean floor was created by injection of molten material from below, at the mid-ocean ridges; that the floor was split along the central axis of the ridge; and that each side moved away toward the distant deep-sea trenches to be carried down, or subducted, into the earth's mantle.† Hess was an

*Especially the works of Holmes (1929, etc.), Vening Meinesz (1930, etc.), Rothé (1954), Ewing and Heezen (1956, preprints, etc.), Bullard et al. (1956), and Girdler (1958, etc.).

†Although he published earlier (1961a) than Hess, Robert Dietz, who coined the term "seafloor spreading," acknowledged in 1968 that the concept originated with Hess. Hess had lectured on and discussed his ideas with Dietz and many others long before 1961 (Heezen, oral comm., VIII-23-76). Both Hess and Dietz were intellectually indebted to Bruce Heezen (1959, 1960), Warren Carey (1958), and Ronald Girdler (1958), who all proposed a tensional origin for the mid-ocean ridges within the context of their belief in an expanding earth. Hess passed through several long stages, all fixist, before arriving at the overview of earth mobility presented in his "geopoetry" paper of 1962. He early (1938) borrowed the broader term "tec-

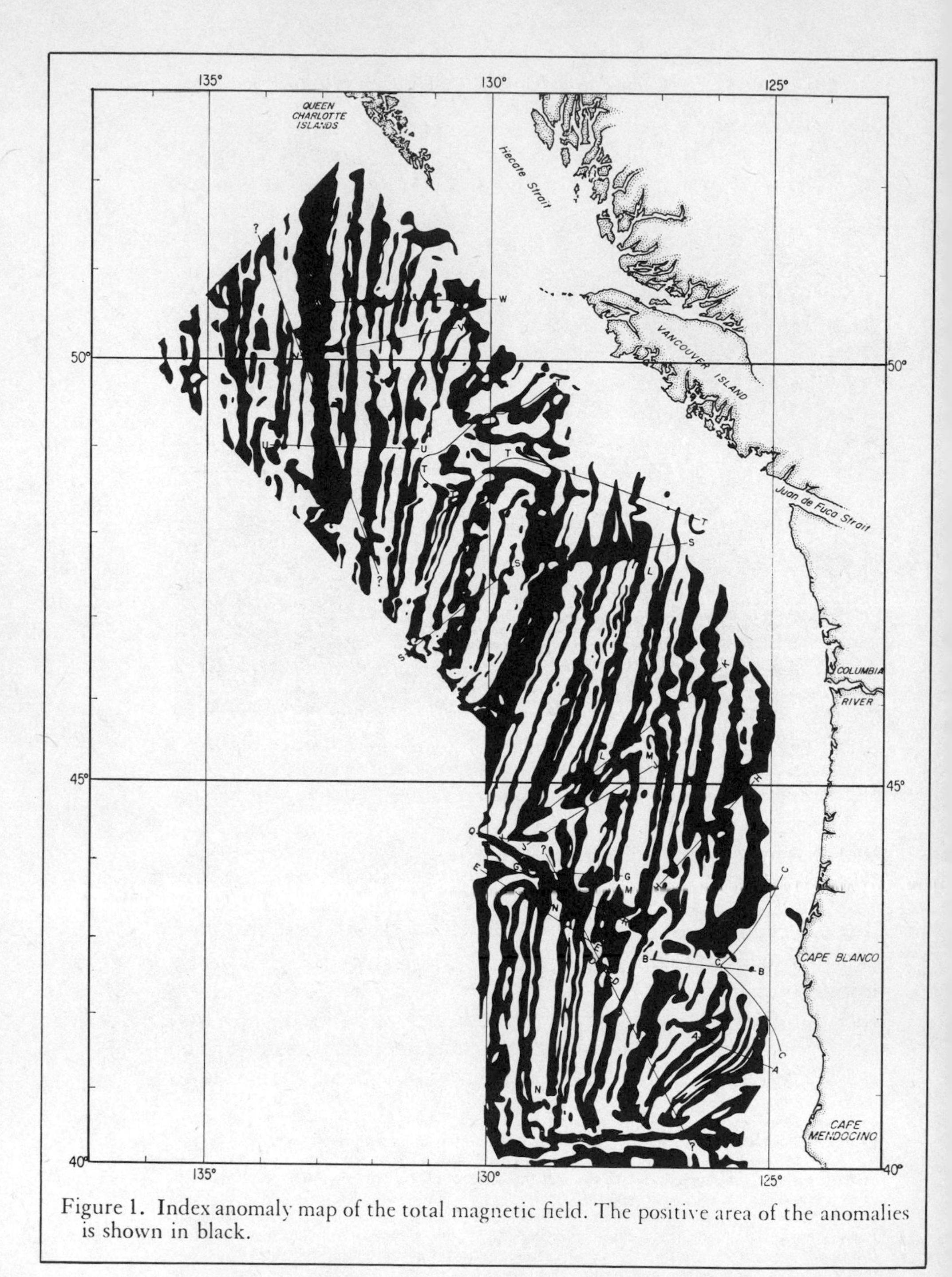

Figure 1. Index anomaly map of the total magnetic field. The positive area of the anomalies is shown in black.

Fig. 7.3. Raff and Mason published a map of total magnetic-field anomalies southwest of Vancouver Island in figure 1 of their August 1961 paper in the *Geological Society of America Bulletin*. The remarkable zebra-stripe pattern had never been found in magnetic surveys on land.

extremely dynamic researcher, and was instrumental in assembling a group at Princeton University who, if not sympathetic to drift, were open-minded in their scrutiny of it.*

Robert Dietz's exposition of Hess's hypothesis, published in 1961, drew attention to the lineated magnetic pattern of the northeast Pacific discovered by Mason and Raff; he surmised that it fit nicely into the concept of seafloor spreading, with the north-south anomalies developing perpendicularly to the direction of spread. Dietz suggested a causal relationship between the striped pattern and linear crustal stress.

Arthur Raff offered a number of explanations for the pattern of magnetic stripes on the seafloor, including parallel bands of rock of high and low temperature, and patterns of electric currents in the earth's crust. Raff's astute observation (which he carried no further) that the magnetic stripes were generally disposed parallel to the mid-oceanic ridges also lacked the essential synthetic element. In 1962, Victor Vacquier published a speculation that deserves mention because of the way it anticipated the Vine-Matthews-Morley hypothesis; more important, it demonstrates the casual and almost random speculative hypothesizing among geologists attempting to explain the striped enigma at that date. Vacquier wrote,

> Because the displacements measured in the ocean do not propagate onto the continent, the oceanic crust must be regenerating at different rates, by

---

togene" from Haarmann (1926, p. 107) and restricted it to linear, narrow downfolds of continental crust related to mountain building; he explained a number of major structural features by the tectogene, and also embraced Holmes's (1929) mantle-convection idea in 1950 to explain the earthquake patterns of island arcs. The crucial element in his seafloor-spreading hypothesis came out of his own professional background—petrology. As early as 1933, Hess was concerned with the origin of serpentinite, a low-temperature, high-pressure metamorphic rock formed from water and parent rock rich in olivine. He first (1938) believed that serpentinite was formed from magma, but was convinced by the demonstration of Bowen and Tuttle (1949) that it can form at temperatures as high as 500°C by the action of water on olivine-enstatite mixtures, or from olivine alone if the water is rich in CO_2. By 1955, he wedded the new data on serpentinization to the discovery, in the late 1940's, that the Mohorovičić discontinuity (the Moho) rises from 35 kilometers depth under the continents to about 5 kilometers below the seafloor; by the attendant increase in volume of sub-Moho material by serpentinization at shallow depths, he explained the mid-ocean ridges and thus removed the paradox of deriving mountains (whose continental analogs demanded compression) at mid-oceanic sites that bore evidence of tension. However, Hess did not accept continental drift until 1960. Then, in a preprint, he joined volume increase by serpentinization at the mid-ocean ridges to Holmes's seafloor conveyor-belt idea and invoked a descending crustal slab at the trenches to complete his seafloor-spreading hypothesis. Henry Frankel (preprint, 1979) details certain parts of Hess's career in "Hess's development of his seafloor spreading hypothesis."

*In 1962 that group included Walter Elsasser, Tuzo Wilson, Jason Morgan, and Harry Hess. Earlier, Hess had gone to the University of Cambridge on sabbatical leave, where he interested the Cambridge group in seafloor spreading. Wilson was at Cambridge in 1965 and was puzzled by the distribution of earthquakes along ridges and fracture zones—a question that led to his 1965 article containing incipient plate-tectonic theory (see Appendix C).

Victor Vacquier, attempting to make paleomagnetic measurements on a spinner magnetometer that refused to work, on board the research vessel *Argo* on the Monsoon Expedition in 1960. A year before the publication of the Vine-Matthews-Morley hypothesis in 1963 Vacquier profoundly anticipated it: "Because the displacements measured in the ocean do not propagate onto the continent, the oceanic crust must be regenerating at different rates, by a process of the kind postulated by Carey for the formation of . . . rift oceans. . . . The north-south lineation of the magnetic anomalies would be the record of the regenerative process. Therefore a band of strong magnetic lineations should run parallel to the great oceanic rises."

a process of the kind postulated by Carey* for the formation of rift valleys and rift oceans, or perhaps by the rise of mantle material along the northern extension of the East Indian rise. *The north-south lineation of the magnetic anomalies would be the record of the regenerative process. Therefore, a band of strong magnetic lineations should run parallel to the great oceanic rises.*[45]

*Vacquier had been greatly influenced in this matter by his contact with Carey during a visit to Hobart, Tasmania in 1960 during the maiden voyage of the *Argo* on the Monsoon Expedition. Vacquier and Robert Warren were exposed to Carey's dynamism and his innovative bent. Vacquier recalled: "I bought the volume *Continental Drift*, which was on the symposium held at the University of Tasmania in March 1956" (t. 1, s. 1). Published in 1958, it contained Carey's notion of the development of a rift ocean by stretching, faulting, isostatic adjustment, and "flow in depth," but the mechanism that Carey proposed at the mid-ocean ridges "fills the gap neither with sediments nor igneous intrusions although both may be involved to a minor or major degree" (Carey, 1958, p. 189). Carey did not discuss seafloor magnetics because there were essentially none known when he wrote.

In the course of an interview, Vacquier did not recall ever having written that profoundly anticipatory note, and remarked that at the time of writing he regarded it as "a low-probability postulate."[46]

It is curious that although Vacquier was vitally interested in the questions surrounding the anomaly patterns and their connection with the great seafloor faults, and speculated on their origin in a form that closely approached the Vine-Matthews-Morley hypothesis, as late as 1964 he wrote that "a theory consistent with the facts is still needed to account for the existence of the north-south magnetic lineation in the northeastern Pacific."* Vacquier had rejected the hypothesis of Vine and Matthews of 1963, mainly because at that time no linear magnetic anomalies were known paralleling the central anomaly over the axes of mid-ocean ridges.

To summarize, it can be said that at the time of its formulation in 1963, all of the major conceptual components of the Vine-Matthews-Morley hypothesis were considered open to serious question in varying degrees. The hypothesis itself was thus destined to a poor reception.

Simultaneous Conception: Morley

The epochal synthesis of Lawrence W. Morley was accomplished in complete ignorance of, but was coincident with and virtually identical to, that of Vine and Matthews.† Morley had done his first rock-magnetism and paleomagnetic studies in 1948 under J. Tuzo Wilson at the University of Toronto, on rocks from the Canadian Shield, acquiring his Ph.D. in 1952. He had flown one of the earliest aerial magnetic surveys for Gulf Corporation in Venezuela with a magnetometer developed at Gulf by Victor Vacquier.‡ He had also promoted widespread aeromagnetic surveys extending into the Arctic, examined reversely magnetized dikes in the Cana-

*Vacquier, 1965, p. 80. Published in the *Philosophical Transactions of the Royal Society of London* on x-28-65, the "Symposium on Continental Drift" (held iii-19–20-64) included manuscripts received up to xi-23-64. Neither Vine and Matthews nor Morley had seen Vacquier's written postulate; nor were they familiar with it in any form (oral comm.: Vine, v-18-79; Matthews, iv-28-79; Morley, iv-24-79).

†James Heirtzler recalled that "before the paper of Vine and Matthews, an Australian graduate student, Geoffrey R. Dickson, strolled into my office [at Lamont] and explained the linear magnetic anomalies in just the way that Vine and Matthews did in their classic 1963 paper. The idea was pretty widely prevalent in the community by then" (oral comm., viii-17-76). Heirtzler, however, did not accept the idea at that early date, nor did he and most others at Lamont Geological Observatory accept it as late as 1966.

‡Vacquier developed an improved three component fluxgate magnetometer in the name of Gulf Research and Development prior to World War II; when war began, the U.S. government used it in submarine detection after its further refinement by Bell Laboratories.

Lawrence W. Morley, while chief of the Geophysics Division of the Geological Survey of Canada, in February 1963 wrote his hypothesis in the form of a letter to *Nature*. They replied (about two months later) that they "did not have room to print" it. He then submitted the hypothesis to the *Journal of Geophysical Research* in April 1963, but received no answer until late September (after the appearance of the Vine-Matthews paper, which *Nature* had promptly published on September 7, 1963). A didactic editor finally responded acerbically, flatly rejecting the manuscript, which contained one of the most important hypotheses in the history of earth science. Morley is presently Science Counsellor with the Canadian High Commission (Embassy) in London.

dian Shield, and extensively sampled rocks for magnetics studies. Morley later came to manage a group of about 100 scientists and technicians as chief of the Geophysics Division of the Geological Survey of Canada.

In February 1963, Morley submitted his hypothesis in the form of a letter to *Nature*. An apologetic reply from the journal about two months later said that they "did not have room to print" his letter. He thereupon submitted the letter to the *Journal of Geophysical Research* in April, but received no answer until late September (after the publication of the Vine-Matthews paper). The editor apologized for the tardy response (which was due to his absence during the summer), explained that the paper had gone to an anonymous referee, and concluded that "such speculation makes interesting talk at cocktail parties, but it is not the sort of thing that ought to be published under serious scientific aegis."*

Following that second rejection, Morley made an oral presentation of his ideas to the Royal Society of Canada in Quebec City at their 1964 annual meeting on June 4; the hypothesis was not published until late in 1964, in an article written with Andre Larochelle.† The original letter to *Nature* was reproduced in the *Saturday Review* in an article by John Lear, "Canada's Unappreciated Role as Scientific Innovator," on September 2, 1967:

Several investigators and authors writing on the subject of continental drift and convection currents in the earth's mantle have referred to the puzzling linear magnetic anomalies in the Eastern Pacific Ocean basin reported by scientists of the Scripps Institution of Oceanography.

If one accepts in principle the concept of mantle convection currents rising under ocean ridges, traveling horizontally under the ocean floor and sinking at ocean troughs, one cannot escape the argument that the upwelling rock under the ocean ridges, as it rises above the Curie point geotherm, must become magnetized in the direction of the earth's field prevailing at the time. If this portion of rock moves upward and then horizontally to make room for new upwelling material, and if, in the meantime, the earth's field has reversed, and the same process continues, it stands to reason that a linear magnetic anomaly pattern of the type observed would result. This explanation has the advantage, over many others put forward, that it does not require a petrologically, structurally, thermally, or strain-banded oceanic crust. It requires a convection cell whose axis of rotation is at least as long as the linear magnetic anomalies and whose horizontal distance of travel stretches from ocean rise to ocean trough. In addition to this, it requires a large number of reversals of the earth's magnetic field from at least the Cretaceous period (which ended 68,000,000 to 72,000,000 years ago) to the present (since no rocks older than Cretaceous have been found in the ocean basins; that is, no rocks older than 140 million years).

R. L. Wilson reported that Mrs. J. Cox, in a recent search of the paleomagnetic literature, was able to find 136 normally polarized cases and 141

*A fire at Morley's home on xii-5-78 destroyed all of his correspondence, including that from *Nature* and the *Journal of Geophysical Research*; the quotes in this paragraph are Morley's recollections (t. 1, s. 1, iv-24-79).

†Andre Larochelle stated explicitly that he played no role in the formulation of the Vine-Matthews-Morley hypothesis (t. 1, s. 1). This seems to contrast with the impression one is likely to gain from a letter of Larochelle to C. G. Winder of x-8-76, which Winder published in 1979 (p. 32).

reversely polarized from the Carboniferous to the present. Since there is no evidence to suggest that the earth's field should have been "normally" polarized for any more periods or for longer periods than it has been reversely polarized, it is entirely possible that there may have been as many as 180 reversals since the Lower Cretaceous. This would be one reversal about every half million years on the average (a figure which T. Einarsson gives from his investigations of Icelandic lavas. He also suggests that the time taken for a reversal of the field is geologically very short—a few centuries to 10,000 years).

From an examination of the Scripps magnetic maps, the width of a complete positive and negative cycle, averaged over the widest part of the available surveyed section, is about 35 kilometers. To travel this distance in 1,000,000 years (time of two reversals), the convection current must have a rate of about 3.5 centimeters per year. This figure is only good to an order of magnitude, *because no accurate data are available on the length of the periods of reversals.* A better way to arrive at the rate of convection travel and the reversal period would be to measure the ages of rocks at widely spaced locations in the Pacific and to count the number of reversals occurring between these points.*

R. G. Mason and A. D. Raff (who made the magnetic survey for Scripps) report that some of the many guyots (flat-topped undersea islands) which were detected on the echo sounder produced magnetic anomalies, while others apparently had little or no effect. It seems unlikely that these guyots would be divided into two classes—those containing magnetite and those containing little or none. A more likely explanation would be that the ones which give little or no effect are negatively polarized to an intensity which nearly equalized their magnetization induced by the present earth's field. If the "non-magnetic" guyots always occur in the negative anomaly bands and the magnetic ones in the positive bands, this would be evidence that they cooled below the Curie point at approximately the same time as the rock surrounding them, because they were magnetized in the same direction. Indeed, since at that time they would have been in the shallow water of the oceanic ridge, they would have protruded above the surface and have had their tops flattened by erosion. As they proceeded along with the mantle convection current, they would pass into deeper water. This is an alternative explanation of origin to that suggested by Darwin for the flat-topped guyots in the deep Pacific.

The purpose of this letter is to point up the possibility of calibrating the frequency and duration of reversals of the earth's magnetic field in geologic history from a study of magnetic surveys of the ocean basins, and the idea presented is considered to add support to the theory of convection in the earth's mantle.

*Spreading rates, discussed by Morley, is a dimension lacking in the hypothesis as presented by Vine and Matthews.

At the time the idea came to Morley, he was engaged in aeromagnetic surveys over the continental shelf off Newfoundland, an outgrowth of the Geological Survey of Canada's dissatisfaction with the slowness of ship-towed magnetometer surveys, which had begun in 1958. Morley recalled,

The thing that got me going on it was the definitive paper done [in 1961] by the Scripps people Mason and Raff, who were the first to outline the magnetic ridges off the coast of California. I was struck by the fact that they frankly admitted that they had no explanation for those ridges. There was just nothing to correlate the magnetic anomalies with. They postulated different types of rocks, of different magnetization, emplaced as vertical dikes. There was no evidence anywhere to point to the vertical-dikes model in the ocean basins. They had no ocean-floor samples and therefore couldn't prove it. I couldn't imagine any geological process that would reproduce dikes a thousand miles long in the ocean, each with different petrological characteristics; I presumed we had a mantle that was fairly uniform. When I came up with that hypothesis, I was not a researcher; I was an administrator with a staff of 100. Research was something I did on the side: that explains why I had not read the papers of Girdler and Peter and Laughton, Hill, and Allan [both in 1960]. I was very familiar with the idea of inverse remanence.[47]

Morley also noted that he was aware of all the publications of Vacquier and the question of the possibly great difference in magnetic properties between oceanic and continental rocks. He emphasized that his background in both geology and physics at Toronto allowed him to understand geological papers. He continued:

I was puzzled until I read the paper given by Robert Dietz in August 1961, at the Pacific Science Congress, in which he presented the concept of ocean-floor spreading.[48] I did not hear Dietz give his paper on seafloor spreading but read it about eight months later. I had the two papers in front of me—that of Mason and Raff and the one by Dietz. It was a natural thing that the two should go together, given my background. I said, "That is the explanation; this dike business is a lot of hooey!" I put that together with that of Mason and Raff. Additionally, it was a matter of having my knowledge of rock magnetism and knowing the Curie point of rock, about 500°C, and here I [went] back to Vacquier: if you go down about 13 kilometers or so, the rock is non-ferromagnetic. It was the Eureka thing! In our laboratory I was the leader of a group doing paleomagnetics, seismic work, and offshore geology. These things were on the top of my mind, and we had been puzzled by negative anomalies.

I had been involved in this big argument between [Keith] Runcorn and

[John] Graham and Jim Balsley about whether reversals were caused by the field reversing or by mineralogy of the rock. I must confess that I was on the fence but leaning toward Runcorn, who pressed his arguments very strongly for field reversals. My own ambition in the early 1950's, after being influenced by a paper by John Graham, was to try to find evidence for continental drift in paleomagnetism.[49]

Morley was influenced by the Icelanders in his belief that a one-million-year periodicity of reversals was likely. He thus reasoned that each of the magnetic blocks on the seafloor should be about 13½ miles wide, which was in keeping with prevailing estimates of continental drift rates and also roughly in agreement with the width of the stripes mapped by Mason and Raff.

Morley's administrative responsibilities left him virtually no time to write further papers and press forward his original ideas. Unlike Morley, Vine was able to extend and refine his theory through sustained research in favorable contexts at Cambridge and Princeton. In a letter to *Geology* in April 1974, Norman D. Watkins, speaking of the fate of Morley's paper, opined that it is "probably the most significant paper in the earth sciences ever to be denied publication."*

An Auspicious Gathering at Madingley Rise

The Vine-Matthews-Morley hypothesis did not gain much attention when first presented in 1963, owing in large measure to valid criticism on several counts. First, the scale of polarity reversals available to Vine and Matthews at the time of publication of their paper (September 7, 1963) was that of Cox, Doell, and Dalrymple of June 15, 1963 (scale number one of this book). It was quite simple and not detailed or accurate enough to be compared with the magnetic profiles available from the seafloor magnetometer surveys invoked by Vine and Matthews from the Carlsberg and Mid-Atlantic Ridges.† Second, the magnetic profiles cited in that

*The question of priority of Vine and Matthews over Lawrence Morley entails not only the matter of publication dates, but also the point that Morley's hypothesis, submitted in letter form, lacked the pertinent data that Vine and Matthews derived and included in their manuscript. That lack of data likely contributed to Morley's twice being denied publication. A persistent rumor appeared in reliable quarters, but could not be definitively corroborated, that a conscious effort was made to delay the publication of Morley's paper by certain editors in order to give priority to Vine and Matthews.

†It bears remembrance that the geophysics community was not instantly converted to belief in polarity reversals, or to the chronology implied, by the publication of the first potassium-

historic first paper were not nearly as regular as the hypothesis predicted; the symmetry of magnetic-intensity values across the ridge axes was not in evidence. Third, the sporadic distribution and timing of terrestrial volcanism seemed to detract from the notion of volcanism being confined to the axes of the mid-ocean ridges in a way that would continuously produce clearly demarcated regular magnetized bands. Such a mechanism was unprecedented and appeared quite improbable.*

Vine felt that "very little happened" concerning the hypothesis "from mid-1963 to early 1965. One was preoccupied with other things. There was so little new information to confirm or deny it."[50] Matthews agreed that "the paper dropped into a sort of vacuum, as we expected it to, and we carried on with other things. Teddy Bullard used to proselytize for it a bit, but American labs wouldn't hear anything of it—thought it was all nonsense. Indeed, there were papers written which continued to flay it forever, including Heirtzler's."†

Vine also recalled that Victor Vacquier was "quite adamantly opposed to the Vine-Matthews hypothesis at the Symposium on Continental Drift," held in London on March 19 and 20, 1964.[51] Vacquier was clear in his surmise that Vine and Matthews's "attractive mechanism is probably not adequate to account for all the facts of observation. . . . Where the East Pacific Rise can actually be seen, no lineated magnetic pattern was found."‡ The same was true of the Mid-Atlantic Ridge between 6° and 30° S. latitude in 13 profiles, "except for a single prominent broad magnetic anomaly," which

---

argon polarity-reversal time scale by Cox, Doell, and Dalrymple in June 1963; Vine himself had lingering doubts at that time.

*The pioneering seafloor investigators were familiar with geological models of volcanism (and most other processes) derived almost exclusively from studies on the continents—but the case of Iceland is an exception treated shortly.

†Heirtzler and Le Pichon's 1965 paper reflected the thinking at Columbia University's Lamont Geological Observatory, discussed in Chapter 8. Although most papers on the subject ignored or criticized Vine and Matthews, a notable exception was George E. Backus's "Magnetic Anomalies Over Oceanic Ridges" in the ii-8-64 issue of *Nature*. In a laconic note sympathetic to Vine and Matthews, Backus suggested a magnetic survey on the flanks of the south Mid-Atlantic Ridge to discover whether magnetic stripes widening toward the south could be found, in keeping with the geometry of the opening of the Atlantic suggested by Carey (1958) and Bullard et al. (1965; then in press). He believed that the presence of such stripes would constitute evidence for Vine and Matthews, but that a negative result "might mean simply that other mechanisms more powerful than [that of] Vine and Matthews were also producing magnetic anomalies." Writing from the University of California in La Jolla, Backus thanked Sir Edward Bullard for "helpful discussions." As Matthews said, Bullard "proselytized" for Vine-Matthews.

‡Vacquier, 1965, p. 80. Ironically, the magnetic anomaly profile from *Eltanin* Leg 19 (discussed in Chapter 8), definitive in proving the Vine-Matthews-Morley hypothesis, was from the East Pacific Rise.

seemed to be the only clearly continuous feature.[52] Vacquier concluded with the suggestion that the magnetic anomalies in the northeastern Pacific "may have been generated by other mechanisms than those associated with ridges in the Atlantic and Indian Oceans."[53] It is noteworthy that in that same proceedings volume (from the Symposium on Continental Drift, published in October 1965), which included papers by two dozen eminent authors from both Europe and North America, including Tuzo Wilson and Edward Bullard, no one other than Vacquier ever mentioned the Vine-Matthews-Morley hypothesis.

The personalities and research products of the Cambridge group, which had so vitally conditioned Vine and Matthews in the development of their hypothesis, were widely known abroad. A notable assemblage of highly regarded, wide-ranging proponents of drift thus came to be gathered at Cambridge in 1965, which contributed to further development of the Vine-Matthews-Morley hypothesis. They included J. Tuzo Wilson from Toronto, Harry Hess of Princeton, and Edward Bullard and Maurice Hill of Cambridge.

The interchange among them further stimulated an already fever-pitched interest in the relationship between magnetic-anomaly patterns, the polarity-reversal scale, seafloor spreading, and related topics, including large-scale faulting, which had been a lifelong interest of Wilson's.

Vine recalled that "Tuzo Wilson arrived first, in January 1965, while virtually all the marine geophysics people were at sea except for me. There was a major cruise going on, and I'd been left behind basically to finish up my thesis." Harry Hess arrived shortly afterward at Madingley Rise on sabbatical leave, and Vine, somewhat overawed at finding himself "the only geologist left for them to talk to," was delighted at such an unusual opportunity for a graduate student. Madingley Rise is located some distance from the main college area, which contributed further to Vine's opportunity to interact with them on a sustained basis.

On arrival, "one of the first things that Hess said was that he thought the Vine-Matthews hypothesis was a fantastic idea." Vine was pleased: "No one had ever said that to me before. I said to Tuzo Wilson—just for something to say—that the equatorial fractures in the Atlantic couldn't be like the transcurrent faults that Bruce Heezen had suggested. I got the impression that Tuzo wasn't listening at all: he started talking about something else. But

J. Tuzo Wilson, whose many important contributions included the idea of hot spots and the pivotal concept of the transform fault as the third type of boundary of a rigid crustal plate; the transform paper of 1965 was the bridge from the concept of seafloor spreading to the theory of plate tectonics.

lo and behold, he came back a few weeks later, in February 1965, telling me about transform faults."*[54]

Matthews also recalled that period:

> Wilson hired a yacht and went sailing with his family on the south coast of Turkey. On his return, I remember him pounding up the stairs, at the stables where we all three lived, taking three steps at a time and saying, "What are you doing? It can't be as important as what I want you to do—Stop!" For the next three weeks or so we all worked for Tuzo; he had come back with superb ideas. One was that the hypothesis that we presented implied that the anomaly pattern must be symmetrical about the ridge crest. We simply had not believed the hypothesis well enough to twig. The second idea was the transform fault. I remember him tearing a piece of paper and showing that the transform would have motion in the direction other than what you might expect. Wilson put those ideas into publication before he left Cambridge at the end of that academic year.[55]

Wilson surmised, on the basis of seismicity and structural analysis, that a segment of mid-oceanic ridge was bounded by a "new

*Wilson first recognized the new class of transform faults between the offset, parallel segments of mid-ocean spreading ridges, where the newly formed crustal plates travel in opposite directions and cause earthquakes along a fault between them. Transform faults also arise between spreading centers and subduction zones at trenches. Where a descending plate is subducted into two trenches, one farther from the ridge than the other, a transform fault will con-

class" of lateral faults (designated transform faults*) near the Strait of Juan de Fuca (see Fig. 7.3, above).

Vine continued:

> As a result of [conceiving of the transform fault], Wilson projected a ridge off the Oregon and Washington area in the [second] 1965 paper.† There was one morning when we were all together and Tuzo said, "Look, there should be a ridge here," and Harry Hess said, "Well, if you're going to put a ridge there, then there ought to be some magnetic expression of it on the Raff and Mason map." He pointed out that it was one place where we had good magnetic data, so shouldn't one apply the Vine-Matthews hypothesis if you've got a ridge and you've got magnetics?
>
> I hustled up to the library and got out the Raff and Mason map, and lo and behold, for the first time, although it had been in print for four years, we realized that there was symmetry in the anomalies. From that flowed the paper of Tuzo describing the morphology of the Ridge and then our joint paper packed with interpretations of magnetics.[56] This was the first time we were seeing symmetry. *The problem was that we couldn't intepret the anomalies in terms of a confident spreading rate. We couldn't do that because we didn't have the right time scale.* Had I taken the big jump and insisted on interpreting it at a constant spreading rate, it would have been considered too outrageous at that time.[57]

Vine and Wilson collaborated in a study (and the naming) of the Juan de Fuca Ridge, which was marked by magnetic stripes easily recognizable, lying parallel to and bilaterally symmetrical to the ridge axis. It is important to recall that at the time of their collaboration in that study, the only detailed magnetic survey in existence of a large region of the seafloor was Mason and Raff's, west of North America (Fig. 7.3, above). The presence of the Juan de Fuca Ridge segment had not been recognized before Wilson described it in *Science* in 1965. Vine and Wilson opened their paper of October 1965 by invoking the most recent polarity-reversal scale of Cox, Doell, and Dalrymple (scale seven), which was greatly refined and included the Olduvai and Mammoth events. By means of a computer, they modeled a magnetic profile that would be expected if normal and reversed bands of rock were produced, as hypoth-

---

nect them with the characteristic line of earthquakes. How Wilson derived both the transform fault concept and his hot-spot theory is discussed in Appendix C.

*A case of partial simultaneous discovery concerning the transform fault is discussed later in this book, in Appendix C.

†Menard (1964) had already identified a young mid-ocean ridge in the same area, but believed it to be the northern end of the East Pacific Rise, with a northwest strike; Wilson thought it would be an isolated ridge with a northeast strike, separate from the East Pacific Rise but linked to it by the San Andreas transform fault.

esized, by spreading of the seafloor from the ridge axis; the rate of spreading, however, was unknown. They tried models based on various rates of seafloor movement in an attempt to derive a theoretical profile that matched closely the ones recorded over the Juan de Fuca Ridge. The model profile produced by assuming a spreading rate of 2 centimeters per year on each side of the ridge bore a very close similarity to the one derived from field data (see Fig. 7.4).

The spreading model proposed by Vine and Wilson was also improved (over that of 1963 by Vine and Matthews) by the growing suspicion that the thick subsea layer of serpentinite (which Harry Hess thought was emplaced as a solid) did not account for the magnetic anomalies. They adopted Hess's new model of 1965 for oceanic ridges, which led them to assume that the "'magnetic' material of the crust would be largely confined to the basalt layer (layer 2)," a layer of extrusive rock approximately 1 kilometer thick lying on a deeper layer of serpentinite.* Their success with the Juan de Fuca Ridge model was almost equaled in their match of an observed profile across the East Pacific Rise (reported by Heirtzler in 1961) with that of the theoretical model and with the Juan de Fuca Ridge pattern (Fig. 7.5).

That demonstration of a clear magnetic symmetry imprinted in the seafloor rocks, which appeared so clearly in keeping with the postulated cause of spreading, attracted much interest in and beyond the paleomagnetic community. The correlation of the two mid-ocean ridges with the theoretical model based on the potassium-argon time scale of the reversalists was good, but it lacked the fineness of detail that would be absolutely convincing, especially to the skeptics, who were still in the majority.

At Cambridge, Fred Vine had established close rapport with Harry Hess as a result of common interests over a broad range of subjects, including petrology, in which they were both formally trained. Vine thus declined an offer from James Heirtzler to work at the Lamont Geological Observatory, at Palisades, New York, and instead joined the staff at Princeton University in September 1965

*Vine and Wilson, starting from the "basic principle" that the steep magnetic gradients "delineate the boundaries between essentially normally and essentially reversely magnetized crust," found "no difficulty in explaining the anomalies but only in deciding on the distribution of magnetization within various layers of oceanic crust. . . . The actual distribution of magnetization within layer 2 [1- or 2-kilometer-thick basalt] and layer 3 [serpentinite, hydrated mantle] . . . is a matter of speculation at the present time." Their preferred model across the Juan de Fuca Ridge showed the basalt (layer 2) as 1.7 kilometers thick.

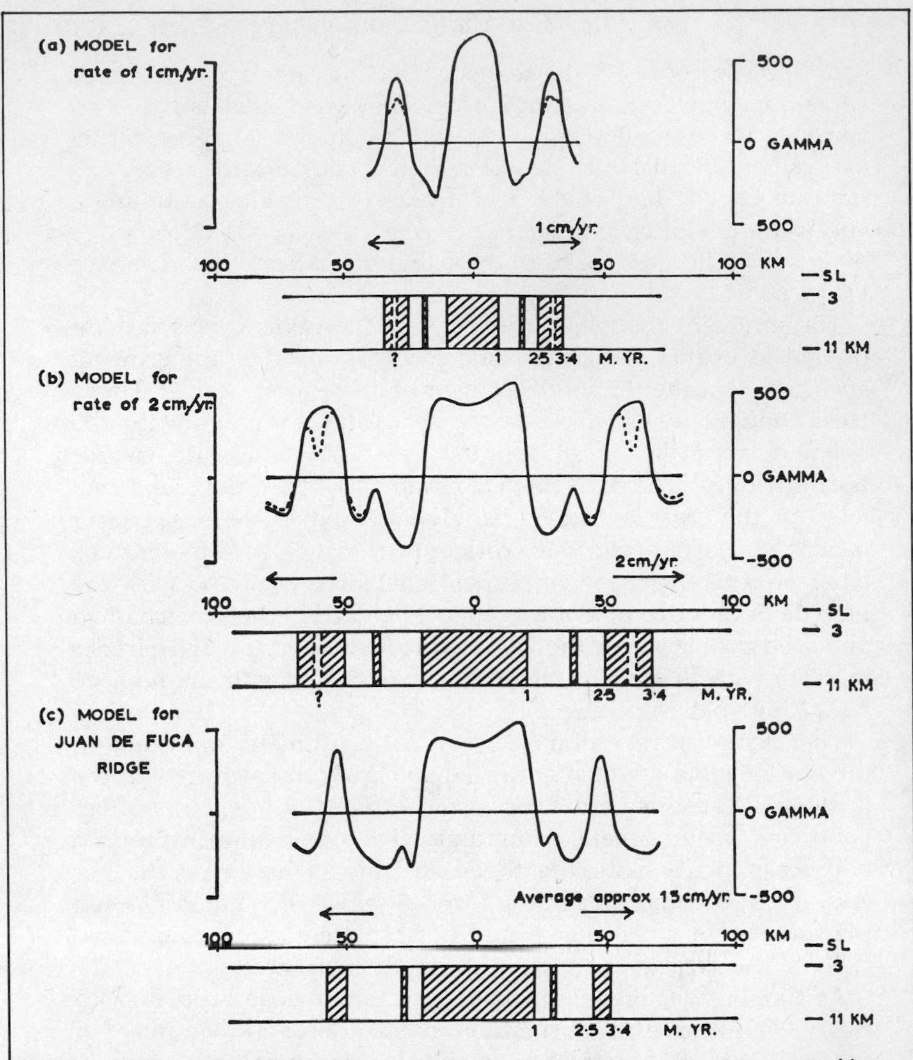

Fig. 1. Models and calculated total field magnetic anomalies resulting from a combination of suggested recent polarities for the earth's magnetic field (*7*) and ocean floor spreading. Normally magnetized blocks are shaded; reversely magnetized blocks unshaded. Portions *a* and *b* assume uniform rates of spreading. Portion *c* was deduced from the gradients on the map of observed anomalies. The dashed parts of the computed profiles show the effect of including the possible reversal at 3 million years (*7*; see also *15*).

Fig. 7.4. Magnetic anomalies across the Juan de Fuca Ridge were compared with two theoretical models based on different seafloor-spreading rates in figure 1 of Vine and Wilson's October 1965 article in *Science*.

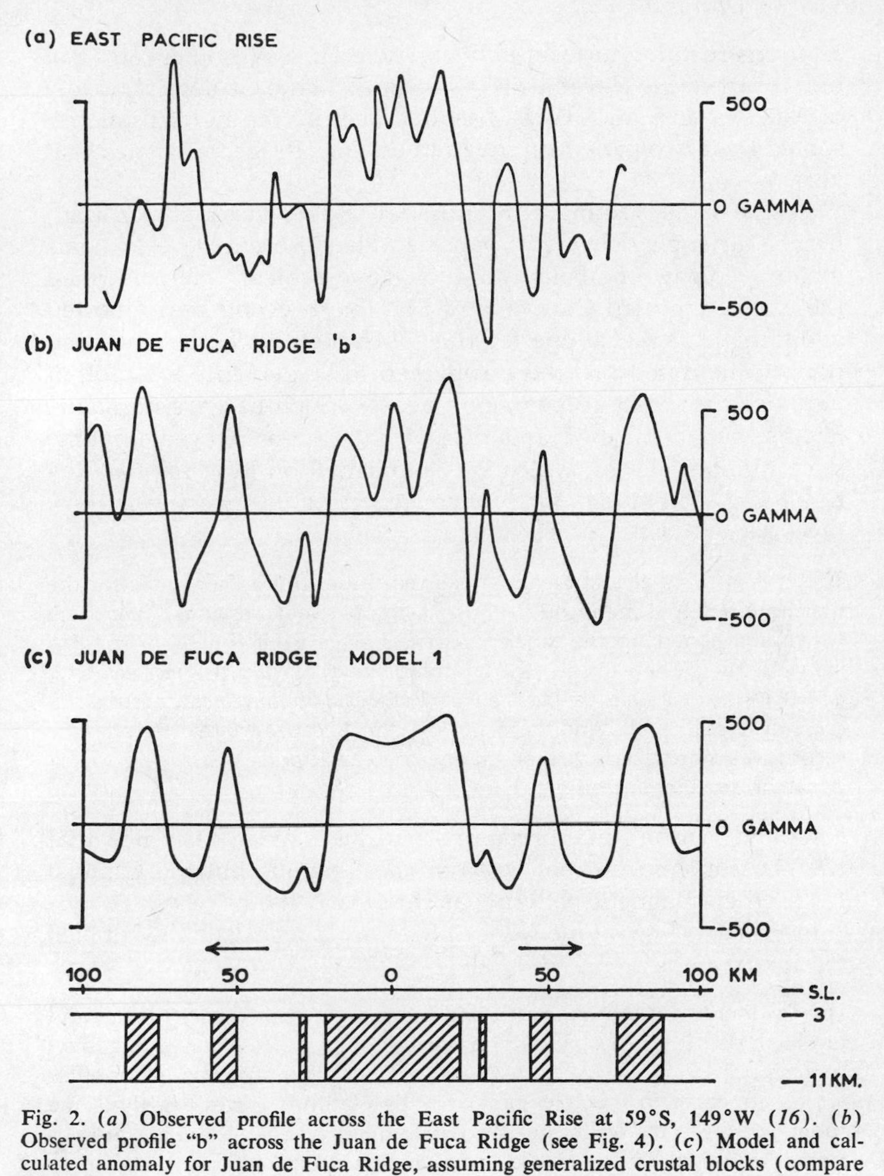

Fig. 7.5. In figure 2 of their October 1965 paper, Vine and Wilson compared two observed magnetic profiles, one across the East Pacific Rise and one across the Juan de Fuca Ridge, with their theoretical model.

as an instructor. Shortly after his arrival, Tuzo Wilson encouraged him to attend the November 1965 meeting of the Geological Society of America, in Kansas City, Missouri. There, Vine and Wilson presented their two papers on the Juan de Fuca Ridge, published that year.[58]

At that same meeting,* on Thursday, November 4, at 9:00 a.m., Brent Dalrymple delivered a paper entitled "Recent Developments in the Geomagnetic Polarity Epoch Time Scale."[59] He confirmed the earlier reported Olduvai and Mammoth events and reported additional data that supported the epoch boundaries; but most important, he noted that there appeared to be evidence for another new polarity event at 0.9 million years, which he gave no name. It was later confirmed and named the Jaramillo event in time scale number eleven, which was published on May 20, 1966, by Doell and Dalrymple. Vine's recollection of that session is most revealing:

> The crucial thing at that meeting was that I met Brent Dalrymple for the first time. He told me in private discussion between sessions, "We think we've sharpened up the polarity-reversal scale a bit, but in particular, we've defined a new event—the Jaramillo event." *I realized immediately that with that new time scale, the Juan de Fuca Ridge could be interpreted in terms of a constant spreading rate*. And that was fantastic, because we realized that the record was more clearly written than we had anticipated. Now we had evidence of constant spreading; that was very important.[60]

Vine and Wilson's interpretations had been made assuming that the latest magnetic event was not the Jaramillo but the Olduvai Event; their magnetic pattern was thus thrown off.

In February 1966, Vine went to Lamont "specifically to visit Neil Opdyke."

> He was literally in the same area and same room with Walter Pitman and Jim Heirtzler. It was a very, very important day. . . . Neil was poring over [papers on] a light table; he must have been drawing up magnetic profiles which appeared in the 1966 paper on high-latitude cores.[61] In the same room, on the wall, Walter Pitman had pinned up all the *Eltanin* profiles from the South Pacific; he was working on interpretation of the magnetic anomalies. While talking to me about research in general, Neil said: "We've discovered a new event and we're just about to publish it. Look, we've got all the detail of the time scale of Cox, Doell, and Dalrymple; we've got an event at about 0.9 million years." [Opdyke obviously

*Earlier alluded to in Chapter 6 in the discussion of scale eleven.

hadn't been at the Kansas City meeting and had not been in contact with Cox, Doell, and Dalrymple.] I said, "Neil, I hate to tell you this, but Cox, Doell, and Dalrymple have already defined it and named it the Jaramillo event." Neil just about fell off his chair—he said they'd given it another name.[62]

Vine's news meant that the magnetics group at Lamont had lost scientific priority in recognizing the new event. They had found signs of the event in both a remarkable magnetic-anomaly profile and in deep-sea sediment cores; both records were part of the peerless data bank that was Lamont. Opdyke had already decided to name the event after the Emperor Seamount in the Pacific Ocean.

But the disappointment of the Lamont group could only have been short-lived, for they were in a state of jubilation over their discovery of almost incredible magnetic-anomaly data that clearly indicated that seafloor spreading was a reality. The activities leading to that singular contribution and the research context in which it was made bear scrutiny, for unlike Cambridge, which disposed its students in favor of mobilism, Lamont was largely dominated by fixists with scant regard for the Vine-Matthews-Morley hypothesis.

Chapter 8

Lamont Observatory: The Data Bank

MAURICE EWING was the prime mover in the founding of Columbia University's Lamont (now Lamont-Doherty) Geological Observatory. Starting in 1945 from a budget of $50,000, Lamont had become, by the 1960's, one of the leading marine geophysical research centers in the world. Its research vessels *Vema* and *Conrad* were kept at sea more than 300 days a year, acquiring the largest store of geophysical data extant at the time Fred Vine was searching for information relevant to his hypothesis. The smallest and youngest of the three major oceanographic institutions in the United States, Lamont had produced geophysical data, and innovations in securing it, that were out of proportion to its size. Ewing's wide-ranging intellect, "intuitive sense of what is important in science," compulsive work habits, seemingly endless work regimen, dynamic administrative leadership (frequently heavy-handed), and forceful personality made him not only Lamont's director but also a persuasive intellectual influence in almost all that was done there.*

"Ewing was an avid anti-drifter and the leader against mobility, both intellectually and emotionally. Jim Heirtzler and Xavier Le Pichon were similarly disposed," recalled Neil Opdyke. "I came from a department of geology at Columbia University that was dead-set against continental drift. Marshall Kay, Walter Bucher, Joe

*Wertenbaker's *The Floor of the Sea* (1974), the most complete treatment of Ewing, fails to convey his adamantly fixist views, which apparently contributed to Ewing's long-term less than amicable relationship with Harry Hess. This personal element may have conditioned Ewing's rejection of seafloor spreading for three years after most others at Lamont and elsewhere accepted it in 1966.

Maurice Ewing, first recipient of the Vetleson Prize and numerous other signal honors, was instrumental in the founding of the Lamont-Doherty Geological Observatory; as director he guided its growth to a position of preeminence in the acquisition of marine geophysical data. He seemed to possess an intuitive sense of what is important in science. Neil Opdyke recalled (as did others): "Ewing was an avid anti-drifter and the leader against mobility, both intellectually and emotionally."

Worzel, and those boys wouldn't hear of it; it was anathema."* Walter Pitman, too, remarked that "Ewing, Heirtzler, and Le Pichon did not believe in drift and seafloor spreading. Ewing had a very strong influence in setting the people around him against it. Walter Bucher had influenced Ewing against drift." (Bruce Heezen was sure that "Walter Bucher was the guiding scientific light at Lamont."[1]) Pitman remarked also that "Joe Worzel was a true believer who actually worshipped Ewing intellectually" and was set against drift as late as 1965.† There were many geophysicists at Lamont, as there were at most institutions, who were not truly familiar with the geological evidence, but were fixists because as far as they knew, physics did not permit an earth of the mobility required by the seafloor-spreading model. Ellen Herron, a graduate student at Lamont in the mid-1960's, was quick to affirm that "Doc Ewing's philosophy that the oceans were permanent features was the party line at Lamont; seafloor spreading was anathema!"[2]

The mobilist views of Bruce Heezen marked him as an exception

*Opdyke, oral comm., vii-25-79; the named group were professors of geology at Columbia University.

†Worzel was clearly against drift at the Symposium on Continental Drift in 1964 in London (Blackett et al., 1965, pp. 137–39).

at Lamont and Columbia; he was an outspoken advocate of an expanding earth and as early as 1955 was converted to a belief in a tensional origin for the mid-oceanic ridge system. He had been recruited by Ewing while still an undergraduate at Iowa and spent his illustrious career at Lamont often publishing papers with Ewing. He was in sympathy with the views of Warren Carey and provided much of the impetus for Hess's formulation of the seafloor-spreading hypothesis; he spoke a number of times in Guyot Hall at Princeton at Hess's invitation in the late 1950's on the origin of the mid-oceanic ridge system by tension and basalt injection. Heezen recalled:

> In 1955, the geophysicists at Lamont didn't come down to earth far enough to think about things like tensional interpretation of rifts. They were thinking of the earth as a physical and mathematical model. It didn't have rocks in it; it had things of different velocities. It didn't have history; it didn't have an age. They were on a different plane: it was the difference between talking to a historical geologist and a solid-state physicist. I wouldn't say they wouldn't buy a tensional origin for the ridges; it just didn't seem like a problem of any importance or interest to them.[3]

Neil Opdyke recalled that William Donn, a research associate at Lamont and professor at Brooklyn College, was another rare exception who had also shown sympathy for continental drift and polar wandering.

The geophysical scene at Lamont was much as it was in many institutions; geophysicists too often lacked the geological training and perspective required to apply their physical data to the construction of broadly encompassing geological models and overviews.*

*The roster of earth scientists who pioneered in the development of plate tectonics is dominated by geophysicists with training or sustained long-term interests in geology, and geologists who, by training or inclination, could utilize geophysical information. Those operating with ease at disciplinary interfaces were predictably most often in possession of diverse and seemingly unrelated elements required for the synthesis of broad conceptual constructs—a pattern not peculiar to earth scientists. The major figures in this book typify the genre. Beyond displaying characteristics in keeping with David McClelland's (1970) generalization about the preoccupation of scientists with analysis, the group seems to fit Liam Hudson's (1966) taxon of "divergers," who prefer to work on, or conceive of, poorly defined problems. Divergers are synthesis-oriented and have wide lateral vision. "Convergers," in contrast, prefer to work on well-defined problems with a high probability of solution; they are analytical and component-centered, and are skilled at disarticulating complicated wholes into understandable parts. Hudson also noted that convergers tend to become scientists while divergers predominate in the humanities and arts. The distinction between the two types is not often simply made, since such attributes of personality are spectrally distributed. Kuhn (1963) regarded it as not essential that an individual possess both divergent and convergent approaches in a research project, merely that both types be members of the team. Mitroff (1974) provides a fine overview of the question and the psychology of scientists in general.

James Heirtzler, the leader of Lamont's magnetics group, had assumed the task of organizing the enormous volume of magnetics data that had been accumulated during the early 1960's. In order to store the information for convenient future use, the navigational records, showing the vessel's changes in speed and direction, first had to be integrated with it. Newly available computers that could be operated by one person allowed Heirtzler's team to store the vast number of data and retrieve them in the form of magnetic profiles plotted at selected scales. The comparison of such profiles, by computer, from many different areas was vital to the success of the marine magnetics program.

Ironically, Lamont, an institution that was a bastion of antidrifters, had gathered more than half of the oceanic magnetic profiles and more than three-quarters of all deep-sea cores. The mass of systematically stored data and the cores would provide evidence that, when made public in 1966, would convince the geological community almost overnight of the truth of seafloor spreading. Heirtzler himself, whose data made this great turnaround possible, controverted the Vine-Matthews-Morley hypothesis in his published work while certain of those working beneath him were unknowingly laying the groundwork for its confirmation.

Objections to Vine-Matthews-Morley

During the year before the confirmation of seafloor spreading, which began to unfold at Lamont in late December 1965 or early January 1966, the Lamont marine magnetics group led by James Heirtzler submitted for publication three papers that clearly demonstrated that he and his coauthors were fixists still at that date. "Magnetic Anomalies Over the Mid-Atlantic Ridge," published by Heirtzler and Xavier Le Pichon in the *Journal of Geophysical Research* in 1965, grew from examination of their unparalleled store of magnetics data from the seafloor.* They predictably proposed an alternative to the Vine-Matthews-Morley hypothesis—one, of course, that would be in accord with the then-faltering but still tenable fixist paradigm.† They traced a continuous central anomaly in the At-

*Heirtzler and Le Pichon's paper was received by the *Journal of Geophysical Research* on iv-23-65 and published on viii-15-65.

†Walter Pitman recalled that "the Vine-Matthews hypothesis was not taken seriously at Lamont. Heirtzler thought it was nonsense" (t. 1, s. 1, iv-24-79). Others, including John Verhoogen, who was sympathetic to drift theory, were similarly disdainful of Vine-Matthews.

lantic more than 6,200 miles long on the basis of 58 profiles lying between 60°N and 42°S latitude. The anomaly was "everywhere associated with the axis of the ridge" and was "continuous over all latitudes except for offsets at the fracture zones." They believed that "a volcanic body of limited width and depth" with "an upper surface at, or very near to, the sea floor" gave rise to the axial anomaly. In order to explain the dual nature of the anomaly over the ridge axis, which was a single one north of 30°S, but consisted of several stronger (higher amplitude) anomalies toward the axis south of 30°S, Heirtzler and Le Pichon combined two ideas: one, that the volcanic rock filling the fractures of the axial zone is more magnetic than the surrounding rock (first suggested by Drake and Girdler in 1964 to explain the Red Sea's axial magnetic pattern), and two, that the difference on either side of 30°S resulted from a "difference in the fracture pattern." Analyzed by a computer program, the body that gave rise to the continuous linear axial anomaly was assumed to be 10 kilometers wide, 10 kilometers deep, and infinitely long. Most significantly, they also assumed that the body was "magnetized only by induction."[4] Although they noted the probability that some ridge rocks are reversely magnetized and that the inclusion of such a variable in their computer program would have permitted smaller apparent values of susceptibility, they eschewed that analytical course because the overall anomaly pattern they mapped, extending for more than 1,000 kilometers from the ridge axis, did not agree with Vine and Matthews's assumption of a progressive decrease in amplitude and increase in wavelength of anomalies with distance from the crest. Vine and Matthews had attributed such an anomaly change to increasing depth of water and depth of the Curie-point isotherm away from the axis. In contrast, Heirtzler and Le Pichon argued that the reduction in amplitude of the axial anomaly in the South Atlantic could not be explained by the effect of changing bathymetry. They also attributed the regional reduction in the magnetic field over the ridge crests to increased heat flow, which naturally implied a shallower depth to the Curie-point isotherm in that area.*

The second paper, "East Pacific Rise; the Magnetic Pattern and the Fracture Zones" by Manik Talwani, Le Pichon, and Heirtzler, appeared on November 26, 1965, in *Science*. The authors did not

*In a lengthy discussion (pp. 4028–31) of heat flow and its effect on the Curie-point isotherm, no mention is made of the question of mantle convection. This is predictable, since that would imply consideration of seafloor spreading—not likely at Lamont.

pursue the question of the origin of the ridge anomalies any further, but on the basis of new data from the East Pacific and also from the Reykjanes Ridge south of Iceland (a paper on the Reykjanes Ridge was in press) concluded that "it is unlikely that the offsets in the magnetic anomaly pattern are caused by transcurrent faulting."[5] They reviewed the "difficulties associated with the hypothesis of large fault displacements," and considered the remarkable linearity and symmetry of the Reykjanes anomaly pattern and the tacit implication that "the long fracture zones of the northeast Pacific Ocean existed before the formation of the rise" (which is precisely what Tuzo Wilson expounded in his transform paper of 1965,[6] to which the authors referred). They concluded, in predictable, fixist fashion, that "the important motion on the fracture zones, due to the formation of the rise, would then be vertical and not horizontal."[7] They noted that the difference in character of the axial and flank anomalies implied different origins: "flank anomalies are *not* axial anomalies at greater depths." They also believed that the flankward reduction in amplitude of the axial anomalies was also not explainable in terms of Vine-Matthews: the flank attack, opened in Heirtzler and Le Pichon's paper, was repeated but not pressed home; it was resumed in the paper on the Reykjanes Ridge, which appeared shortly.

The third paper was written by Heirtzler, Le Pichon, and J. Gregory Baron* in the fall of 1965 but not published until June 1966 in *Deep-Sea Research*. "Magnetic Anomalies Over the Reykjanes Ridge" had its origins in Heirtzler's early interest in the ridge. He had fruitlessly encouraged Ernst Zurflueh to do a magnetic survey of the ridge in 1961; instead Victor Goldsmith did it from a U.S. Navy aircraft in 1963, and Heirtzler "redigitized" Goldsmith's data and processed them by computer.[8] It was the first study that detailed the magnetic pattern over a large area of a mid-ocean ridge that was identified as such beyond doubt (see Fig. 8.1). The paper quite clearly sets out the conclusions of the Lamont group concerning seafloor magnetic anomalies and their rejection of the Vine-Matthews-Morley hypothesis. It was submitted for publication less than two months prior to the discovery of the remarkable *Eltanin* Leg 19 profile.

The Reykjanes Ridge was selected for the study because it was known to have a very large axial magnetic anomaly, as well as a

*Baron was then at the U.S. Naval Oceanographic Office in Washington, D.C.

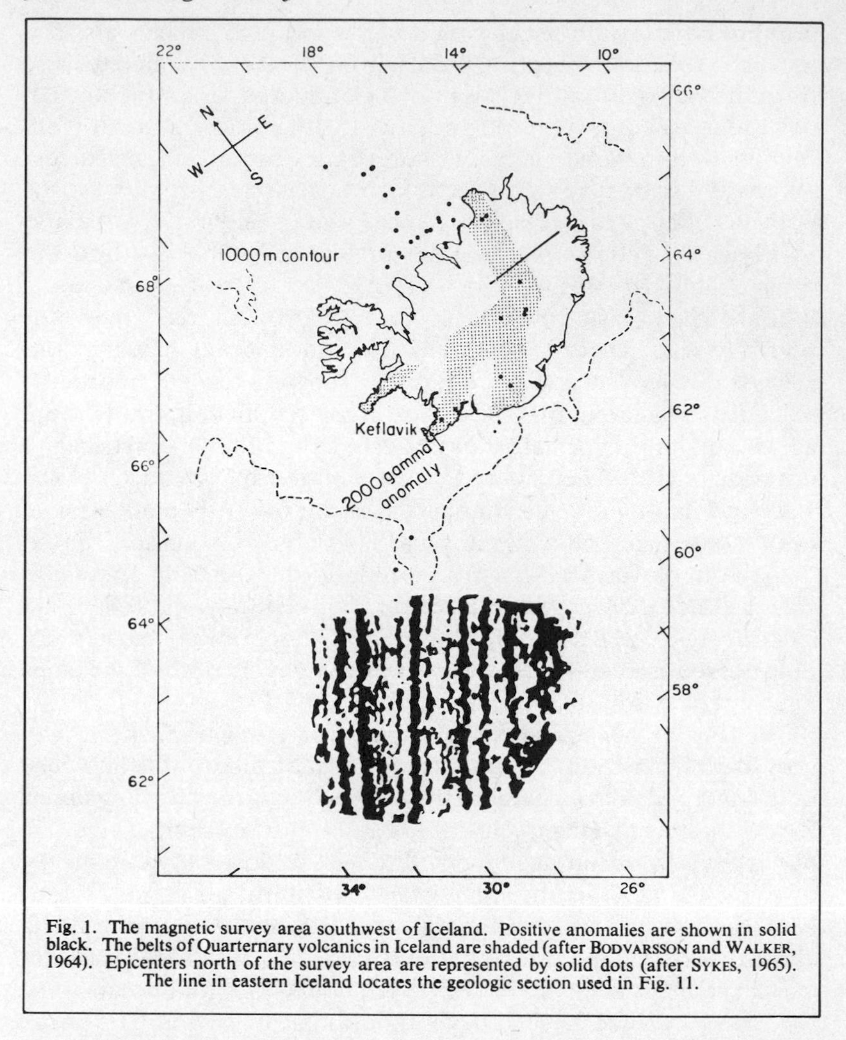

Fig. 1. The magnetic survey area southwest of Iceland. Positive anomalies are shown in solid black. The belts of Quarternary volcanics in Iceland are shaded (after BODVARSSON and WALKER, 1964). Epicenters north of the survey area are represented by solid dots (after SYKES, 1965). The line in eastern Iceland locates the geologic section used in Fig. 11.

Fig. 8.1. The magnetic pattern over the Reykjanes Ridge was shown as figure 1 of Heirtzler, Le Pichon, and Baron's June 1966 paper in *Deep-Sea Research.*

nearby airfield and magnetic observatory (essential for adjusting the data according to changes in the field strength); most important, the submarine Reykjanes Ridge is part of the Mid-Atlantic Ridge and continues into Iceland as the Reykjanes Peninsula. Geologic and seismic work had indicated that Iceland is mainly a large plateau of basalt topped with a pile of volcanic rock 10 kilometers thick; in contrast, the Reykjanes Ridge is a narrow structure with a much thicker basalt cap. The structural differences between the two were seemingly reflected in their respective magnetic patterns. Although the survey did not include the Icelandic continental shelf, a large 2,000-gamma anomaly about 20 kilometers wide was detected over the Reykjanes Peninsula on flights from Keflavik airfield (on the peninsula) to the main survey area. The anomaly was similar to but smaller than that over the ridge crest. The Reykjanes Ridge displayed extremely linear anomalies parallel to the ridge axis and strong axial symmetry. The axial area contained a central 3,000-gamma anomaly believed to be "caused by a normally magnetized body, 20 km wide"[9] coinciding with the crest and seismic zone, and six shorter (1,000-gamma) wavelength anomalies on each side of the crest. On each side of the axial area, the anomaly pattern changed sharply. Wide, weak (300-gamma) anomalies of irregular spacing appeared along with narrower, still weaker (100–200-gamma) anomalies.

Heirtzler, Le Pichon, and Baron thus spoke of the "axial magnetic zone," composed of strong, narrow linear anomalies with strong bilateral symmetry about the crest, and a "flank province" of "longer wave-length, smaller amplitude, irregularly spaced anomalies, on which is superimposed small wave-length 'magnetic' noise."[10] Although the pattern over the Reykjanes Ridge was narrower, it was basically like that over the Mid-Atlantic Ridge, described by Heirtzler and Le Pichon in 1965, and that over the East Pacific Rise, described by Talwani, Le Pichon, and Heirtzler in 1965. Fig. 11 of the June 1966 paper, reproduced here as Fig. 8.2, is remarkable in that it shows a "simplified" geological cross section of eastern Iceland (prepared by Hans Wensink, who had published on Icelandic magnetostratigraphy with Rutten in 1960). The section shows a "wide block of recent, normally magnetized lava with, on each side, flows of alternately reversed and normal magnetization dipping toward the graben [center of the central block]." Lava flows, labeled R_1, N_2, R_2, and so on, were depicted as continuous features that dipped toward and under the central block to emerge at distances of tens of kilometers on the other side of the axis. The resulting picture was of a trough, or syncline, with a symmetrical pattern of outcrops across it. The Lamont group had mapped a reversal pattern, bilaterally disposed to the central rift—apparently strong evidence for the Vine-Matthews-Morley hypothesis—but also explicable, in fixist terms, as a syncline. Besides Wensink,[11] Gunnar Bödvarsson and

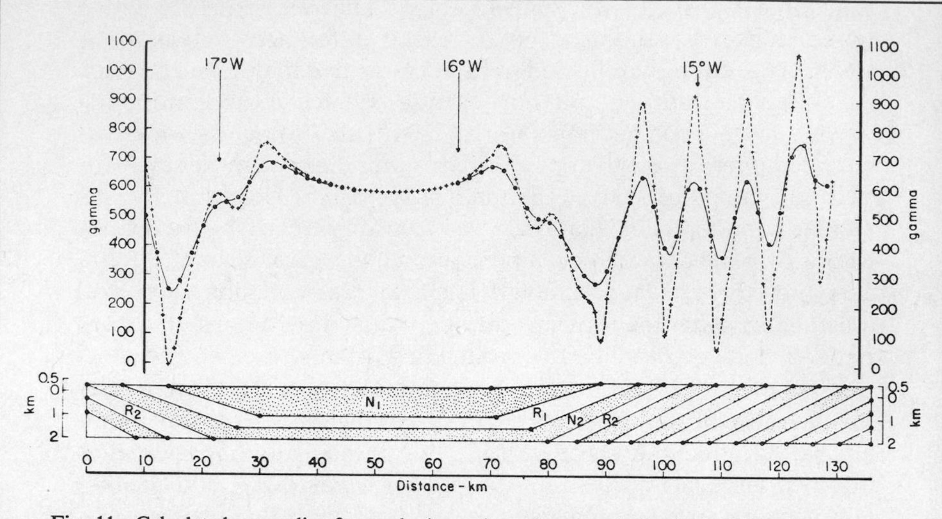

Fig. 11. Calculated anomalies for geologic section of eastern Iceland. "N" and 'R" refer to normal and reversely magnetized material. Anomalies are calculated for heights of 1 km (dashed line) and 2·5 km (solid line) above sealevel, or 0·5 km and 2 km above the eastern plateau. The value of magnetization assumed is + 0·006 c.g.s. for the normally magnetized material and — 0·006 c.g.s. for the reversely magnetized material.

Fig. 8.2. Hans Wensink prepared this geological cross section of eastern Iceland, which appeared as figure 11 of Heirtzler, Le Pichon, and Baron's June 1966 article. The cross section depicts a syncline-like structure with lava flows of alternately reversed and normal magnetization dipping toward the graben (a crustal block dropped between faults produced by tension).

George P. L. Walker, to whom we shortly turn, had examined such a possibility and rejected it in 1964, but not forcefully.

Heirtzler, Le Pichon, and Baron emphasized in rebuttal to Vine-Matthews-Morley that although the large axial anomaly over the ridge continues within the belt of Quaternary volcanics over the Reykjanes Peninsula, "the magnetic pattern over the survey area [the ridge] breaks down before the edge of the Icelandic shelf." They reiterated in concluding that "while a geological structure [such as the main Icelandic graben] . . . could explain the axial magnetic pattern observed over the ridge, the differences between the Icelandic Plateau and the Reykjanes Ridge are so large that it is not clear to us how their geologic structures can be similar."[12] They thus discounted the likelihood of structural continuity, and in the process militated against spreading.

In an article then in press, Heirtzler, Le Pichon, and Baron referred to Bödvarsson and Walker and also cited Matthews, who had "assumed" a process of crustal extension similar to that of Bödvarsson and Walker.[13] The latter two believed that the rocks of extreme eastern and western Iceland "may have been carried apart by 400 km or more." Such crustal "drift" was thought to result "mainly from crustal extension through the injection of dykes." They further believed the structure of Iceland to be "closely related to the world-wide rift system."[14] Although Bödvarsson and Walker presented that conclusion forthrightly in their introductory summary, the field and geophysical evidence was in fact not conclusive. In the body of the paper, they had carefully weighed two hypotheses: (1) that "the structure of Iceland is to be taken to be synclinal" within the context of "a contracting state hypothesis," and (2) the steady-state hypothesis, which involved a continuously uniform, wide main belt of active volcanism and implied that the oldest rocks in the extreme east and west of Iceland had a common origin in the central volcanic zone and had been carried apart 400 kilometers by crustal drift.[15] In supporting the syncline hypothesis, they said, "On purely geophysical ground"—certainly an appealing expression to the physics-centered Lamont group—"the first possibility . . . appears to provide the most attractive interpretation of the seismic data."[16] Further, after considering the evidence, they concluded that Tuzo Wilson's evidence of 1963 for transport of Atlantic islands away from the mid-ocean ridge by crustal drift implied a spreading rate of a few centimeters per year, "which is one order of magnitude higher than the rate envisaged on the dyke-model suggested in this paper."[17] The discrepancy detracted from the strength of their major conclusion, as set forth above. The doubt of the Icelandic geologists about many important aspects of the geological structure of their island appears to have contributed to the way in which the Lamont group chose to interpret their data. Beyond that, the breakdown in the magnetic pattern between Iceland and the Reykjanes Ridge implied that there might well be some structural discontinuity between them.

Heirtzler, Le Pichon, and Baron concluded: "Vine and Matthews (1963) . . . , following Dietz [1961a], suggested that this pattern was due to a spreading ocean floor, originating at the ridge axis, and alternately normally and reversely magnetized. However, this hypothesis in its present form does not explain the characteristic change in magnetic pattern from the axial zone to the flanks and the difference between the axial anomaly and the adjacent ones."[18] It was an important question, which the Vine-Matthews-Morley hypothesis in fact did not answer—but neither did the Lamont group.

Thus the marine magnetics group under Heirtzler was uniformly fixist in its published and spoken interpretations of data from the seafloor during the year prior to the confirmation of Vine-Matthews-Morley. However, within Lamont, a recently arrived young product of the English heresiarchs was sympathetic to a mobilist interpretation of the magnetics data. In his new surroundings, he was suitably deferential, but his day was dawning.

Opdyke and the Deep-Sea Cores

Although Maurice Ewing had a basic distrust of the magnetic directional data and rejected the conclusions drawn from it by the English, he did feel it important to have someone at Lamont who was versed in what had become a growing area of inquiry: paleomagnetics was attracting increasing attention and lay clearly at the center of the reemerging continental-drift controversy of the late 1950's. Prompted by the efforts of John Graham, who began pioneering paleomagnetic studies in the late 1940's (discussed in Chapter 4), and also Ellis Johnson, Thomas Murphy, and Oscar Torreson's studies of paleomagnetism of deep-sea cores in 1948, Ewing had decided that an attempt should be made to determine the magnetism of the rapidly accumulating store of deep-sea cores at Lamont, which he thought were "underutilized."* Although no particular research goal or plan of attack was formulated, he encouraged Manik Talwani to undertake the investigation.†

The magnetic intensities of sediments in cores are many times less than those of basalts; what was required, then, was an extremely sensitive magnetometer that could also operate without disturbing the column of soft sediment in a core. Because Lamont lacked such an instrument, Talwani visited John Graham's laboratory in 1958 in an attempt to measure some cores on a magnetometer but was unsuccessful. Little was attempted during the next five years, but by 1963, M. J. Keen published in *Nature*, demonstrating that deep-sea cores from the Iberian Abyssal Plain had magnetic

*The core work of Johnson, Murphy, and Torreson was part of their ongoing program of studies in secular variations (discussed in Chapter 5).

†Formerly director of the Lamont-Doherty Geological Observatory, Talwani had written a doctoral dissertation on gravity under J. L. Worzel at Columbia University.

Neil D. Opdyke at the Lamont-Doherty Geological Observatory in 1980. Walter Pitman recalled: "Opdyke was most important; he believed in drift and was a very important guy to have around. Because he was our age he was a colleague; he was encouraging and kept on pushing. . . . I remember staying there all night long one day, running out magnified [magnetic anomaly] profiles. . . . I pinned up all the profiles of *Eltanin* 19, 20, and 21 on Opdyke's door, and went home for a bit of rest. When I came back the guy was just beside himself! He knew that we'd proved seafloor spreading. . . . Opdyke stimulated us by knowing intellectually exactly what it meant and also pushing us."

inclinations congruent with the ambient dipolar field. About that time Ewing recruited Hans Wensink, on sabbatical from Holland, to develop a paleomagnetic capability at Lamont, but as late as 1964 the program remained unsuccessful. That was the year in which Christopher Harrison and Brian Funnell, while working at Scrips on leave from Cambridge, made the first report of polarity reversals in deep-sea cores.*

*The Rock Magnetics group had discussed the possibility of using reversals in young lake sediments and deep-sea cores before the first recognition of polarity reversals in cores by Harrison at Scripps in 1964. "We thought it virtually impossible that such sediments would contain a reliable magnetic record, and the idea was never pursued" (Dalrymple, oral comm., xii-21-78). David Hopkins noted: "Throughout most of the 1960's, the Rock Magnetics group's equipment was unsuitable [insufficiently sensitive] for work on sediments. When I was in the Marine Geology Branch of the U.S.G.S. in 1968 or 1969, Parke Snavely and I discussed with Doell a plan to work on Bering Sea cores. . . . Doell was not yet set up to work on cores" (written comm., i-5-78).

Other reports quickly followed Harrison's; in 1965, T. I. Lin'kova showed clearly that reversals were present in cores from the Arctic Ocean, and in 1966 Michael D. Fuller, Harrison, and Y. Rammohanroy Nayuda found reversals in sediment cores recovered by the Mohole drilling project. Victor Vacquier had suggested magnetic study of ocean cores to Johnson, Murphy, and Torreson, and also to Harrison and Funnell.

In spite of Lamont's earlier difficulties in securing a capability in paleomagnetics, Ewing demonstrated his sustained belief in the promise of such a program; he had James Heirtzler offer Neil D. Opdyke a post funded by the Office of Naval Research, which Opdyke assumed in January 1964.

Neil Opdyke, born in New Jersey, got his undergraduate training in geology at Columbia University in New York City. There he met and served as field assistant for Keith Runcorn, who subsequently invited him to graduate study at the University of Cambridge. Opdyke arrived at Cambridge in the fall of 1955. He recalled:

Shortly afterward I moved to the University of Newcastle-on-Tyne with Runcorn, because the money I was being supported by went with Runcorn. I completed my Ph.D. jointly in the departments of geology and physics with a dissertation on "Paleoclimatology and Paleomagnetism in Relation to Polar Wandering and Continental Drift." I was introduced to the subject of paleomagnetism, about which few people in North America knew anything, the minute I got into the automobile to go out west with Runcorn as his field assistant. I was tutored for days at a time in the field in the U.S. and in England. We were looking for evidence of polar wandering. At that time Runcorn was *not* in favor of continental displacements. I remember arguing with [John A.] Clegg, [Mary] Almond, and [Peter

Christopher G. A. Harrison in the late 1970's. While a student at Cambridge, Harrison was attracted to paleomagnetism through a talk by Patrick Blackett. John Belshé advised him to work on deep-sea sediments; he arrived at Scripps in July 1961 to gather data for his doctoral studies at Cambridge. Brian Funnell, also from Cambridge, and versed in radiolarian biostratigraphy, urged Harrison to collaborate with him in a joint microfossil and magnetic study of oceanic sediment cores. In 1964 the two made the first report of polarity reversals in deep-sea cores.

H. S.] Stubbs in London, and told them that they were bananas looking for continental drift.*

During the late 1950's, the accumulating magnetic directional data acquired mainly by the English quite convinced them not only of polar wandering but also of continental displacement. Opdyke was thus an early convert to drift as a student at Newcastle, which was a major center for mobilist studies. He also spent a Fulbright postdoctoral year in Canberra in 1959 at the Australian National University, at the invitation of Edward Irving, who was collecting much of the important directional data that supported drift. While Opdyke was in Australia, John Belshé visited the A.N.U. and lectured on the magnetics of deep-sea sediment cores, which Christopher Harrison had studied under Belshé at Scripps for his thesis problem. Opdyke was thus informed of core work by 1960. He recalled:

When I came to Lamont [in January, 1964] I knew there were problems in doing cores, although you could do it. I'd seen results [from] cores—Harrison's paper came out in 1964.† [But] I wasn't in the magnetic stratigraphy game, and so I pursued the Silurian and White Mountains series of New Hampshire with Hans Wensink of Utrecht, who spent a year here at Lamont. We spent much time in putting the lab together and finally went down to Princeton and saw that they had an exact copy of the magnetometer we were trying to build. It was a Joe Phillips design; you could buy the electronics off the shelf, which was a big step forward. Princeton supplied plans for the spin magnetometer, which we built and put in operation by [June 1965].

Although Ewing didn't tell me what I had to do, he did say I should look at cores. Wally Broecker, a geochemist, had gone to sea and brought back red-clay cores from the central Pacific in the fall of 1965 on the *Vema* 21 leg from Hawaii to Japan. I was not anxious to do it. I told John Foster, my graduate student, that the boss wanted some cores looked at—why don't you do it? John Foster didn't do anything with the project and tried to give it away. Billy Glass, another graduate student, decided to do the cores. . . . He was in the same office as Jim Hays, who was just finishing his Ph.D. on the zonation of Radiolaria in the Antarctic; both were students under Bruce Heezen. Glass [and Foster] and Hays began to measure the cores and got very good results. This was precisely the time that Walter Pitman was running out his magnetic-anomaly profiles.

*Opdyke, t. 1, s. 1, v-11-79. The 1954 paper of Clegg, Almond, and Stubbs provided the earliest significant magnetic directional evidence for drift.

†Harrison and Funnell, 1964, "Relationship of Palaeomagnetic Reversals and Micropalaeontology in Two Late Cainozoic Cores from the Pacific Ocean."

Opdyke, interested mainly in directional studies based on terrestrial rocks, was without a magnetometer at Lamont when he first arrived in 1964. He thus sent his specimens to the Dominion Observatory in Canada for magnetic determination in the course of his collaborative studies with Irving on the Silurian Bloomsburg red beds of Pennsylvania.[19] Opdyke also collaborated with Hans Wensink (a coauthor with Martin Rutten in 1959 and 1960) in an attempt to build a paleomagnetics laboratory, and also on a study of the White Mountain Series of New Hampshire.

As he had done earlier with Talwani, Ewing suggested that Opdyke examine deep-sea sediment cores, but Opdyke was "not anxious to do it." Opdyke and Wensink were both geologically oriented paleomagnetists, and neither was highly skilled in the technical aspects of instrumentation. Opdyke had arrived at Lamont with plans for a paleomagnetics laboratory modeled after that of Ian Gough at the University of Rhodesia and Nyasaland (now Zimbabwe) in Salisbury, where the equipment had been developed by Gough, Michael McElhinny, and Dai Jones. Opdyke requested that Jim Cottone, one of James Heirtzler's engineers, build a copy of the Gough spin magnetometer (which was evolved from an early model of John Graham's). After more than a year in construction, the Gough spinner was not able to detect a useful signal from a sample of the highly magnetic Clinton iron ore.*

Opdyke asked his graduate student John Foster to correct the machine. Foster instead offered to build a new one. His education had included a diploma in electronic technology as a result of a long-term interest in instrument building and ham radio work. He had subsequently completed a B.S. degree in geophysics at McGill University and entered the graduate school there to pursue a master's degree involving studies in micropulsations and magnetotellurics.† The magnetotelluric apparatus he built at McGill led to an invitation to Lamont to work with Heirtzler, who was then interested in studies in micropulsations. Foster arrived at Lamont in May 1964; by the spring of the following year he decided to study with Opdyke.

*The Gough spinner (Gough, 1967) was a high-speed compressed-air turbine machine built for study of the Pilansberg dikes (Gough, 1956); it was therefore not suitable for the study of soft sediment cores since the associated high centrifugal force disturbs the sediment and its magnetism.

†Magnetotellurics entails the measurement of the earth's electric and magnetic fields. The two horizontal electric field components and the three magnetic field components are usually included. A micropulsation is a short-period (0.2–600 seconds) variation in the geomagnetic field that normally shows an oscillating waveform.

The instrument that Foster built, now referred to as his Mark I spin magnetometer (one of a series of three models built between spring 1965 and late 1966), was completed in June 1965. It was a slower (5 cycles per second) modified version of a high-speed instrument designed by Joseph Phillips at Princeton University.* Foster had decided from his experience in developing low-noise receivers that a slow spinner was "the best approach."[20] It consisted of a pick-up coil matched to an input transformer, a preamplifier, and a two-channel phase-sensitive voltmeter. The instrument was originally intended not for use in the study of deep-sea cores, but rather for use in Opdyke's and Wensink's continuing studies of terrestrial rocks and for Foster's thesis study on the Silurian red beds of the Eastport, Maine, quadrangle, a problem assigned to him by Opdyke.

It was fortuitous that the extremely low spin rate, selected in order to produce a low signal but with very much lower noise, lent itself perfectly to later use in spinning soft sediment cores without destroying their magnetism. Opdyke was not originally interested in marine core studies, and did not regard them as a fruitful area of inquiry before Foster and Bill Glass first recognized reversals in oceanic sedimentary cores at Lamont in the summer of 1965. Opdyke was a product of the English school of rock-magnetic directionality studies, which was mainly concerned with resolution of the drift question. As such, he was likely to order research priorities much as did Keith Runcorn and Edward Irving, with whom he had worked and studied. Neither emphasized the polarity-reversal data acquired in the course of their studies, and Irving had been openly dissuasive of the efforts of Cox, Doell, and Dalrymple to formulate a polarity-reversal time scale. Opdyke remarked: "My own scientific judgment at the time was that the continental drift question was more important, and I was pursuing that because it became clear that the most important thing was to shake people loose from their fixist attitudes or the rest of it could go hang. It's a question of what you perceive as most important."[21]

"Foster, more than anybody at Lamont, had pushed the building

*One of Phillips's thesis advisers was Michael McElhinny, who had trained under Edward Irving at the A.N.U. and spent some time at Princeton.

On the spin magnetometer, see Foster, 1966. By September 1966, the spinners were used in over 10,000 magnetic measurements at Lamont. An important design criterion in the magnetometers was that they could be built of commercially available components and with a minimum of technical facilities. The Mark II was similar to the Mark I, but the Mark III consisted of a fluxgate gradiometer and a two-channel phase-sensitive voltmeter.

of magnetometers that could measure deep sea cores," said one informant; another remarked, "Foster was the catalyst of the core work."[22] Although not broadly trained in geology, and apparently not fully aware of how useful and significant a successfully derived body of data from deep-sea cores could be, Foster had an early "intuitive" sense of their potential importance. He was mainly concerned with the instrumental and technical aspects of the unfolding core program: his almost compulsive drive in building a series of successful slow-speed magnetometers was crucial in the early stages of the research. "Foster was mainly interested in the instruments," according to Dragoslav Ninkovich, "not the analysis of the cores."[23] Neil Opdyke, in fact, viewed Foster's early preoccupation with the instruments and measurement of deep-sea cores as detracting from his proper concern with completion of his doctoral dissertation on the red beds of the Eastport, Maine, quadrangle.* Furthermore, the magnetic work done elsewhere on deep-sea cores had not been very promising; thus Opdyke "didn't believe the sediments would show anything and were probably a waste of time."[24]

After completion and satisfactory operation of the Mark I spinner in June 1965, Opdyke and Wensink collected a number of cores from the White Mountain intrusives (of New Hampshire) in the fall. The day after the spinner was completed, Wally Broecker, a Lamont geochemist, met Foster in the cafeteria and was excited to learn that the spinner was working. Broecker and Cesare Emiliani of the University of Miami were rivals in dating things in the ocean; both were looking at oxygen-isotope data and were interested in any adjunct to dating processes. Broecker had visited Scripps and seen the magnetic core work of Christopher Harrison and Brian Funnell.†

Broecker went to sea on the *Vema* (cruise 21) to secure cores of red clay, which he peppered with one-inch-diameter push-through vials and returned to Foster for magnetic determination in mid-1965.[25] The red clays presented an intractable problem. They found that in the purest red clays, those with the fewest microfossils in them, the magnetometer signal became unintelligible below about

*A number of respondents commented on the wide range of Foster's interests and activities, which appeared out of keeping with the single-minded pursuit of dissertation studies expected of graduate students.

†Broecker's enthusiasm and sense of the potential usefulness of ocean-core magnetostratigraphy gave impetus to Foster in his early instrument-building efforts.

1 meter of depth in the core. Broecker grew discouraged, and Opdyke suggested that Foster return to his thesis work on the Eastport quadrangle. Foster continued working at night on Broecker's cores, but nothing came of it: "I did not want to play with the deep-sea sediments; I wanted to get a Ph.D. degree and do the Eastport quadrangle. Helping Wally Broecker was fine, but that petered out. I didn't know a thing about deep-sea cores and couldn't have cared less."[26] During that time, only Foster and Hans Wensink were running the magnetometer.

Donald Corrigan, a student in magnetics under Dale Krause at the University of Rhode Island, had come to Lamont to acquire a copy of the Foster spinner in order to determine the magnetism of deep-sea cores for Cesare Emiliani at Miami. Foster recalled:

> In a discussion among [William B. F.] Ryan, Billy Glass, and myself, with Ryan contributing the bulk of the ideas, we sketched out Corrigan's approach to the magnetic stratigraphy business. Corrigan was to look for reversals in deep-sea sediments. Ryan sought cores with faunal control and knew that the cores already worked up in James Hays's thesis were likely candidates; Hays had done a radiolarian stratigraphic study for his Ph.D. Bill Glass, who shared an office with Hays, said, "Some of those cores in Jim Hays's thesis are down in the core shed." I didn't even know where the core shed was and had never taken a course in marine geology. My entire education on how to do this came from Bill Ryan.* Glass took out a segment of core; we carefully marked it and cut out a sample and a meter down the core another sample, and another meter still further down. We shaved the three specimens, preserving their orientation. We measured the first one; [its magnetization] was strong and easily recognizable. Then we did the next one and it was reversed; then we did the next one and it was normal. This was much farther down the core than we'd been able to get with the red clays.[27]

In an excited state, Foster called Bill Ryan, Neil Opdyke, and Bruce Heezen. Opdyke was not favorably responsive at that point, but Heezen immediately drove to Lamont, examined the data in great excitement, and within a day supervised the preparation of many more samples for analysis. Foster continued:

*On a Lamont cruise to the eastern Mediterranean on which Ryan acquired data for his doctoral dissertation, he attached a compass in the field of the corehead camera in an attempt to orient cores by laying a string down the side of the core to produce a fiducial mark. He wanted to be able to orient any directional feature in the photographs. Maurice Ewing ordered him to discontinue the practice and threw the equipment for magnetic orientation overboard, citing Manik Talwani's earlier lack of success in deep-sea core magnetostratigraphy at John Graham's laboratory. Ryan had been an associate of Graham at Woods Hole Oceanographic Institution,

Billy Glass was going to cut up the Pacific-Antarctic cores, and I was going to measure them. Lamont was on a fire-drill basis: as soon as something interesting happened—Bang! Pour on the coal. Heezen grabbed all of his graduate students and put them to work; Eric Schneider was going to do the North Atlantic, Ruddiman another ocean, Ninkovich the North Pacific, and Ryan the eastern Mediterranean. My role was to take my machine and do the measurements. They were going to select the cores from as high a latitude as possible in order to meet the orienting criterion to make it work.* Within three hours I was committed to doing work for Ninkovich, Ruddiman, Schneider, and Glass, and the next day Geoff Dickson got into the act.[28]

The precipitous occupancy and use of the paleomagnetics laboratory by Heezen's group led to a request from Opdyke that they curtail their use of the laboratory; an "acrimonious discussion" between Opdyke and Foster also ensued over the use of the magnetometer built by Foster. Heezen appealed to the director, Maurice Ewing, to permit Glass, his doctoral student, to continue the work on his thesis area, which had been under way before the first successful magnetic results on deep-sea cores. Opdyke recalled:

I told Bruce Heezen that he could not be an author on the first paper,[29] which caused a big blow-up because I'd been doing paleomagnetism—it's my business—and his very presence on the paper would have been enough to lead people to think he had been the prime mover. He was very encouraging after it started, but he did not conceive of it; I said he could be [an author] on the second paper we wrote with Ninkovich.[30] That was the end of our collaboration.

Glass and Foster really got down to figuring out that things were really interesting and important in January [1966]. It was all geared up to go in February. The important thing about the Antarctic cores was that they had an independently derived frame of reference within the cores themselves: they had the fossils in them, whereas the cores from the Pacific had no fossils. We would have got there sooner or later, but it just took Billy Glass to say that these are the cores that should be done for this purpose. James Hays had zoned the cores for his Ph.D. thesis; they turned out to have very good [biostratigraphic] control. There were very few cores in the world that had that kind of control. Those cores had long sections that went back into the Pliocene; they were the best at the time that I knew of. I didn't know about that [core] work until it was brought to my attention by Billy Glass. . . .

---

and was apprised of efforts in sediment-core studies before he came to Lamont (W. B. F. Ryan, oral comm., viii-20-79).

*Foster vividly recalled that the important idea to select cores from high latitudes, in order to get clearly recognizable polarity determinations and thus avoid the difficulties inherent in low-latitude cores, was contributed by Ryan. Ryan could not clearly recall that event.

Bruce Heezen, one of the foremost investigators of the seafloor, in his miraculously crammed office and research quárters at Lamont-Doherty Geological Observatory in August 1976. Heezen's ability to acquire space and shipboard time was greatly curtailed after a dispute with Lamont's director, Maurice Ewing, in the mid-1960's. Heezen died of a heart attack while aboard a submarine off the coast of Iceland in June 1977.

In the autumn of 1965 we found [what was later named] *the Jaramillo event.** *We couldn't understand that then. We couldn't interpret the data*; the reversal wasn't 180 degrees, it was 120, which was caused by overprinting mud. There were technical problems, but we did have data and reversals on those cores which had been collected on the *Vema* 21 leg from Hawaii to Japan. At precisely this time, Walter Pitman was running out his [*Eltanin* 19] profile, so suddenly the whole thing fell together. I could see that the magnetic stratigraphy of the cores could be used to verify the pattern that Pitman was getting in the magnetic anomaly profiles. He was sitting in the next office to me, and what he knew, I knew within a day. It became apparent that this was exciting and important and we could do the job. *That is when I decided to pursue the deep-sea core problem*—that was in February 1966.[31]

*Opdyke recalled that the event as recognized in the cores at Lamont had tentatively been named the Emperor event; Foster noted that it had been referred to as the Chinook event. No other members of the Lamont group interviewed could recall that episode.

The Clincher: Pitman's Eltanin 19 Profile

Unlike Opdyke, who was a student of the English directionalist school, a mobilist, and fully informed of the Vine-Matthews hypothesis, Walter C. Pitman III was trained at Columbia University and thus had not acquired a background that might have disposed him toward a search for data, nor ready recognition of any, that would have been applicable to the Vine-Matthews-Morley hypothesis.

As a graduate student under James Heirtzler, Pitman traversed the Pacific-Antarctic Ridge aboard the National Science Foundation vessel *Eltanin* from September to late November 1965. Ellen Herron, a new graduate student who had recently left studies in seismology to enter magnetics at Lamont, worked with Pitman aboard the *Eltanin*, and upon their return to Lamont the two examined the data gathered on the cruise, which included Legs 20 and 21. During an earlier cruise, Leg 19, the *Eltanin* had perpendicularly crossed the East Pacific Rise, where it curves west at about 51°S latitude, and obtained remarkably intelligible data. Herron noted that when aboard the *Eltanin* with Pitman, "we spent most of the time keeping the seismic gear going and didn't know what we were doing with the magnetic data. None of us was aware of the Vine-Matthews hypothesis at the time."[32] Opdyke remarked that there was "*no mention of the Vine-Matthews hypothesis in the research proposals for those cruises*."[33]

The data were gathered in analog and digital form showing total magnetic intensity. Pitman and Herron mathematically removed the ambient field and thus derived the remaining, anomalous field along the seafloor profiles of Legs 19, 20, and 21. Although the Leg 19 data had been gathered earlier, Pitman and Herron decided to process the data from Leg 20 first because they had jointly gathered them. The first step involved editing the digitized data from the ship, to remove any spurious readings, and plotting the total intensity. Herron noted: "The magnetics data were recorded on paper tape, which was played directly into the computer. First you'd [convert] the data into total field and then ship all the cards, containing the data points, down to the big computer at Columbia, where the earth's main field would be removed, and then we'd plot up the residual or local anomaly field. That was part of our routine chores; as graduate students we were expected to reduce data."[34]

Pitman recalled:

Walter C. Pitman III, as a graduate student at Columbia University under James Heirtzler, traversed the Pacific-Antarctic Ridge aboard the research vessel *Eltanin* from September to late November 1965 gathering magnetics data with Ellen Herron. Pitman recalled: "I took that *Eltanin* 20 [magnetics] profile down to [a staff member who] had written on the subject. . . . He was talking to [another researcher] and when they came out I said, 'Look at this, this is very puzzling; these magnetic anomalies that we got . . . look almost the same as the profile over the Juan de Fuca Ridge.' He said, 'Ha! Ha! I suppose you're going to prove Vine and Matthews are right.' . . . That was in December of 1965."

At exactly the time that we got back from sea, we saw the Vine and Wilson paper [in *Science*], and the short magnetic-anomaly profile that they had across the Juan de Fuca Ridge. [It] was very similar to the axial portion we could see in the south Pacific from *Eltanin* 20: the big Brunhes normal, the Jaramillo, and the Olduvai events. Whatever you could see on the Juan de Fuca Ridge you could see on the *Eltanin* 20 profile. At that point in December 1965, Ellen Herron and I were just getting data out and had not yet selected our doctoral dissertations and didn't know what the *Eltanin* 20 profile really meant. I'd read Dietz's paper casually; I didn't know the Vine-Matthews hypothesis very well at all. I was not aware of it in detail. I had not read the paper and studied it; it was not something that seemed important to me at the time. I was unaware that Le Pichon and Heirtzler had made that strong comment against Vine and Matthews in print.*

*Dietz, 1961, "Continent and Ocean Basin Evolution by Spreading of the Sea Floor." Heirtzler and Le Pichon, 1965, p. 4028: "The pattern of the anomalies is such that we had no basis for

I took that *Eltanin* 20 profile down to [a staff member's] office, because I knew he had written on the subject and was interested in the problem. He was talking to [another researcher], and when they came out I said, "Look at this, this is very puzzling; these magnetic anomalies that we got on the *Eltanin* 20 look almost exactly the same as the profile over the Juan de Fuca Ridge." He said, "Ha! ha! I suppose you're going to prove Vine and Matthews are right." I questioned him about that and he said something about seafloor spreading. I had been doing broad reading while searching for a thesis problem and remembered Dietz. They also said something about seafloor spreading and Harry Hess. That was the first time Hess's name came into this thing. That was in December of 1965. The conversation and laughter about Vine-Matthews, Dietz, Hess, and seafloor spreading stuck in my mind. His laughter didn't bother me at all; I must say I began to smell something at that time. I didn't understand it and I didn't know what it was. I went back and continued to process the data with Ellen Herron, but what problems we were mutually or exclusively interested in I had only the faintest idea.[35]

About a month later, when Pitman and Herron first noticed the similarity between the Juan de Fuca Ridge and *Eltanin* Leg 20 profiles, and "long after" they had submitted an abstract (which said "nothing important"), concerning the *Eltanin* magnetic data processed by that time, to the American Geophysical Union for the Spring 1966 meeting, Pitman was able to process the data from *Eltanin* Legs 19 and 21. He recalled:

One of the very important factors in all this was the presence of Neil Opdyke and a graduate student named Geoffrey Dickson, who, because he was Australian, was more receptive to the idea of continental drift. Opdyke was most important; he believed in drift and was a very important guy to have around. Because he was our age he was a colleague; he was encouraging and kept on pushing. He talked about the paleoclimatological evidence for drift and believed in field reversals and that the polar wander paths could be explained by drift. As we processed all those data, the similarity between the Juan de Fuca [Ridge] and those profiles became obvious. I remember staying there all night long one day, running out magnified projected profiles, simply taking the magnetic anomaly data and making slight adjustments to them so they look as though you've run perpendicular to a ridge axis. I pinned up all the profiles of *Eltanin* 19, 20, and 21 on Opdyke's door and went home for a bit of rest. When I came back the guy was just beside himself! He knew that we'd proved seafloor spreading! It was the first time that you could see the total similarity be-

---

including reversed magnetization in our models. . . . It is clear from this study that most of the profiles do not follow the pattern assumed by Vine and Matthews." Although Pitman here refers to the "Jaramillo," he did not know of the Jaramillo event by name at that time.

tween the profiles—the correlation, anomaly by anomaly. The bilateral symmetry of *Eltanin* 19 was the absolute crucial thing.* Once Opdyke saw that he said, "That's it—you've got it!" Opdyke was a very important catalyst at that point. He saw immediately what it meant and became our advocate. Having him around was a very important thing. Heirtzler took a while to come around; at first he tried to explain it away by electrical currents.† Opdyke knew the significance of the profile immediately; he knew all the ramifications. Heirtzler was not a geologist and didn't appreciate it.[36]

During several weeks of interaction with Opdyke, Pitman, Dickson, and others at Lamont, Heirtzler came to recognize not only that the *Eltanin* 19 profile contained all of the polarity epochs and events of the past 4 million years that had been ingeniously and laboriously discovered by the reversalists, but that the full 2,000 kilometers of magnetic stripes on both sides of the East Pacific Rise recorded a vastly longer interval of polarity behavior.‡

*The bilateral symmetry is apparent to some extent in *Eltanin* profiles 20 and 21, but much clearer in profile 19, which was acquired from a place where the magnetic record of seafloor spreading had been faithfully captured and well preserved. The perfection of the *Eltanin* 19 profile was surprising. Studies that supported a tensional interpretation for the origin of the rifts (and led to the notion of seafloor spreading) had suggested that mid-ocean ridges with well-developed axial valleys were most active and spreading at the highest rates. Ironically, it was found that the most perfect magnetic profiles are produced where rapid spreading gives rise to gentle, low topography lacking a rift valley.

†Opdyke recounted: "Heirtzler said the *Eltanin* 19 profile was too perfect and caused by electrical currents in the upper mantle in order to get out of the Vine-Matthews hypothesis. It was a process of about a month getting Heirtzler to change his mind about it. The reason was, the summer before we had had a hell of an argument on three papers that were published by Heirtzler and Le Pichon on the Mid-Atlantic Ridge. Hans Wensink and I had reviewed the paper on the magnetics [Heirtzler and Le Pichon, 1965, "Magnetic Anomalies Over the Mid-Atlantic Ridge," where they argued that the asymmetrical anomaly pattern militated against the Vine-Matthews-Morley hypothesis]. We told Heirtzler and Le Pichon that they were probably wrong, that they could not disregard the magnetic reversals as a possibility in the ordering of the magnetic anomalies. They wouldn't buy it, and it didn't make a bit of difference what we said" (t. 2, s. 1, v-11-79). Heirtzler later called in Joe Worzel and showed him the profiles, reversing them to emphasize the bilateral symmetry. Pitman recalled: "Worzel looked at it for a while and finally said, 'Well, that knocks the seafloor-spreading nonsense into a cocked hat.' I said, 'What do you mean, Joe?' He said, 'It's too perfect,' and walked out of the room" (Pitman, t. 1, s. 1; t. 2, s. 1, iv-24-79). H. William Menard related a similar episode (1979, p. 28): "I had occasion to talk to Francis Birch [of Harvard, an outspoken opponent of drift] at a meeting sometime in the mid fifties, and he remarked that if convection really existed, there would be a systematic pattern of regions with high and low heat flow. Our *Downwind* [oceanographic] expedition was accordingly designed to test this hypothesis. . . . After harrowing months [Richard] von Herzen got his heat-flow data, and they were in a pattern with very high values on the crest of the East Pacific Rise and low values on each flank [which was in keeping with Arthur Holmes's convection theory in support of continental drift]. I showed a map of our results to Birch and you can imagine my reaction when, after inspecting it quite carefully, he said, 'Wrong pattern.'"

‡Recall that the chronology of reversals based on terrestrial rocks did not extend beyond approximately 4 million years due to the limits of resolution of potassium-argon dating.

Heirtzler called in Jack Oliver and Lynn Sykes to examine the *Eltanin* 19 profile and asked why, if the seafloor was in motion, didn't "we hear more seismological noise as the seafloor moves." The profile apparently prompted Sykes to set aside his research at the time to pursue first-motion studies, by which he confirmed the character of transform faults as Tuzo Wilson

Walter Pitman remembered of this period that "Heirtzler was very supportive; he got the money and the data to begin with and pushed us hard. It took a long time for us to realize the magnitude of what was happening. During that period Le Pichon stimulated us by forcing us to compete with him; Opdyke stimulated us by understanding exactly what we had in hand and knowing intellectually exactly what it meant and also pushing us."[37]

Fred Vine's arriving at Lamont in February 1966 was like a starving man let loose in the kitchen of the Cordon Bleu. When Pitman and Heirtzler showed him the *Eltanin* Leg 19 profile, he immediately knew that in terms of proving the Vine-Matthews-Morley hypothesis "it was all over but the shouting."[38] Vine, perhaps better than anyone else, realized the implications of that profile and how it might best be articulated with other available data in support of his hypothesis. He did so brilliantly in the December 16, 1966, issue of *Science*.

Walter Pitman recounted the way in which Fred Vine acquired the powerfully persuasive *Eltanin* 19 profile:

> Vine saw the *Eltanin* 19 profile in February 1966 and went away. Heirtzler and I began to write our paper up in about May or June of 1966; we had it pretty much in the shape in which it appeared in *Science* [Dec. 2, 1966]. Vine was coming up to Lamont from Princeton to give a Friday afternoon colloquium talk in early May or June 1966, after the April 1966 A.G.U. meeting; Heirtzler was away, and I was a little worried that Vine might possibly scoop us. So I finished off a redraft of the paper Heirtzler and I were writing and sent it to Vine by special delivery, so that he got a copy of the paper before he came up here to talk—which was a smart thing, because when he gave the talk, there it was, the paper that he eventually published in *Science* [Dec. 16, 1966]. I had not known if Heirtzler had given Vine the *Eltanin* 19 profile, which he saw early in the year, or not. I went back and looked through the files and found a letter from Heirtzler to Vine giving him the data, but he was very vague about it. Heirtzler said, "Here is the data, but please keep in mind that several of my students are using it in their theses."* Vine was obviously preparing a large paper based on the *Eltanin* 19 profile.

---

had postulated in 1965 (Heirtzler, t. 1, s. 1, viii-16-76; Pitman, t. 2, s. 1, iv-24-79). Sykes reported his results at both the G.S.A. meeting in San Francisco and the Goddard Symposium in New York in November 1966 (Sykes, 1967, 1968); he showed tensional movement at the ridges, lateral motion on the active parts of the transforms, and seismic inactivity along the transform inactive extensions.

*Pitman's account was essentially corroborated by Heirtzler (t. 1, s. 2, viii-16-76), who recalled his concerns about the use to which Vine intended to put the *Eltanin* 19 data and how soon he planned to publish. Heirtzler had not consulted with Pitman before he gave Vine the profile.

So there we were, up a tree. Heirtzler was coming back very shortly, so I said to Vine, "Maybe we ought to think of publishing together," but he declined. Very shortly after that, we went down to Princeton to confer with Vine about how we were going to settle all of this. I think basically Vine didn't want to be a coauthor with us for a lot of good reasons. His paper went into much more depth about the implications of this than we did; and again, it was because the guy had been thinking about it all these years. If [the Vine-Matthews-Morley hypothesis] were true, what did this [*Eltanin* 19 profile] mean? Well, God, *he* knew what it meant—why should he share that with anybody? After some arm-wrestling with Vine and with *Science*, it was agreed that we would publish [two weeks] before he did; his would be a major article. *Science* wanted to publish both papers together in one issue, which would have been very logical from a purely editorial point of view. But on the other hand, the guys that run *Science* understand geopolitics, and once that was explained they said "We don't like it, but we'll do it."

As a general rule in this business, you never give data away with strings attached to it, you just don't give it away, and you don't have to apologize to people. The *Eltanin* 19 was the whole ball of wax. You can't give an intelligent man a piece of data and tell him not to work on it. You just can't do it, particularly if the guy is a scientist; particularly a young guy like Vine, who had had his brains beaten out by that time. He was really searching around. I must say that I learned something from it: Data is either given away with no strings or not given at all.[39]

Thunder at the A.G.U.: Trilogic Triumph

At the April meeting of the American Geophysical Union in Washington, D.C., Allan Cox was chairman of the session in which James Heirtzler first publicly displayed the *Eltanin* 19 magnetic anomaly profile. Heirtzler showed the profile next to its mirror image, to show the bilateral symmetry clearly, and compared both with the computed profile made by assuming that the rocks were magnetized in accordance with the potassium-argon/polarity-reversal time scale, as it was then known (see Fig. 8.3). A spreading rate of 4.5 centimeters per year on each side of the East Pacific Rise was indicated.

Cox thought seeing the *Eltanin* 19 profile "a truly extraordinary experience." Shortly before the meeting, while having a beer with Heirtzler in the hotel bar, Cox had seen the profile and immediately recognized the Jaramillo event. He recalled:

The *Eltanin* 19 profile really has everything on it that we found in all our work on reversals. When it came out, it had things I knew were there and

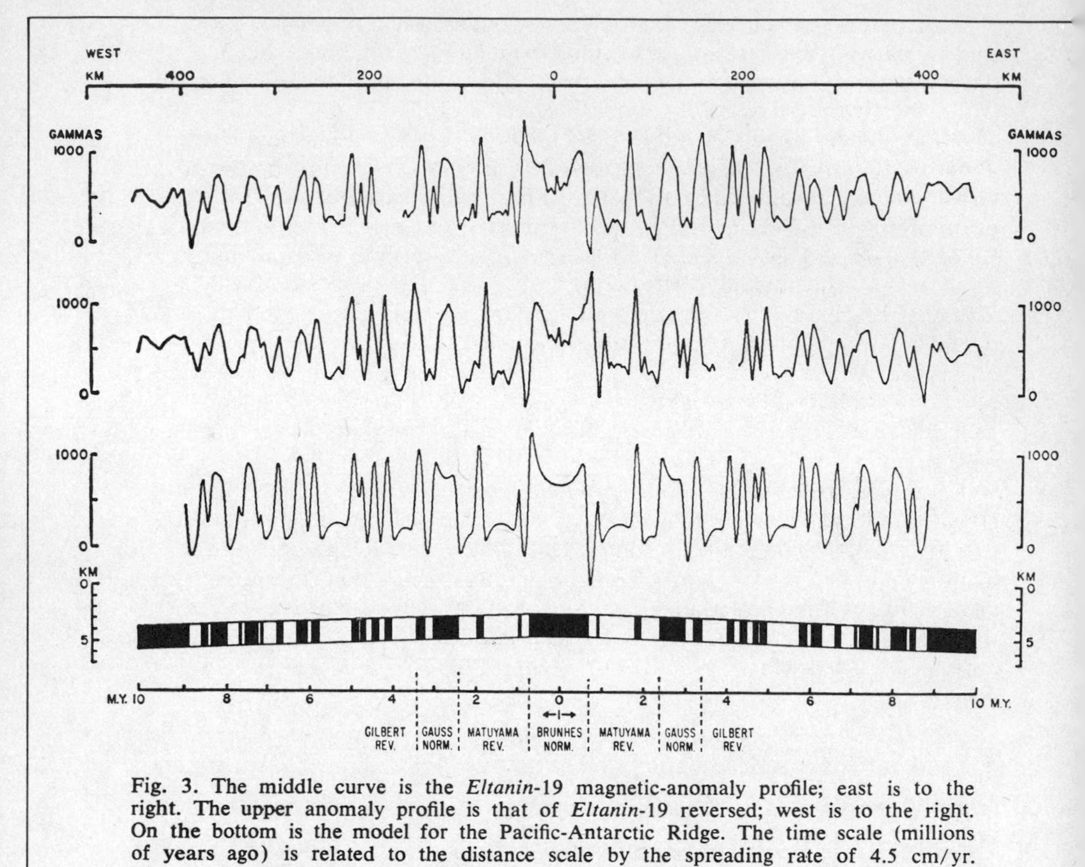

Fig. 3. The middle curve is the *Eltanin*-19 magnetic-anomaly profile; east is to the right. The upper anomaly profile is that of *Eltanin*-19 reversed; west is to the right. On the bottom is the model for the Pacific-Antarctic Ridge. The time scale (millions of years ago) is related to the distance scale by the spreading rate of 4.5 cm/yr. The previously known magnetic epochs since the Gilbert epoch are noted. The shaded areas are normally magnetized material; unshaded areas, reversely magnetized material. Above the model is the computed anomaly profile.

Fig. 8.3. The *Eltanin* 19 magnetic-anomaly profile, as it appeared in figure 3 of Pitman and Heirtzler's December 2, 1966, paper in *Science*.

things I thought were probably there, including short polarity intervals slightly older than the Olduvai event. The potassium-argon dates at the beginning of the Olduvai event are more inconsistent than they should be, in view of what we know of the dating errors. This led me to suspect that one or more short events, slightly older than the Olduvai, were fuzzing up the boundary. *Eltanin* 19 shows a big event, the Olduvai, and then slightly older than that on both flanks of the Rise, two little blips come in—I think they're both real. I said so in an article shortly afterward. There was so much happening all at once. That was the most exciting year of my life, because in 1966, there was just no question any more that the seafloor-spreading idea was right.[40]

Neil Opdyke recalled: "I saw the stunned look on Dick Doell's face; he was sitting in the lab outside my office that April in 1966 when we gave those talks in Washington. Doell looked at the magnetic stratigraphy in the cores [soon to be discussed] and at the *Eltanin* 19 profile and said, 'It's so good it can't possibly be true, but it is.'"[41]

In their short paper of December 2, 1966, which followed from that presentation at the A.G.U. meeting, Pitman and Heirtzler applied the Vine-Matthews model, which assumed that the alternately polarized crustal blocks matched the observed positive and negative peaks, and also the assumed spreading rate of 4.5 centimeters per year on the East Pacific Rise. Such a procedure permitted the polarity-reversal scale to be extrapolated across the entire survey area. The resulting model then became the standard against which they examined the profile across the Reykjanes Ridge; a good match between the two spreading centers resulted when a spreading rate of 1 centimeter per year was assumed for the Reykjanes Ridge.

Fred Vine, like Pitman and Heirtzler, wrote his landmark paper of December 16, 1966, with two key components in hand: the refined polarity-reversal scale (see Fig. 8.4), which included three polarity events (the Olduvai, Mammoth, and newly found Jara-

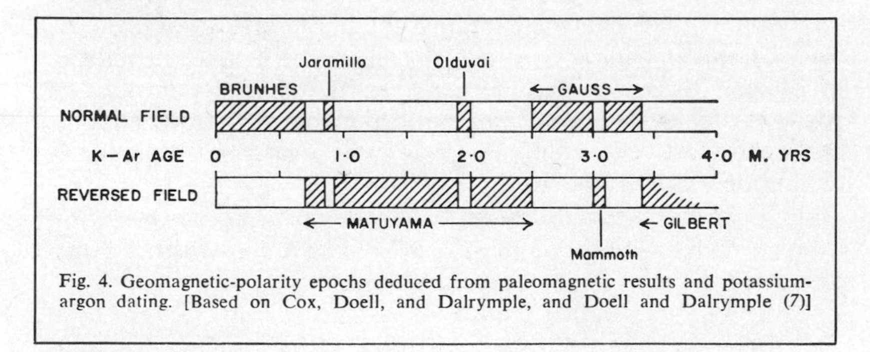

Fig. 4. Geomagnetic-polarity epochs deduced from paleomagnetic results and potassium-argon dating. [Based on Cox, Doell, and Dalrymple, and Doell and Dalrymple (7)]

Fig. 8.4. Geomagnetic polarity-reversal scale eleven, containing the newly discovered Jaramillo event, as depicted by Vine in his landmark paper of December 16, 1966, in *Science*.

millo), and the *Eltanin* 19 profile. As earlier agreed upon, Vine's paper appeared in *Science* two weeks after the much briefer one of Pitman and Heirtzler.

Vine attended the April A.G.U. meeting with a preprint of that paper. In it he compared the magnetic anomalies from various mid-ocean ridges and rises (Juan de Fuca, Gorda, East Pacific, Reykjanes, Carlsberg, Mid-Atlantic, and Red Sea) with anomalies calculated from models of the seafloor that assumed rocks of alternately normal and reversed polarity that matched the geomagnetic reversals dated in the potassium-argon/polarity-reversal time scale (see Figs. 8.5–8.9). In a masterful integration, he touched on all of the factors that contributed to or were the consequences of the assumption that

> the entire history of the ocean basins, in terms of ocean floor spreading, is contained frozen in the ocean crust. The hypothesis is supported by the extreme linearity and continuity of oceanic magnetic anomalies and their symmetry about the axes of ridges. If the proposed reversal time scale for the last four million years is combined with the model, computed anomaly profiles show remarkably good agreement with those observed, and one can deduce rates of spreading for all active parts of the mid-oceanic ridge system for which magnetic profiles or surveys are available. The rates obtained are in exact agreement with those needed to account for continental drift.[42]

The completeness and detail of the *Eltanin* 19 profile led Vine to suggest, in answer to the long-standing central objection to his hypothesis, that the change in character of the anomalies from the axial zone to the flanks might reflect a higher frequency of reversals before about 25 million years ago. Thus, narrower blocks would be produced, with a resulting decrease in the bulk magnetization of adjacent blocks. The transition from the axial zone to the flanks could therefore reflect both a change in intensity and a change in the frequency of reversals.

The polarity-reversal time scale was indeed the key to the meaning and interpretation of a very great, global array of magnetic data preserved in ocean floor crust and sediment.

Walter Pitman recalled that "the push really didn't begin until after that A.G.U. meeting in April 1966; at that time we began to see the anomalies [number] 25 to 32 at the oldest end of the time scale, which we could recognize all through the Pacific. We returned from that meeting and began to put the whole Pacific to-

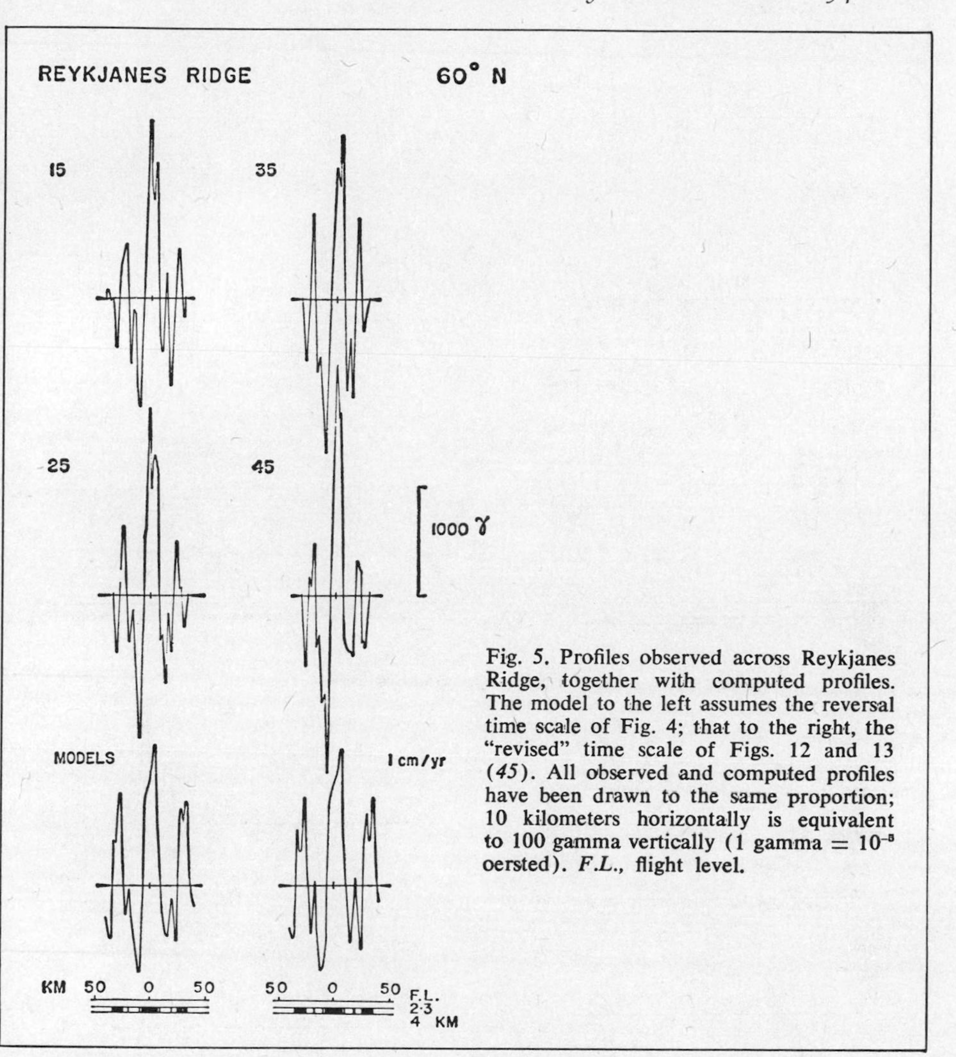

Fig. 5. Profiles observed across Reykjanes Ridge, together with computed profiles. The model to the left assumes the reversal time scale of Fig. 4; that to the right, the "revised" time scale of Figs. 12 and 13 (*45*). All observed and computed profiles have been drawn to the same proportion; 10 kilometers horizontally is equivalent to 100 gamma vertically (1 gamma = 10^{-5} oersted). *F.L.*, flight level.

Fig. 8.5. Observed and computed magnetic-anomaly profiles across the Reykjanes Ridge, shown as figure 5 of Vine's December 16, 1966, paper in *Science*.

gether. We took a cut at the South Atlantic and Indian Oceans and began to recognize the pattern everywhere."[43]

A series of papers by the scientists at Lamont followed shortly;[44] they demonstrated repeatedly that the symmetrical magnetic anomaly pattern along the mid-ocean ridges (both active and dor-

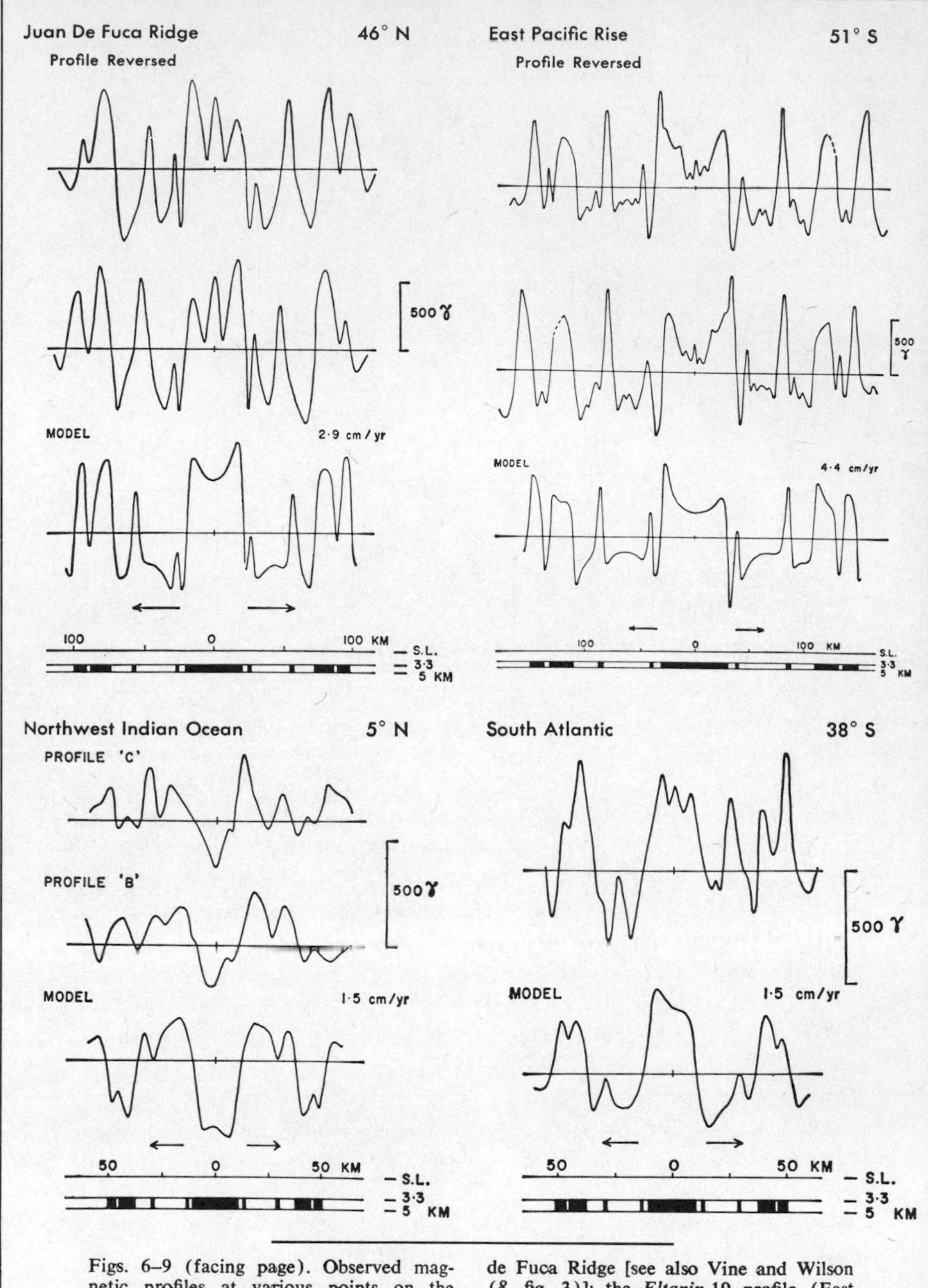

Figs. 6–9 (facing page). Observed magnetic profiles at various points on the midocean ridge system are compared with simulated profiles based on the reversal time scale of Fig. 4, a constant rate of spreading, and the model outlined in the text (*47*). The observed profiles are taken from Raff and Mason (*27*, pl. 1) for Juan de Fuca Ridge [see also Vine and Wilson (*8*, fig. 3)]; the *Eltanin*-19 profile (East Pacific Rise) of Pitman and Heirtzler (*23*, fig. 2); the Owen profiles (northwest Indian Ocean) of Matthews, Vine, and Cann (*46*, fig. 2); and the Zapiola-2 profile (South Atlantic) of Heirtzler and Le Pichon (*15*, fig. 1). *S.L.*, sea level.

Fig. 8.6. Vine's figures 6–9 compared magnetic profiles across the Juan de Fuca Ridge, the East Pacific Rise, the Northwest Indian Ocean Ridge, and the South Atlantic Ridge (from *Science*, December 16, 1966).

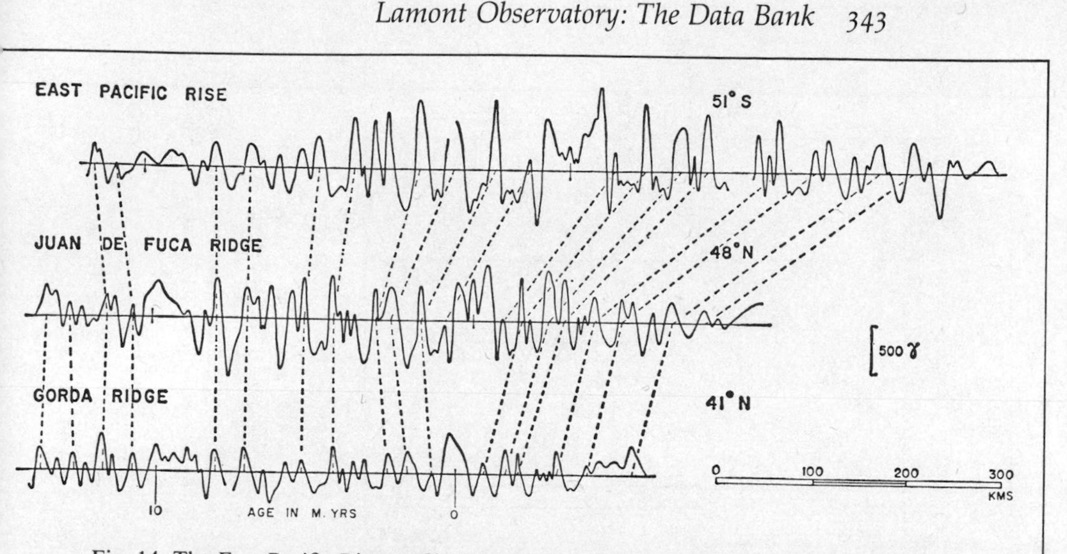

Fig. 14. The East Pacific Rise profile *Eltanin*-19 [Pitman and Heirtzler (*23*)] compared with a composite profile across and to the northwest of Juan de Fuca Ridge, and with a profile normal to the strike of the anomalies across and to the west of Gorda Ridge. [The last two profiles from Raff and Mason (*27*) and Vacquier *et al.* (*50*)]

Fig. 8.7. Vine's figure 14 correlated the *Eltanin* 19 profile from the East Pacific Rise with profiles from the Juan de Fuca and Gorda ridges (*Science*, December 16, 1966).

mant seafloor-spreading centers) was present over vast portions of the South Atlantic, Indian, North Pacific, and South Pacific Oceans. These papers also made clear that the pattern is essentially identical in all of the different oceanic regions, and that it is produced by the same mechanism that generates the alternate strips of normally and reversely magnetized basalt. Furthermore, although each spreading center has opened at a unique rate, thus producing a stretched pattern at a fast spreading center and a compressed pattern at slower ones, the various magnetic profiles could be matched perfectly. *By the use of the terrestrially derived polarity-reversal time scale, the Lamont group could determine accurately the rates of crustal spreading beneath the sea*. By extrapolating that rate, with the implicit assumption (open to great question at that time) that spreading had occurred at a fixed rate during the last 80 million years, and using the profiles from the South Atlantic Ocean as the most trustworthy standard, Heirtzler, Dickson, Herron, Pitman, and Le Pichon proposed, in 1968, the remarkable polarity-reversal scale shown as Fig. 8.10.

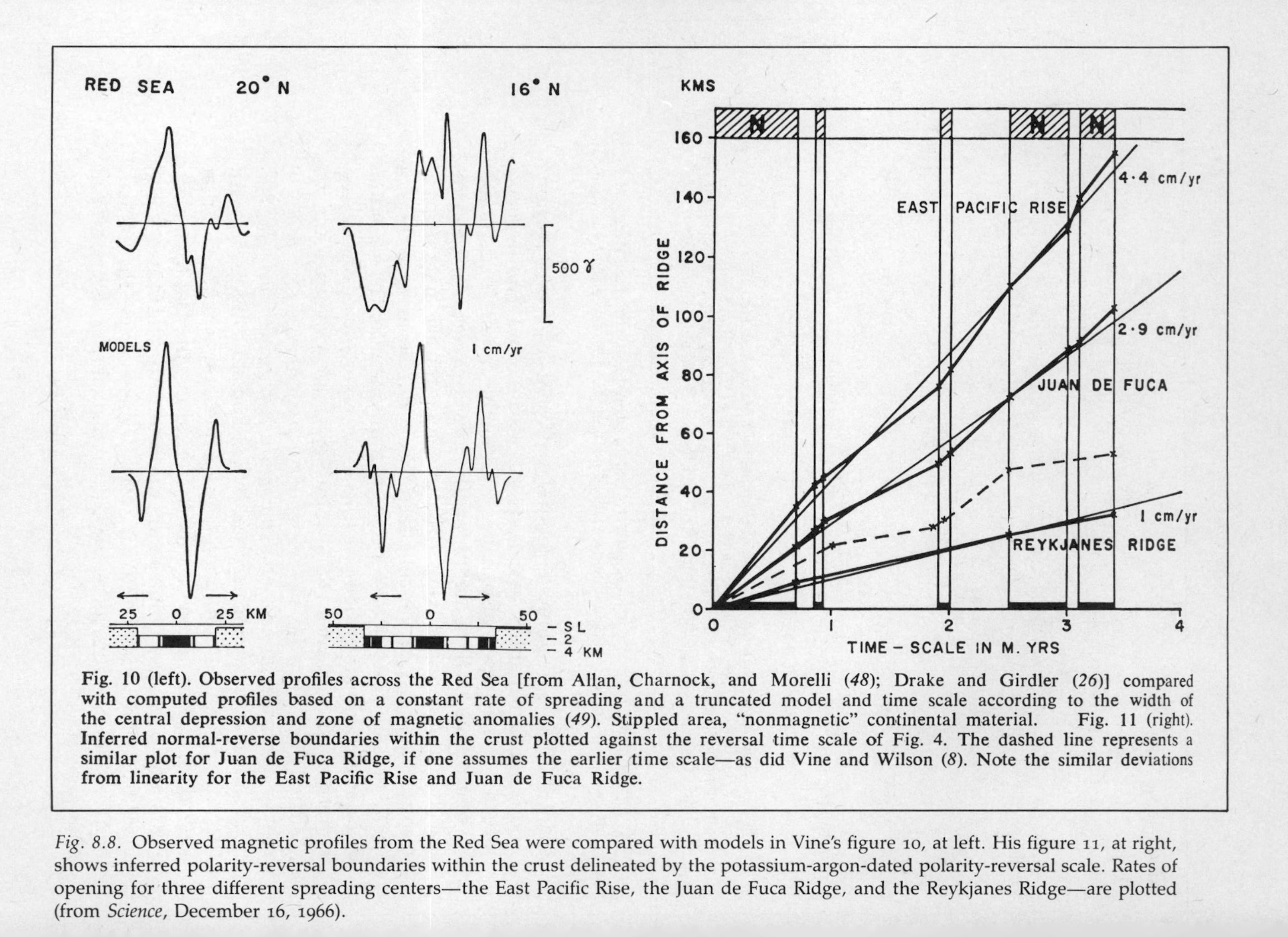

Fig. 10 (left). Observed profiles across the Red Sea [from Allan, Charnock, and Morelli (*48*); Drake and Girdler (*26*)] compared with computed profiles based on a constant rate of spreading and a truncated model and time scale according to the width of the central depression and zone of magnetic anomalies (*49*). Stippled area, "nonmagnetic" continental material. Fig. 11 (right). Inferred normal-reverse boundaries within the crust plotted against the reversal time scale of Fig. 4. The dashed line represents a similar plot for Juan de Fuca Ridge, if one assumes the earlier time scale—as did Vine and Wilson (*8*). Note the similar deviations from linearity for the East Pacific Rise and Juan de Fuca Ridge.

Fig. 8.8. Observed magnetic profiles from the Red Sea were compared with models in Vine's figure 10, at left. His figure 11, at right, shows inferred polarity-reversal boundaries within the crust delineated by the potassium-argon-dated polarity-reversal scale. Rates of opening for three different spreading centers—the East Pacific Rise, the Juan de Fuca Ridge, and the Reykjanes Ridge—are plotted (from *Science*, December 16, 1966).

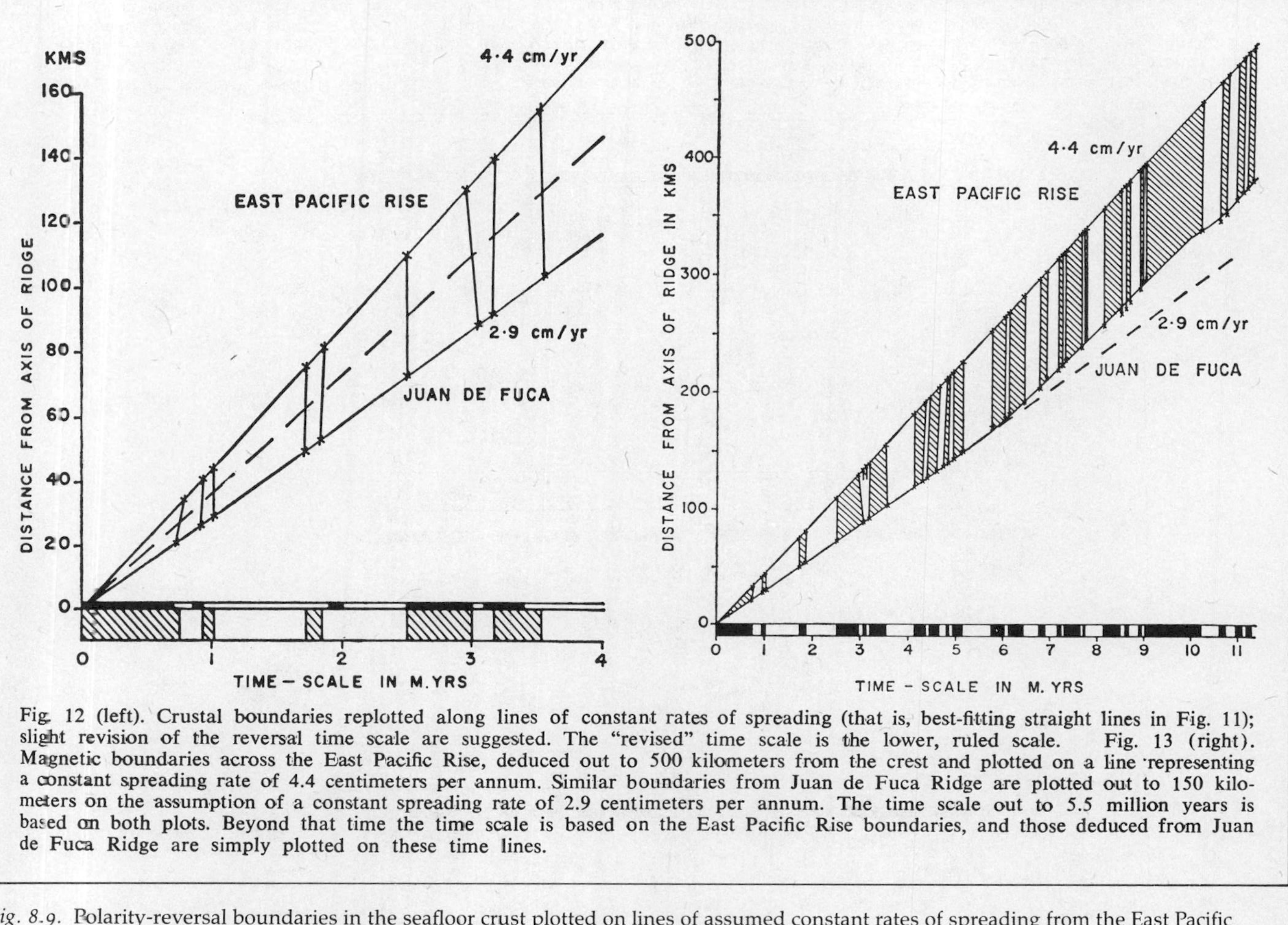

Fig. 12 (left). Crustal boundaries replotted along lines of constant rates of spreading (that is, best-fitting straight lines in Fig. 11); slight revision of the reversal time scale are suggested. The "revised" time scale is the lower, ruled scale. Fig. 13 (right). Magnetic boundaries across the East Pacific Rise, deduced out to 500 kilometers from the crest and plotted on a line representing a constant spreading rate of 4.4 centimeters per annum. Similar boundaries from Juan de Fuca Ridge are plotted out to 150 kilometers on the assumption of a constant spreading rate of 2.9 centimeters per annum. The time scale out to 5.5 million years is based on both plots. Beyond that time the time scale is based on the East Pacific Rise boundaries, and those deduced from Juan de Fuca Ridge are simply plotted on these time lines.

Fig. 8.9. Polarity-reversal boundaries in the seafloor crust plotted on lines of assumed constant rates of spreading from the East Pacific Rise and the Juan de Fuca Ridge (figure 12 from Vine, *Science*, December 16, 1966).

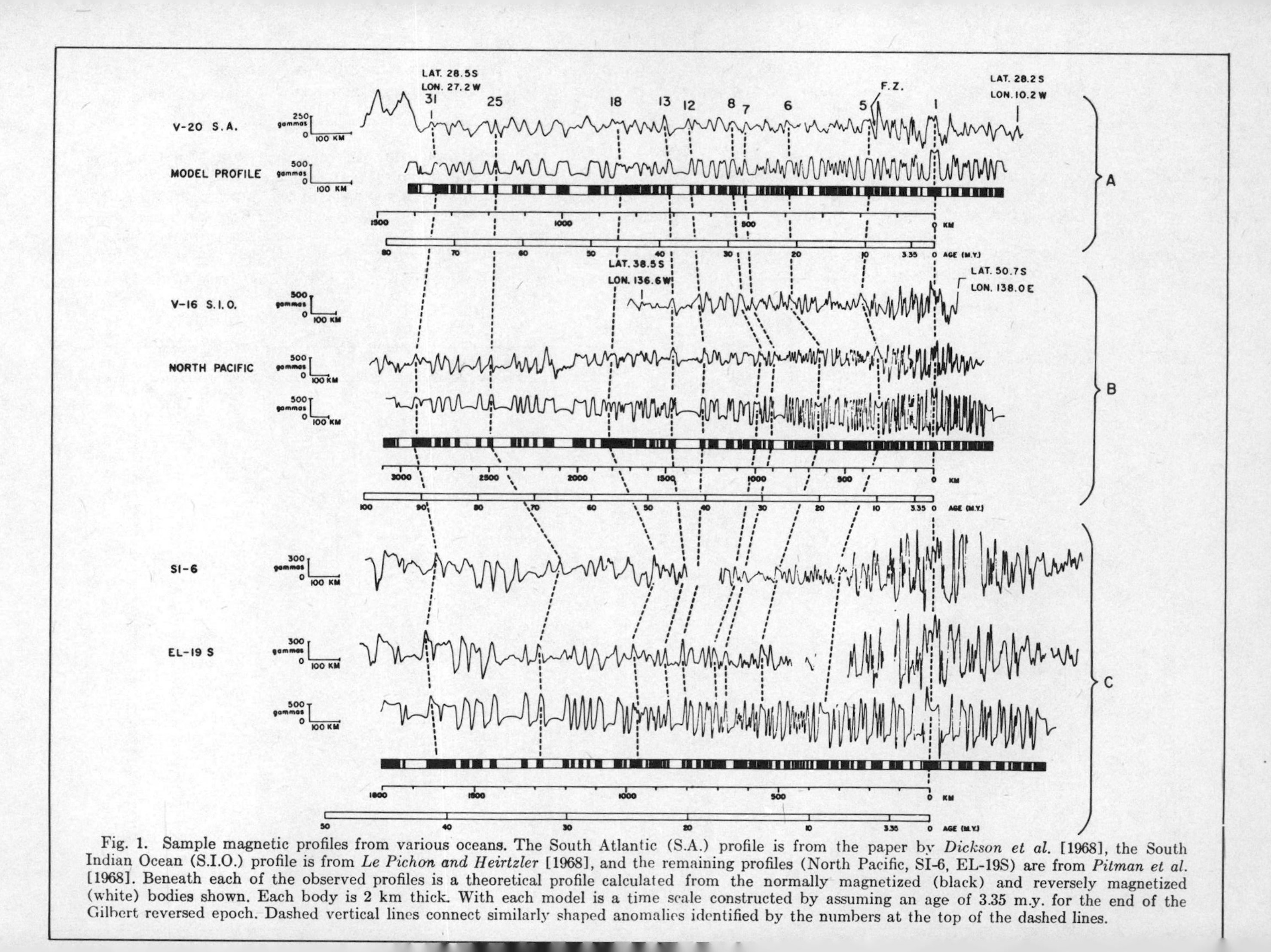

Fig. 1. Sample magnetic profiles from various oceans. The South Atlantic (S.A.) profile is from the paper by *Dickson et al.* [1968], the South Indian Ocean (S.I.O.) profile is from *Le Pichon and Heirtzler* [1968], and the remaining profiles (North Pacific, SI-6, EL-19S) are from *Pitman et al.* [1968]. Beneath each of the observed profiles is a theoretical profile calculated from the normally magnetized (black) and reversely magnetized (white) bodies shown. Each body is 2 km thick. With each model is a time scale constructed by assuming an age of 3.35 m.y. for the end of the Gilbert reversed epoch. Dashed vertical lines connect similarly shaped anomalies identified by the numbers at the top of the dashed lines.

The essential accuracy of this time scale has been verified by several subsequent studies. The extended reversal scale interpreted from seafloor magnetic patterns has become a standard by which newly surveyed ocean-floor areas can be dated from magnetic profiles. This revolutionary magnetic tool also permits refined analysis of the structural evolution of the ocean basins, and establishes a more precise schedule for continental drift. More was learned in a few years about the structural evolution of the oceans by magnetic surveys than was learned over two centuries about the structural history of the continents from geologic mapping. No one had ever dreamed that oceanic geology could be so simple. The generation of new oceanic crust at spreading centers and the destruction of old in the deep-sea trenches have produced an endlessly rejuvenated seafloor, generally less than 100 million years old. That crust is magnetically banded in chronologic sequence and covers much of the earth's surface.

Heirtzler's display of the astonishing *Eltanin* 19 profile and Vine's masterful use of it in the interpretation of observed anomalies from several different, widely separated spreading centers had indeed overwhelmed the audience at the A.G.U. meeting; seafloor spreading was suddenly ascendant. A sense of the import of that moment pervaded the audience—any further speaker could only have been anticlimactic, but Neil Opdyke was not!

Although informed in February by Fred Vine on his visit to Lamont that the magnetic event to be named the Jaramillo had been discovered by the Menlo Park group, Neil Opdyke excitedly told Brent Dalrymple and Sherman Grommé in Norman Watkins's hotel suite that his group had discovered a new polarity event. (Recall that the paper describing and naming the Jaramillo was not published until May 20, 1966; thus it was not yet formally named, and Opdyke did not know at that time if further study had been done by the Menlo Park group or if their results were yet in press.) Dalrymple replied, "Too late—I can tell you the name. We've called it the Jaramillo." Opdyke was encouraged to speak at that meeting by the session chairman, Allan Cox.[45] He reported that the Lamont

---

Fig. 8.10. The extended polarity-reversal scale interpreted from seafloor magnetic patterns has become a standard by which newly surveyed ocean-floor areas can be dated from magnetic profiles. Figure 1 of Heirtzler, Dickson, Herron, Pitman, and Le Pichon's 1968 paper in the *Journal of Geophysical Research* presents profiles from various ocean areas, compared with each other and with theoretical profiles; correlation of magnetic reversals among them is clear.

group had successfully determined the magnetism of deep-sea cores and identified polarity reversals, including the Jaramillo event. Thus, the time scale of the reversalists was corroborated from a third independent source.

Although the Lamont group had successfully correlated reversals in deep-sea cores with the potassium-argon/polarity-reversal time scale, they were not the first to attempt to do so. Christopher Harrison, while at Scripps on leave from Cambridge, published the first such attempt in June 1966 (see Fig. 8.11). And just as Vine had been misled by the absence of the Jaramillo event at that early date, in the case of the seafloor magnetic anomalies, so too was Harrison misled in his work with cores. He correlated the second normal interval from the surface (the present) with the Olduvai event instead of with the Jaramillo, the Jaramillo having not yet been named or published; his diagram thus gave the impression that the rate of sedimentation was very sporadic. Although Dalrymple had reported evidence for the Jaramillo on November 4, 1965, at the Kansas City meeting of the G.S.A., Harrison was apparently not aware of it. Harrison's paper was submitted to the *Journal of Geophysical Research* on January 10, 1966, and revised on March 18, only a month before the Jaramillo event was to become a "household" term at the April 1966 meeting of the American Geophysical Union.

Immediately after that momentous meeting of the A.G.U. in April, Opdyke returned to Lamont, and after several months of fever-pitched activity, he, Glass, Hays, and Foster published a most significant and influential paper entitled "Paleomagnetic Study of Antarctic Deep-Sea Cores" in *Science*, on October 21, 1966. They demonstrated that (1) magnetostratigraphy, coupled to biostratigraphic zonation, permitted reliable dating and correlation between widely separated sites; (2) cores generally held a more complete and finely detailed magnetic record of the time interval they represented than was obtainable from other studies; and (3) the cores provided a third independent source of polarity-reversal data; the core data could be matched with the reversals defined by the potassium-argon/polarity-reversal scale and also with the seafloor magnetic-anomaly profiles.

On a sequence of seven cores taken in the Antarctic region, they performed magnetic measurements and biostratigraphic subdivision (based on Radiolaria) demonstrating that the "magnetic rever-

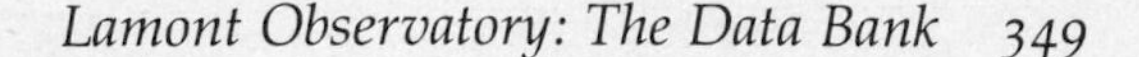

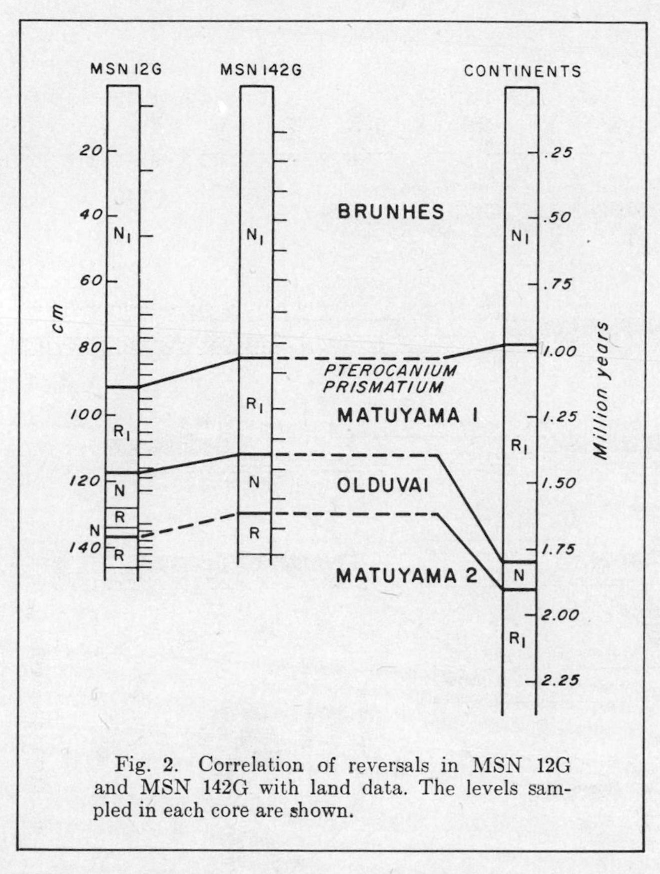

Fig. 2. Correlation of reversals in MSN 12G and MSN 142G with land data. The levels sampled in each core are shown.

Fig. 8.11. Harrison's June 16, 1966, paper in the *Journal of Geophysical Research* presented in figure 2 a remarkable schematic of the first attempt to correlate reversals in deep-sea cores with the polarity-reversal scale.

sals and faunal boundaries are consistently related to each other" (Fig. 8.12). Although the Jaramillo event had early been detected in the cores, it was at first mistaken for the much earlier Olduvai event;* only after the Jaramillo was identified in the *Eltanin* 19 profile, and correlated with the newly announced polarity scale (number eleven) of Doell and Dalrymple, did its position within the

*In one sense, it might be said that Johnson, Murphy, and Torreson (1948) or Harrison and Funnell (1964) recorded the Jaramillo event since their tabulated data of polarity against depth in the cores show proportions that indicate its presence. Of course, that early they, like the Lamont group, were without a temporally framed polarity-reversal scale, and thus their reversal data lacked contextual significance.

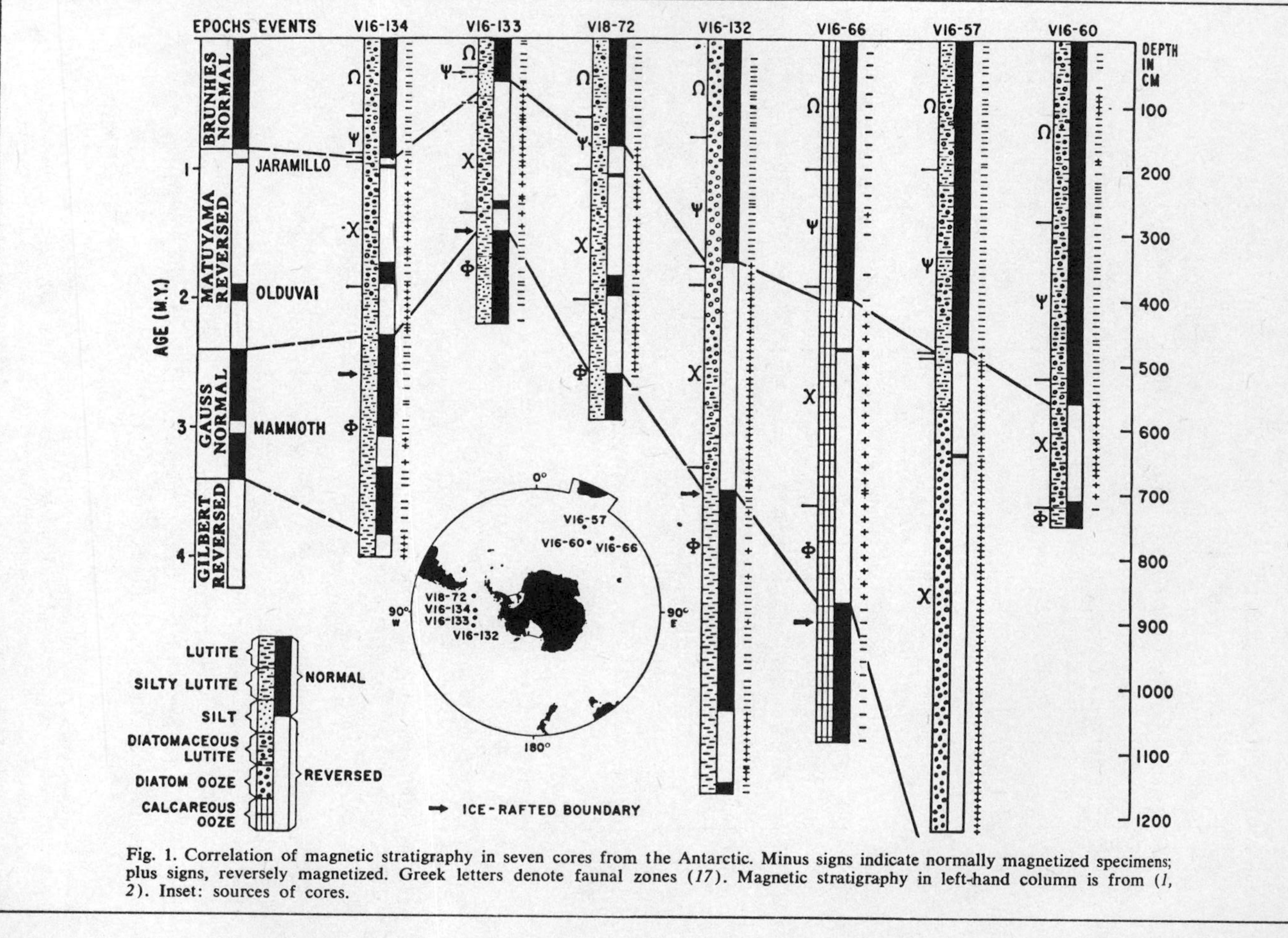

Fig. 1. Correlation of magnetic stratigraphy in seven cores from the Antarctic. Minus signs indicate normally magnetized specimens; plus signs, reversely magnetized. Greek letters denote faunal zones (*17*). Magnetic stratigraphy in left-hand column is from (*1*, *2*). Inset: sources of cores.

Fig. 8.12. A correlation among seven deep-sea cores from the Antarctic, showing that polarity-reversal and microfossil boundaries have a consistent relationship, was presented as figure 1 of Opdyke, Glass, Hays, and Foster's October 21, 1966, paper in *Science*.

cores make sense and provide an important additional measure of proof for seafloor spreading.

Lest their readers be misguided concerning the intrinsic value and potential limitations of their "revolutionary method of dating events in Earth's history" (the subtitle to the paper), the section on "magnetic stratigraphy" opened as follows: "A zone of normally magnetized sediment is present at the top of all cores studied; we believe this represents the Brunhes normal epoch *which has been defined on the basis of K-A dating by Cox et al. . . . We emphasize that the magnetic stratigraphy obtained from deep ocean cores is based on the classical methods of stratigraphy—the law of Superposition.*" *

Twelve days later, on November 2, 1966, Ninkovich, Opdyke, Heezen, and Foster reinforced the Antarctic core results with a second important announcement (in *Earth and Planetary Sciences Letters*) of successful magnetic correlations encompassing 12 deep-sea cores from the North Pacific, work that was alluded to in the paper of October 21. Only four of the cores penetrated the Matuyama-Brunhes boundary: one went into the Gauss epoch, two into sediments representing the Olduvai event, and one reached to the middle Matuyama. They were able to accurately calculate rates of sedimentation and date a volcanic eruption 1.2 million years ago that had left a brown ash bed in three of the cores (Figs. 8.13 and 8.14).

The newly emergent deep-sea core-research program, besides providing a third, independent record of reversal behavior of the Earth's magnetic field, also provided, because of the uniquely complete and detailed record represented by certain cores, data that facilitated refined estimates of seafloor spreading rates.

J. Tuzo Wilson, in reviewing the extraordinary discoveries that confirmed that the record of reversals is preserved in terrestrial igneous rocks, seafloor basalts, and deep-sea sediments, has remarked that these data constitute a revolution in earth science. "Three different features of the Earth all change in exactly the same ratios. These ratios are the same in all parts of the world. The results from one set are thus being used to make precise numerical predictions about all the sets in all parts of the world."[46]

For decades, from 1915 onward, continental-drift theory seemed

*Opdyke et al., 1966, p. 350 (emphasis added). Opdyke and his colleagues mistakenly cited Cox, Doell, and Dalrymple's scale of vi-26-64 (scale seven) as the basis for their placement of the Matuyama-Brunhes boundary. In scale seven, the boundary was still at 1 million years, and Opdyke et al. showed it at 0.7 million years.

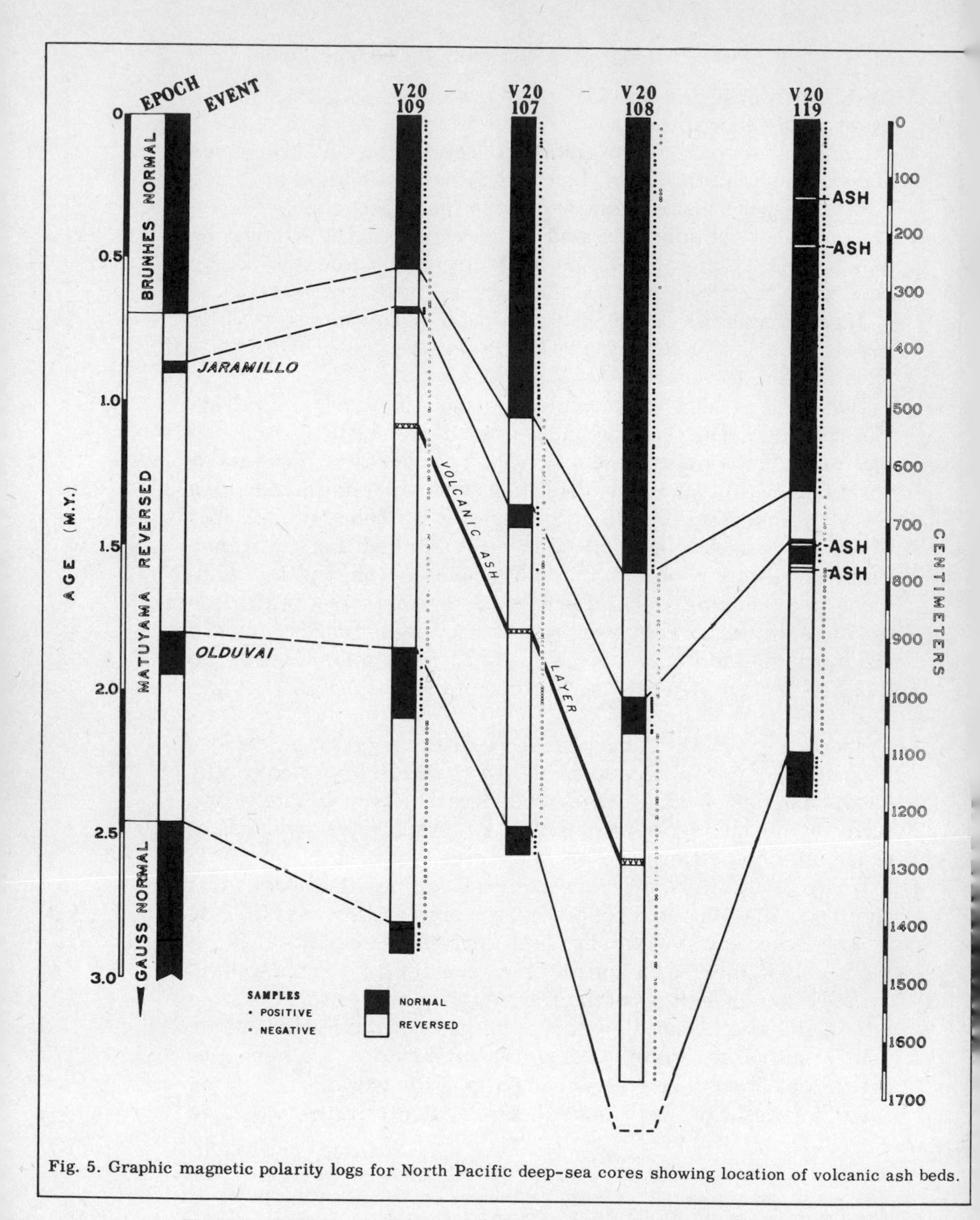

Fig. 5. Graphic magnetic polarity logs for North Pacific deep-sea cores showing location of volcanic ash beds.

Fig. 8.13. Magnetic polarity reversals correlated among four deep-sea cores, as shown in figure 5 of Ninkovich, Opdyke, Heezen, and Foster's November 2, 1966, paper in *Earth and Planetary Sciences Letters*; the potassium-argon-dated polarity-reversal scale at the left serves as the standard of reference.

Table 2
List of piston cores studied.

Core	Lat. N	Long.	Length (cm)	Depth (m)	Earliest Paleomagnetic epoch or event penetrated	Number of volcanic ash layers	
						colorless	brown
107	43°24'	178°52' W	1282	5872	Olduvai		1
108	45°27'	179°14.5'W	1671	5625	Matuyama		1
109	47°19'	179°39' W	1452	5629	Gauss		1
119	47°57'	168°47' E	1170	2739	Olduvai	4	
120	47°24'	167°45' E	1632	6216	Brunhes	6	
121	46°58'	164°16' E	1604	5859	Brunhes	7	
122	46°34'	161°41' E	1575	5563	Brunhes	6	
123	46°15'	157°55' E	1360	4903	Brunhes	10	
124	45°50'	154°30' E	854	5534	-	5	
125	43°29'	154°22' E	948	5545	-	7	
126	42°00'	155°52' E	1050	5515	Brunhes	7	
127	40°17'	156°55' E	1150	5583	Brunhes	5	
128	38°47'	157°24' E	1063	5612	-	3	
129	37°41'	156°35' E	1277	5766	Brunhes	4	
130	36°59'	152°36' E	1039		-	3	
131	36°20'	151°00' E	1034	5858	Brunhes	6	

Fig. 8.14. In their table 2, Ninkovich and his colleagues presented the data on which the correlations in Fig. 8.13, above, were based.

a vessel barely afloat battered by storms of criticism from almost every point of the geologic compass. But from among the favorably disposed minority a theory of seafloor spreading emerged by the early 1960's, grown mainly from new data contributed by oceanographers. Within that exciting time from 1966 to 1968, seafloor spreading was proved by the evidence of rock magnetism, and that proof immediately catapulted Wilson's transform fault concept of 1965 (containing incipient plate theory, discussed in the Appendix C) to the forefront. The additional inventive components were then advanced to flesh out the theory of plate tectonics—which had grown from, and now subsumed, seafloor spreading.

Plate theory is an overarching conceptual scheme of heuristic power unprecedented in earth science; it became a mercurially ascendent paradigm. But had there been no potassium-argon/polarity-reversal time scale, how and when might seafloor spreading have been confirmed?

Epilogue

MUCH OF what was undertaken on the road to Jaramillo had nothing to do, initially, with the question of continental drift. In science, quantum leaps are seldom anticipated—and often depend on the coincidence of more modest increments of progress. This study has repeatedly demonstrated progress, breakthroughs, and applications almost wholly incongruent with the early expectations of the researchers.

Among the kinds of invention or discovery often regarded as less than important and thus dismissed are those that entail the extension of knowledge by gradual changes both in existing kinds of instruments and in the established concepts and theoretical influences underlying the instruments.* Although such contributions in science are not likely to directly challenge the ruling paradigms, they often lead to breakthroughs that reveal new areas of ignorance and open investigation into unexpected channels. The work of the young-rock radiometry group appears to be of this kind. John Reynolds's contribution, for example, was largely instrumental. He conceived of no new theories or principles, but he recognized prospects and possibilities, and by incorporating several existing technical features with ingenuity and daring, he built a static-mode mass spectrometer that counted isotopes at an undreamed-of level of refinement.†

The Reynolds mass spectrometer was first applied to geological

*Kuhn, in *The Structure of Scientific Revolutions* (1970), sees such activities as unimportant.

†Early in his mass-spectrometry efforts, Reynolds was viewed askance by his physics colleagues; the dating of rocks was a "boundary" problem but considered mainly geological, and he was urged to pursue studies more clearly within physics (unattributed by request).

problems with the aid of Robert Folinsbee at Berkeley. Soon the work of Garniss Curtis and Jack Evernden allowed the dating of increasingly younger rocks. Their refinements in techniques and instruments enabled them to make profound contributions in dating; and their contributions prompted conceptual upheavals in both geology and anthropology. Although they were part of an academic center where research in paleomagnetism was being actively pursued and the magnetic reversal question discussed, Curtis and Evernden were not at first interested in applying their skills, then unique in potassium-argon dating, to little-known and unpromising studies of polarity reversal. When the first polarity-reversal time scale had at last been formulated by the Rock Magnetics group in June 1963, they recognized the challenge; but until then, and although the range of application of the dating tool had been seen by all as almost limitless, no one at Berkeley or anywhere else had envisioned the role that potassium-argon dating was to play.

And that was the picture generally. Mason and Raff, who had produced the first large-area magnetic anomaly maps of the seafloor in the 1950's, were not concerned with the resolution of the drift question in the formulation of their studies. And Vacquier, whose structural interpretation of Mason and Raff's maps led to his pioneering demonstration of seafloor mobility, was not directly concerned with attempts to decide the reversal question. His early, casual speculation on the possible origin of the magnetic stripes on the ocean floor strongly hinted at the Vine-Matthews-Morley hypothesis, but when later confronted with their idea, Vacquier rejected it.

Ironically, even the definitive act in the confirmation of seafloor spreading took place in the most unlikely, hostile theatre. That most of those in the Lamont group controverted drift is now well known. Maurice Ewing, whose influence within and beyond Lamont was profound, was largely responsible for those views. Whereas most others who had shared his fixist views were dislodged by the powerful magnetics evidence in 1966, his dedication to the fixist cause precluded his conversion to seafloor spreading for three more years. Bruce Heezen was one of the rare mobilists at Lamont, but although he provided components for and impetus to the derivation of Harry Hess's seafloor-spreading model, he was not conceptually disposed to the early acceptance of subduction-zone kinematics because of his subscription to expanding-earth

Many of the figures in this story were reunited in 1970 at a field trip on Hawaii during the 3rd U.S.–Japan Conference on Paleomagnetism. A new order prevailed—no longer were young geophysics aspirants advised, as they had been for so long, to eschew paleomagnetism in favor of the more practical and rewarding subjects of seismology and gravity. Paleomagnetism had launched the revolution—its community of practitioners was growing rapidly. Among the conferees were Neil Opdyke (standing, extreme left), Richard Doell (3rd from left, seated), Takesi Nagata (4th from left, seated), John Verhoogen (top row, seated, 2nd from right), Brent Dalrymple (seated in front of Verhoogen), and Sherman Grommé (bearded, seated, at right).

theory. He nonetheless immediately recognized the value of (and moved, perhaps indiscreetly, to capitalize on) the first success of John Foster and Billy Glass in their magnetic determinations of oceanic sediment cores.

James Heirtzler was accumulating and systematically storing the crucial magnetics data sought by Vine, but was clearly against drift and openly disdainful of the Vine-Matthews-Morley hypothesis. Just before the revelation of the *Eltanin* 19 profile, he had interpreted his magnetics data as militating against Vine-Matthews-Morley, and when shown the *Eltanin* 19 profile by Walter Pitman and Neil Opdyke, he did not come to accept its true significance until three weeks later. Pitman, a student at Columbia who was neither informed of the hypothesis nor disposed toward a purposeful search for data applicable to it, came into possession of the *Eltanin* 19 profile only fortuitously. But Opdyke, trained as a directionalist, immediately saw its significance and urged Pitman to make the most of it. Opdyke had been a sustained, outspoken champion of continental drift and knew about the Vine-Matthews-Morley hypothesis; the stamp of his English mentors was manifest. By the same token, he had no high regard for magnetics studies of deep-sea cores because polarity-reversals were not at the time given priority in directionalist research. And in any case, the record of such studies was not encouraging. He thus held little hope for Foster's early efforts in deep-sea cores and did not enter into them himself until February 1966, and then only because he recognized that "the magnetic stratigraphy of the cores could be used to verify the pattern that Pitman was getting in the magnetic anomaly profiles."[1]

Vine and Matthews and Morley had been audacious in combining questionable, unproven components (the hypotheses of seafloor spreading and magnetic field reversals among others) into a still more questionable hypothesis. Vine and Matthews in fact did not at first have much faith in the validity of the hypothesis. They regarded it as more likely only after discovering evidence from the Juna de Fuca Ridge, off Vancouver Island, but were not wholly convinced until the appearance of the *Eltanin* 19 profile. The hypothesis had indeed been poorly received and largely ignored, and Vine and Matthews received encouragement from very few; those few were of course sympathetic to drift and seafloor spreading, but not all those sympathetic to drift regarded the highly questionable hypothesis with favor. Lawrence Morley related his version of the hypothesis privately to Harry Hess and Keith Runcorn at the international geophysical meeting in August 1963 at Berkeley. Hess was "fascinated" and took Morley's name and address; but Runcorn, the principal figure among the directionalists, was "completely uninterested."[2] John Verhoogen, long sympathetic to continental drift (and published in support of mantle convection) did not initially regard the Vine-Matthews-Morley hypothesis as probable. And although Edward Bullard received the hypothesis with enthusiasm and spoke favorably of it to many, he was not moved to accept Fred Vine's offer of coauthorship at the time Vine wrote the first draft at Cambridge.

Some of the reversalists had taken very faltering first steps. As early as 1957 (six years before the first scale by the Rock Magnetics group), the exciting possibility arose in Edmonton to begin reversal time scale experiments there. Robert Folinsbee's radiometry group was producing reliable dates on young rocks and Jan Hospers, a Cambridge-trained Dutch paleomagnetist, voiced interest in reversal studies to Folinsbee. But nothing came of it and Hospers did not again pursue the question.

In 1959, Martin Rutten, another Dutch geologist, broadly accomplished, published the first attempt to define a polarity-reversal time scale by radiometry. Very astutely, he had matched a series of potassium-argon dates (indirectly provided by Evernden and Curtis) with his own simple (and seemingly hurried) field determinations of polarity. However, his pioneering effort was not followed up, likely because he lacked and could not gain access to the capabilities in radiometry and rock magnetism prerequisite to success in such an effort.

The directionalists, meanwhile, had been making impressive progress. During the middle and late 1950's the polar wander data of the paleomagnetic directionalists had increasingly suggested continental displacements and were clearly the most important of the several lines of evidence that were attracting converts to the small but growing drift community. Expectations quite justifiably grew that these studies, begun and mainly conducted by Englishmen, amongst whom Keith Runcorn came to play a dominant role, would likely decide the question of continental drift; but they never did.

The English directionalists had long been convinced of field reversals, but generally did not view efforts to base a time scale on them as deserving priority. Still, their work gave impetus to the development of paleomagnetic research abroad, including that at Berkeley under Verhoogen. Allan Cox and Richard Doell were in large measure the intellectual heirs of those pioneering Englishmen. Berkeley, like most other institutions in the United States during the 1950's and early 1960's, had no research programs directed toward resolution of the drift question until Verhoogen implemented rock-magnetics studies with Doell as his first student. But Doell's earliest efforts were not in reversals; he was interested mainly in directionality, chemical remanence, and secular variations. Cox began directionality work shortly after Doell, but was discouraged by his experience with the Siletz River volcanics of Oregon; because they had been rotated on a structural block, they yielded data incongruent with directionalist conclusions. That experience in fact served chiefly to heighten his dedication to reversal studies.

In the "Review of Paleomagnetism" of 1960, Cox and Doell were sympathetic to drift but surmised that the directionality data did not constitute conclusive supporting evidence for it; that surmise appears to have furthered their receptivity to reversal studies. Why and how the two, imbued with the mainly directionalist values of the tiny paleomagnetism community, alone chose instead to address the reversal question has already been shown in part. Cox had the essential "intuitive sense of the importance of the reversal problem and was the major driving force behind its pursuit,"[3] but Doell's contribution to the program of experiments appears no less important. He served a broad range of functions including selection and planning of experiments, derivation and calculation of data, and designing and building instruments; not least important, he was highly regarded as an administrator by the diverse group of the Rock Magnetics Project who served under him. Brent Dalrymple's efforts were also varied: selecting samples for dating and magnetic examination, designing experiments, acquiring and analyzing data, and writing a great number of papers. His consistently reliable isotopic determinations of young rocks were clearly vital to the success of the program.

Ian McDougall's start in reversal studies was prompted by the work of the Rock Magnetics group at Menlo Park and thus was necessarily later than theirs, but once in his own laboratory (assembled for John Jaeger by Jack Evernden) he provided leadership and continuity of effort among the Australians and engendered a spirit of competition with the Californians that appears to have heightened productivity in both research centers. He was among the pioneers who successfully dated seemingly intractable, whole-rock basalt samples. McDougall and Don Tarling provided two of the crucial early versions of the polarity-reversal time scale, and it was of course scale number eleven, formulated by the Menlo Park group and containing the Jaramillo event, that was used with such astounding success in the interpretation of the marine magnetic anomaly profiles.

The contributions of both the directionalists and reversalists were extolled at the dinner given at Columbia University on April 8, 1971, on the occasion of the award of the Vetleson Prize. Allan

After the revolution in geology was well under way, in 1969, the contract to determine the thermoluminescence and paleomagnetism of the first rocks brought from the moon was awarded to the Rock Magnetics Group at Menlo Park by NASA. Those moon rocks, from the Apollo XI mission, had been sent directly to the main vacuum chamber at the Lunar Receiving Laboratory near Houston, Texas; there, Doell, Grommé, and Dalrymple (left to right, above) and also Edward Mankinen had installed a 3-axis fluxgate magnetometer (designed by Doell and built by Nate Sherrill) for immediate magnetic measurements. Later, other studies were done on the moon rocks at Menlo Park.

Cox, Richard Doell, and Keith Runcorn were the recipients. The development of paleomagnetics was reviewed by Maurice Ewing, who in 1955 had been the first recipient of the prize:

[Cox and Doell] began their studies of the magnetic properties of rocks at a time when two decades of rapid acceleration of work in the subject had culminated in interpretation of unexpected directions of remanent magnetization in rocks being made in terms of recurrent geologic hypotheses of continental drift and polar wandering. . . . On the one hand there was evidence that some natural lavas and synthetic rocks are magnetized in the

Allan Cox, Richard Doell, and Keith Runcorn (from left to right), receiving the Vetleson Prize from the late Maurice Ewing at Columbia University on April 8, 1971.

direction opposite to that of the earth's field in which they are formed. On the other hand there was the radical suggestion that the main magnetic field had reversed from time to time—radical because, as I said before, at this stage no one knew why the earth was a magnet anyway. . . . Cox and Doell established the reality of the reversals of the magnetic field in the earth. By measuring remanent magnetization in the series of volcanic flows in widely separated parts of the earth; by enlisting the aid of Brent Dalrymple to measure the age of each sample by analysis of the potassium and argon content; by showing that the alternations between normal and reversed magnetization occurred at the same time in widely separated places of study, thereby they established an absolute time scale for geologic events that is valid for the entire planet. This removed the doubts that had restricted many scientists, and the studies in paleomagnetism surged ahead. . . . Cox and Doell removed the last barrier to progress.[4]

Polykarp Kusch, who shared the Nobel Prize in 1955 with Willis E. Lamb, Jr., for the determination of the magnetic moment of the electron, presented the prize, remarking of Cox and Doell that:

Your pioneering work on the magnetic properties of rocks, your careful choice of significant rocks for study, [and] your demonstration of the reality of the reversals of the main magnetic field of the earth . . . have earned you a permanent place in the annals of earth science. Your timed sequence of reversals . . . provided strong support for the hypothesis of seafloor spreading once it was recognized that the ocean lithosphere was impressed with the record of field reversals. That the influence of your discoveries was felt in so many directions is [borne out by] the rapid evolution of earth science from an immature, largely observational status to that of a mature science with a substantial body of basic theory.[5]

Kusch then remarked of Runcorn that:

Your fundamental work on the earth's magnetic field and its past configuration, your skill and persistence in paleomagnetic measurement and interpretation when others despaired of making order of a great complexity of data, and your demonstration that polar wandering tracks are consistent with continental drift, have helped to free the imagination of earth scientists and set them on a fruitful course.[6]

The prize, awarded on behalf of the Vetleson Foundation and the Trustees of Columbia University, consisted of a gold medal, a bronze replica, a certificate, and cash in the total amount of $30,000; Runcorn received half the cash award and Cox and Doell the other.

Cox and Doell and their reversalist colleagues were sympathetic to drift, and Runcorn and the directionalists worked assiduously to demonstrate its reality; but none of them envisioned the application of reversal studies to decide the questions of drift and seafloor spreading. They were amazed at the way in which the reversal scale was put to use in deciphering the oceanic magnetics data, an application that triggered a scientific revolution.

Cox has remarked: "It was a perfect case of serendipity. . . . No central committee planning the future of earth science could conceivably have guessed that this would happen."[7]

Appendixes

Appendix A

A Note of Concern

G. B. DALRYMPLE wrote me the following note on April 17, 1981:* "Whether or not the reversal experiment could be done in the present-day Geological Survey, I don't know, of course, but I strongly feel that it probably could not be. The attitudes and environment that allowed and even encouraged Allan [Cox], Dick [Doell], and me to pursue basic research full-time for five or six years at what was then a considerable cost, generally don't exist today. We are caught up in the 'mission' and forced to seek short-term 'relevant' results. We still do basic research, of course, but on a smaller scale and often as a sideline—we tend to pick at it more than plunge in, and the results show it. If today I proposed devoting the entire laboratory and staff to a single experiment like the reversal experiment for a period of five years, I would almost certainly be told that it is not possible. I can't think of any current program that would fund such a 'useless' and costly experiment. I think that this is a consequence of the change in the national attitude toward science that started about a decade or so ago. Relevance. 'Let's stop fiddling around and put science to work doing something useful.' In my view, this arises from a mistaken impression of what science is and does best, from a lack of understanding that basic research always pays off economically in the long run, and from a failure to recognize that serendipity plays a major role in nearly all important scientific discoveries. We, the Survey and the nation, are spending the principal, and in a few decades there may not be any interest to collect. It is a horrible, horrible mistake. It doesn't surprise me at all

*Dalrymple has since June 1, 1981, been Assistant Chief Geologist, Western Region, U.S. Geological Survey.

that the U.S. is rapidly relinquishing the world leadership in new patents.

"That was predictable ten years ago given the change in attitude toward research. Unfortunately, the time constant between basic research and economic benefits is quite long, and by the time the politicians and administrators figure out what's happening, the quick fix will not work. It will take several decades to recover.

"I doubt very much if Allan, Dick, or I would have been hired by the Survey had the present practices and attitudes toward research prevailed in the late 1950s and early 1960s. Indeed, I'm not sure any of us would have applied—I'm quite sure I wouldn't have. As you well know from your interviews and history, administrators like Jim Balsley and Bob Moxham had an instinct for encouraging good basic research and either the license to do so or the willingness to do it anyway. That's what Allan, Dick, and I were hired for, that's what we did, and you know the result. 'Andy' (Charles) Anderson, who was Chief Geologist when I was hired and approved my hiring, told me that he felt it was very important that the Survey hire and support five or six good young researchers each year whether they were needed or not! Today, people are hired for a program or mission and usually work on that program or on other programs. As a result, the Survey is gradually becoming mediocre as a scientific research organization. I don't think that our present Survey administrators are necessarily bad. They are helplessly trapped in a system brought about by a change in public attitude toward science, and are not bold enough to do anything about it. Perhaps there is nothing they can do—that's what I'm told—but I choose to believe that basic research is a demonstrably worthwhile endeavor and could be sold even in today's climate." Dalrymple's opinions were shared by a great number of senior geologists and former administrators within the Survey.

Appendix B

Parody on Privilege

THE U.S. GEOLOGICAL SURVEY has long held "Pick and Hammer Shows," at which individuals and groups are lampooned in song; elaborate printed programs contain diverse, high-spirited parodies. Such shows originated in Washington, D.C., early in this century and were begun at Menlo Park in March 1956. They served to vent rebellious feelings, publicize complaints, puncture pomposity, roast administrators and notable individuals, and often to affirm the values of the old-guard majority in the face of encroaching new research programs. David Hopkins (xi-16-79) and others, told me that "there has been, up until recently, the mystique that getting away from field mapping was getting away from reality." The widely prevalent opinion regarding the Rock Magnetics Project may be read in the following selections from the program of the show in March 1963 at Menlo Park (Document No. 37). Cox and Doell were treated at great length; virtuous field geologists trumpeted their disdain for "black-box" practitioners. (Authors are always anonymous.)

DOIN' FIELD GEOLOGY
(Reprise)

Doell and Cox and their black box
Have been exposed as phonies.
This will end their perfidy.
No more magnetology.
No more magnetology.
Wand'ring poles don't fill the holes
In geologic mappin'.

Lab men all have come to grief.
Dibblee is the new big chief,
Dibblee is the new big chief. [Thomas Dibblee
was eminent for field mapping.]

Oh, you don't have to know where the poles have been
Just to split the Triassic from the Pleistocene.
Oh, you don't have to have a field-free space
Just to map the position of a thrust-fault trace.
That's geology,
Field geology.
Sell off surplus equipment
Like magnetometers.
Don't even keep a remnant
Like a gradiometer.

We'll reclaim the Survey's fame,
But not with these black boxes.
We'll map it slick with lens and pick,
Doin' field geology,
Doin' field geology.

AROUND THE WORLD FOR FREE
(Around the World)

DI POLE [Doell] and SMALL POX [Cox]:
Around the world we search for core.
A sample here, a sample there, is what we're looking for
We scarcely care where 'ere we go—
To Timbuctoo, Kalamazoo, or even Kokomo.
We might have been in isotopes,
Or geo-chem, or beating brush along Pacific slopes.
From old Runcorn we took a pregnant cue,
And I found my world with you.

DI POLE:
I've just returned from Waikiki.
Like all the brass, I went first class, as you can plainly see.
I lolled upon the sunlit sands,
I drank my lunch of Planter's punch, and heard the hula bands.
I rode the surf and dived for pearls,
And in between I had some time to spend with hula girls.
Rem'nant magnetism is the key!
Oh, it's the Survey life for me!

SMALL POX:
You think you've had it really grand;
I flew Pan-Am, to old Japan, that lovely geisha land!
And then I pulled another trick:
I sold old Andy [C. A. Anderson] on a junket up to Reykjavik!
I think if I put up a fuss,
What do you know! I'll get to go to the Galapagos!

Rem'nant magnetism is the key!
It surely beats geology.

BOTH:
Rem'nant magnetism is the key!
To get around the world for free!

I DON'T WANT TO BE A FIELD MAN
(After an old English army ballad)

I don't want to be a field man.
I don't want to hike the hills.
I'd rather lounge around,
My feet up off the ground,
Twisting on the dials of weird black boxes.
I don't want to bruise my little tootsies
Walking up a hillside in the heat.
I'd rather be a lab man,
A rich and jolly lab man.
In Doell's lab my life will be complete.
De-Gauss me!

I'll learn a lot of mathematics.
With formulae my brain will be replete.
I'll learn to orient
The magnetic remanent
In the leaning tower of Pisa.
After I have passed their course at Stanford,
I'll get on full-time with Cox and Doell.
Then I'll be on the wagon
With no one's feet a-draggin'.
Their gravey train's my fondly cherished goal.
De-Gauss me!

When I get to be a full-time lab man,
I'll theorize with symbols on a net.
I'll build a big machine
That costs a lot of green
(No one'll know it's my hi-fi set).
Oh! How I want to be a lab man.
I hope that next year's budget has the loot
For me to take a junket
To Egypt or Nantucket
And at a higher grade, to boot.

Appendix C

Tuzo Wilson Conceives of Transforms and Hotspots

J. TUZO WILSON'S contributions to the scientific revolution are numerous. Although two of those contributions lie to the side of the road to Jaramillo and could not reasonably be treated within the main body of this book, they are profound ideas that helped to shape the course of earth science. His concept of transform faults was the bridge by which he took us from seafloor spreading into plate tectonics theory. The idea of hotspots was an important element in corroborating plate movement and in assessing the direction and speed of plate movement. The idea has also triggered new lines of inquiry into processes within the mantle.

After an interview with him in 1980, I felt compelled to include here his remarks on the origins of those two concepts—hence this appendix.

Transform Faults

Wilson postulated the transform fault in his epochal paper of July 24, 1965, as the third boundary of a rigid lithospheric plate; in that postulate clearly lies the origin of plate tectonics theory:

> Seismic activity along [mountains, mid-ocean ridges, or major faults with large horizontal movement] often appears to end abruptly, which is puzzling. . . . This article suggests that these features . . . are connected into a continuous network of mobile belts about the Earth which divide the surface into several large rigid plates. Any [large structrual] feature at its apparent termination may be transformed into another feature of one of the other two types. . . . At the point of transformation the horizontal shear motion along the fault ends abruptly by being changed into an expanding tensional motion across the ridge or rift with a change in seismicity.

Recall that in 1966, about one year after Wilson's contribution of the transform concept, Lynn R. Sykes at Lamont-Doherty Geological Observatory demonstrated by first-motion studies of seismic waves that movement on transforms was as had been suggested by Wilson, thus validating the transform hypothesis and lending additional support to the seafloor-spreading construct.

In another important paper, D. P. McKenzie and R. L. Parker in 1967 carried the application of Euler's theorem beyond that of Bullard[1]; with regard to motions on a sphere, they showed that Wilson's 1965 concept of transform faults articulated with their "essential additional hypothesis . . . that individual aseismic areas move as rigid plates."* Further evidence for spreading was provided at Lamont by Oliver and Isacks in 1967; their study of anomalous seismic zones suggested that the mantle held underthrust slabs of lithosphere beneath island arcs. In 1968 several inventive contributions extended Wilson's concept by treating plate number, geometry, and kinematics in detail: these included papers by Morgan, by Le Pichon, and by Isacks, Oliver, and Sykes.

In 1980 Wilson recalled that he had the "idea of plates very firmly and the idea of boundaries, and of [the various types of transform faults]"; but in answer to my question if he had considered "an Eulerian solution in order to move a rigid slab about the surface of a sphere," he replied, "No, I don't think I did."[2] Wilson was a staunch fixist, and was not converted to drift until 1959 by

> the papers of Irving, Creer, Runcorn, and others on paleomagnetism. Toronto, Cambridge, and Princeton [where he had been a student] were dogmatically against Wegener. . . . They said it was nonsense. I did field work in the Precambrian, on the Canadian Shield, every summer that I was not in Europe, in Nova Scotia and Montana. I could see that faults moved, and the plates on either side were quite rigid. . . . I was born in Ottawa, where you can see the Ordovician [rocks] overlapping the Precambrian. I'd been at many other places where the Precambrian is overlapped by Paleozoic and Mesozoic rocks. [They] show every evidence of complete stability: there are very few faults that cut across that boundary. . . . I was aware that shields are stable and have been stable for immense lengths of time. They are not disturbed or deformed at all; they are absolutely rigid. I'm sure that would have influenced my thoughts on plates. . . . I was aware of how extremely stable the Canadian Shield is; that is largely what prevented me from accepting the idea of continental

*Although Morgan published in 1968, after McKenzie and Parker, he appears to have independently arrived at that important conclusion by at least the same time, or possibly earlier (unattributed by request).

drift, apart from the fact that I was trained not to be a drifter. [I was] also helped by Udintsef's map of the Pacific, which Bullard had pasted on the wall of the lab [at Cambridge during Wilson's sabbatical in 1965]. It was a much better map than had existed before. . . . It showed the Pacific Rise with some offsets. I was trying to rewrite Jacobs, Russel, and Wilson [*Physics and Geology*, the first edition written during his fixist stage] . . . and was interested in great faults [Wilson and Scheidegger, 1950; Wilson, 1962a, b, c, etc.]. I'd also read E. M. Anderson's [1951] book on faulting; he had tried to explain how transcurrent faults petered out by horsetailing. . . . I put together lists of all the big faults of the world and looked in books on structural geology at great length. [Walter] Bucher [at Columbia] had said once, when I was a student at Princeton, that the earthquakes indicated that the San Andreas fault of California went some distance beyond Cape Mendocino and stopped abruptly, and then he said that it was odd that the quakes stopped abruptly. By this time it was clear, from Runcorn's book of 1962 on continental drift, that there were also faults and earthquakes off the Queen Charlotte Islands, and there was a gap between the two. When I plotted the magnetics and earthquakes together . . . on a simplified version of the Raff and Mason magnetics map . . . it was apparent that in the gap there was a pattern of [magnetic] anomalies that were at an angle to the other anomalies, and it [the pattern in the gap] was symmetrical. This later tied in with the ridge on [H. William] Menard's map [of 1955]. I recognized that the Juan de Fuca Ridge must be a spreading ridge, and that the Queen Charlotte Islands and San Andreas faults were shear faults. This then would explain the area of the transform fault: this is where the big faults turn into a spreading center.

Now there are some complications. The Gorda Rise [a ridge lacking an axial rift valley] I recognized also; it's just a repetition of the same idea. I showed this to Vine, because I was aware that he was working on ocean ridges. I didn't know about his and Matthews's paper [of 1963], but he immediately first recognized it. He said we can take that the same way that he and Matthews had done it in the paper in *Nature*. So we wrote a second paper in *Science* in 1965. . . .

[Based on] magnetics, it was already known [as shown by Vacquier] in '59 that there was a 900-mile offset [on the seafloor]. There was the problem of this huge fault, the Mendocino escarpment, coming in and hitting the coast. People like Gilliland [1964] had tried tracing [seafloor] faults across the United States. I was interested in solving that problem. . . . I remember I started, not with a folded piece of paper, but had two L-shaped pieces, which I played with and recognized . . . that you get a shear with opening and spreading. You get a shear [in two places] when you move the L-shaped pieces apart and together: this is a spreading ridge, with a shear at either end. [That] led to the idea of the transform fault. I wrote the paper on transforms about three weeks before it was published in *Nature*."[3]

Within months of Wilson's transform paper, in 1965, Alan M. Coode published a short "Note on Oceanic Transcurrent Faults" in the *Canadian Journal of Earth Science*. Using Sykes's report of 1963 that "earthquakes in fracture zones and epicenters are almost always contained wtihin the limits of the separated ridges," Coode correctly defined a ridge-to-ridge "transcurrent" fault, with the correct sense of motion, as did Wilson. However, Coode did not give it a new name, nor did he treat the notion of plates or carry the transform concept further than a refutation of Gilliland's (1964) view that simultaneous activity has occurred along the thousands of miles of the Murray and Mendocino faults to produce displacements of hundreds of kilometers. Here Coode clearly recognized quiet, extinct fault scars as the extensions of the seismically active ridge-to-ridge faults.

Hotspots

Wilson, although like Hess a fixist until the end of the decade, when once converted, energetically advanced other ideas concerned with mobilism that are still alive. In two papers in 1963, he used volcanoes of the Hawaiian chain as evidence for seafloor spreading. He suggested that Hawaii sits atop a cylindrical thermal plume or rising column of hot material from deep within the mantle. Such plumes, Wilson thought, remain fixed in position for 100 million years or more; as tectonic plates drift over them, they produce age-graded lines of volcanoes, called thread ridges or nemataths. Wilson's hotspot idea prompted Jason Morgan to propose in 1971 that the thermal plumes beneath hotspots were a part of convective systems "in the lower mantle, which provides the motive force for continental drift." Subsequent papers by Morgan and others have bolstered Wilson's original idea. The subject is now both credible and hotly debated, especially as regards the number of extant hotspots (Young summarized the development of hotspot theory in 1980). Wilson recalled:

> I was impressed by the idea of island arcs. I substituted for Merle Tuve, who'd had a heart attack, on a team of the National Academy of Sciences going to the Antarctic in 1960. On the way back, we stopped in Hawaii and visited the Upper Atmosphere Observatory on Mauna Loa at 12,000 feet. We were sitting on a volcano, and I thought about this and knew about Hess's paper. . . . All [the islands] had the same geology: the ABC pattern was repeated in each island, but it didn't mean that all A's and all B's and all C's were concurrent; it meant that every island had the ABC history.

Each island had a similar history, but they had not necessarily occurred at the same time. That was [James] Dana's idea, but [Harold] Stearns [1946] and the majority felt that this had all occurred at once, and Hess had a fault [running] through all of them. You could see the [islands] disappearing into shoals and into atolls, so obviously [the islands were] getting worn down and older.

So I had this idea and I published it. The *Journal of Geophysical Research* turned it down, so I sent it to the *Canadian Journal of Physics*; then I moved on to something else. Morgan revived the idea eight years later; nobody had written anything in between. The hotspot paper [the term was coined in 1968 by Eric Christofferson, now at Rutgers] was written primarily about the Hawaiian Islands, but also about that time I had gotten interested in ocean islands and realized that two-thirds of the earth's surface is covered by ocean, with just these little islands that show you what's going on out there that anybody knew about, because there was very little dredging or coring in those days. One day when I had a cold, I got the index of the *Bibliography of Geology Exclusive of North America*, published by the Geological Society of America. I read the index for all the volumes from about 1930 on up to that date, about 30 years. I discovered that, in the case of the Azores for example, you might find them under "Atlantic Ocean," or under "Portugal," or under "Santa Maria," because [the G.S.A.] had not been uniform in indexing ocean island references. The references were not easily found. I found everything I could about every ocean island, and put together two mimeographed volumes on what was known of their geology and geophysics. I did that in 1963, under a grant from the U.S. Air Force. It was never published [but] went through a number of mimeographed copies—circulars. Menard and other people had copies. [That] compilation gave me all the information available about Tristan da Cunha, and it got me on the ridges and Hawaii and all the other islands. With the exception of one or two, like the Falkland Islands (which were Precambrian), the Seychelles (which were Precambrian), and some larger islands like Ceylon or Madagascar, all the others were very young and tended to get older away from the ridges. I didn't know about [the bends in the island chains] then; I just had the straight lines, but I did have the idea about the bends going two ways from Tristan da Cunha, if they were on a ridge, or one way if not on a ridge.[4]

Reference Matter

Notes

Full bibliographic information on all sources cited is given in the References, p. 393; complete data on all tapes cited are given in the Table of Interviewees, p. 435. See the Note on Citation, p. xvii.

Chapter 1

1. Dalrymple, 1972; Watkins, 1972; Cox, 1973; McDougall, 1977.
2. Dalrymple, 1972, p. 108.
3. Oral comm.: R. R. Doell, iv-78; A. V. Cox, v-78; G. B. Dalrymple, iii-78.
4. See, for example, those of Hospers, Roche, and Einarsson listed in the References.
5. Vetleson Prize Award Dinner, Ewing, t. 1, s. 1.
6. Becquerel, 1896.
7. Rutherford and Soddy, 1902a, b.
8. Rutherford, 1905.
9. Holmes, 1911.
10. Kelvin, 1899.
11. Aston, 1919; Dempster, 1918.
12. J. H. Reynolds, oral comm., ix-77; G. J. Wasserburg, oral comm., iii-78.
13. Rittenburg, 1942.
14. Inghram, 1954.
15. Carr and Kulp, 1957.
16. Dalrymple and Lanphere, 1969.
17. Harper, 1973.
18. Houtermans, in Schaeffer and Zähringer, 1966, p. 5.
19. Pahl et al., 1950; Smits and Gentner, 1950.
20. Gentner et al., 1953, 1954.
21. Marble, 1954, p. 15.
22. Lipson, 1958, p. 139.
23. Marble, 1957.
24. Aldrich and Nier, 1948; Gerling et al., 1949; Gerling and Titov, 1949; Mousuf, 1952; Wasserburg and Hayden, 1954; and others.

25. Oral comm.: J. H. Reynolds, ii-78; G. J. Wasserburg, iii-78; G. H. Curtis, iv-78; and others.
26. F. J. Turner, t. 1, s. 1, ii-1-78.
27. Curtis, t. 1, s. 1, iv-12-78.
28. Reynolds, t. 1, s. 1, i-30-78; Hayden et al., 1949; Inghram, Mess, and Reynolds, 1949; Inghram and Reynolds, 1949, 1950; Reynolds, 1950.
29. Reynolds, t. 1, s. 1, i-30-78.
30. Inghram's double-beta decay experiment: Inghram and Reynolds, 1949. Experiment on xenon: Reynolds, t. 2, s. 2, ii-3-78.
31. Reynolds, t. 1, s. 1, i-30-78.
32. Reynolds, t. 5, s. 1, vi-12-78.
33. May 1951: Reynolds, *Berkeley Lab. Notebook No. 1.*
34. Document No. 1A.
35. Reynolds, t. 5, s. 1, vi-12-78. See also Balestrini, 1954.
36. Cover letter to the proposal of vii-68-55 is Document No. 1.
37. Reynolds, t. 5, s. 1, vi-12-78.
38. Wasserburg, t. 1, s. 1; Reynolds, t. 5, s. 1, vi-12-78.
39. Reynolds, t. 5, s. 1, vi-12-78.
40. Wasserburg, t. 1, s. 1.
41. Alpert, 1953.
42. Reynolds, t. 1, s. 2, i-30-78.
43. Complete letter from Reynolds to Russell is Document No. 2.
44. Reynolds, t. 4, s. 1, iv-12-78.
45. Lipson, t. 1, s. 1, ii-4-78; Kistler, t. 1, s. 2, ii-10-78; J. H. Reynolds, oral comm., vi-12-78; M. Corbett, oral comm., iii-20-78.
46. Document No. 2X.
47. Lipson, t. 1, s. 1, ii-4-78.
48. Lipson, t. 1, s. 2, ii-4-78; Reynolds, 1955 and oral comm., ii-3-78. The failure with the silver coating is described in Document No. 2A, letter from Reynolds to R. J. Hayden of iii-24-55 and Document No. 2B, letter to C. Patterson.
49. Reynolds, t. 5, s. 2, v-14-78.
50. Wasserburg, t. 1, s. 1.
51. Reynolds and Lipson, 1954.
52. Reynolds, 1956a, b; Kistler, t. 2, s. 1, iv-14-78.
53. R. W. Kistler, G. H. Curtis, J. F. Evernden, oral comm., iv-78.
54. Reynolds, t. 1, s. 2, i-30-78; J. J. Lipson, oral comm., ii-4-78; G. H. Curtis, oral comm., iv-19-78.
55. Reynolds, t. 5, s. 1, vi-12-78.
56. Announcement is Document No. 4.
57. Lipson, 1958; Reynolds, t. 5, s. 1, vi-12-78.
58. Reynolds, t. 5, s. 1, vi-12-78.
59. Wasserburg, t. 1, s. 1.
60. Wasserburg and Hayden, 1956.
61. Curtis, t. 1, s. 1, iv-12-78.
62. Letter of iv-29-53, U.S. Dept. of the Interior, National Park Service

(Document No. 4C) and letter of v-26-53 to G. H. Curtis (Document No. 4D).

63. Article from *Berkeley Daily Gazette*, ix-16-53, is Document No. 5.
64. Folinsbee, 1955.
65. Folinsbee, t. 1, s. 1; J. H. Reynolds, oral comm., vi-15-78; J. Verhoogen, oral comm., ix-20-78.
66. Folinsbee, Lipson, and Reynolds, 1955; the abstract appeared on p. 13 of the official program, but only the title, "Further Applications of the Potassium-Argon Method of Age Determination," was published in the proceedings volume (p. 210).
67. Wetherill, Aldrich, and Davis, 1955.
68. Reynolds, 1957.
69. Folinsbee, Lipson, and Reynolds, 1956.
70. Folinsbee, Ritchie, and Stansberry, 1957; Folinsbee and Ritchie, 1957; and others.
71. Folinsbee, t. 1, s. 2; J. H. Reynolds, oral comm., ii-3-78; J. J. Lipson, oral comm., ii-4-78.
72. Reynolds, 1967; Reynolds, Alexander, Davis, and Srinivasan, 1974.
73. Folinsbee, t. 1, s. 2; Lipson, t. 1, s. 1, ii-4-78.
74. Folinsbee and Ritchie, 1957; Folinsbee, Ritchie, and Stansberry, 1957; Folinsbee, t. 1, s. 2; J. J. Lipson, oral comm., ii-4-78.
75. Folinsbee, t. 2, s. 1.
76. Folinsbee, t. 1, s. 2.
77. Folinsbee, Baadsgaard, and Lipson, 1958, 1961; Folinsbee and Baadsgaard, 1958; Baadsgaard and Folinsbee, 1959; Baadsgaard et al., 1959; Folinsbee, 1961; Folinsbee, Krouse, and Sasaki, 1966; and others.
78. Folinsbee, t. 1, s. 2.
79. Folinsbee, t. 1, s. 2.
80. Folinsbee, t. 1, s. 2.
81. Folinsbee, t. 1, s. 2.

Chapter 2

1. Curtis, t. 1, s. 1, iv-19-78.
2. Curtis, t. 2, s. 1, iv-19-78.
3. Curtis, t. 2, s. 1, iv-19-78; Curtis, 1968.
4. F. J. Turner, t. 3, s. 1, vi-21-78; Evernden, t. 1, s. 1, iii-10-78.
5. Evernden, t. 1, s. 1, iii-10-78.
6. Curtis, t. 3, s. 2, iv-26-78.
7. Evernden, t. 1, s. 2, iii-10-78.
8. Evernden, t. 1, s. 1, iii-10-78.
9. Lipson, t. 2, s. 1, ii-4-78; G. H. Curtis and J. F. Evernden, oral comm., iv-78.
10. J. H. Reynolds, oral comm., vi-14-78.
11. Lipson, t. 2, s. 1, ii-4-78.
12. F. J. Turner, oral comm., ii-1-78; G. H. Curtis, oral comm., vi-22-78.
13. Curtis et al., 1956.
14. F. J. Turner, t. 1, s. 1, ii-1-78.
15. F. J. Turner, t. 3, s. 1, vi-21-78.
16. Curtis, t. 2, s. 1, iv-19-78.
17. Curtis, t. 2, s. 2, iv-19-78.

18. Curtis, t. 2, s. 2, iv-19-78.
19. Kistler, t. 1, s. 1, ii-10-78 (quoted). The publication was Evernden, Curtis, and Kistler, 1957.
20. Durham, t. 1, s. 1.
21. Wood et al., 1941.
22. C. A. Repenning, written comm., vii-5-78.
23. Wetherill et al., 1956; Wasserburg et al., 1955a; and others.
24. Wetherill et al., 1955; Folinsbee, Lipson and Reynolds, 1956; and others.
25. Evernden, Curtis, and Kistler, 1957; Evernden, Curtis, and Lipson, 1957.
26. Evernden, t. 3, s. 1, iii-10-78.
27. Curtis, t. 2, s. 2, iv-19-78.
28. Reynolds, 1956b.
29. R. W. Kistler, oral comm., vi-28-78; J. F. Evernden, oral comm., vi-28-78. On the first Pleistocene samples, see Evernden, Curtis, and Kistler, 1957.
30. Carr and Kulp, 1957; Curtis, t. 2, s. 2, iv-19-78.
31. Evernden, t. 3, s. 1, iii-10-78; R. W. Kistler, oral comm., vi-28-78; Curtis, t. 3, s. 1, iv-19-78.
32. Evernden, Curtis, and Lipson, 1957; Kistler, t. 1, s. 2, iv-14-78.
33. R. W. Kistler, oral comm., vi-28-78.
34. Evernden, Curtis, Kistler, and Obradovich, 1960.
35. Curtis, t. 6, s. 1, v-10-78; Evernden and Curtis, 1965.
36. Reynolds, 1956b, p. 139.
37. G. H. Curtis, oral comm., vi-29-78.
38. G. H. Curtis, oral comm., vi-29-78.
39. Evernden, t. 3, s. 1, iii-10-78.
40. Curtis et al., 1956; Evernden, Curtis, and Lipson, 1957.
41. Evernden, t. 5, s. 1, vi-28-78; Curtis, t. 4, s. 2, v-3-78.
42. Curtis, t. 4, s. 2, v-3-78.
43. Keevil et al., 1942; Hurley and Goodman, 1941, 1943.
44. Hart et al., 1960.
45. Erickson and Kulp, 1961.
46. Hart et al., 1960; I. McDougall, oral comm., viii-15-78 (quoted); McDougall, 1961.
47. Bassett and Kerr, 1961; Miller et al., 1962; Miller and Mussett, 1963.

Chapter 3

1. The inauguration of the Miller Institute and award of its first Research Professorships is noted in Documents No. 5A and 5B.
2. Document No. 6, p. 2.
3. For a supporting letter from C. M. Gilbert, Acting Chairman, Geol-

ogy Department, to Professor W. R. Dennes, Institute of Basic Research in Science, Univ. Calif. Berkeley, x-5-56, see Document No. 7; for notice of the award, see Documents No. 5A and 5B.

4. Evernden, t. 3, s. 2, iii-10-78; G. H. Curtis, oral comm., v-10-78; Curtis, t. 2, s. 1, iv-19-78.
5. On the California granitic and sedimentary rocks, see Curtis et al., 1958; on the volcanics, Evernden, Curtis, and Kistler, 1957.
6. Evernden, Curtis, Kistler, and Obradovich, 1960.
7. Evernden, t. 2, s. 2, iii-10-78.
8. Evernden, t. 3, s. 1, iii-10-78.
9. J. F. Evernden, oral comm., vi-26-78.
10. Evernden, t. 3, s. 1, iii-10-78.
11. Curtis, t. 2, s. 2, iv-19-78.
12. S. Washburn, oral comm., vi-26-78.
13. R. W. Kistler, oral comm., vi-27-78.
14. Curtis, t. 3, s. 1, iv-19-78.
15. Letter recommending continuation of appointment as Miller Institute Associate Research Professors from vii-1-59 to vi-30-61 is Document No. 10.
16. E. M. Irving, oral comm., ii-7-78; Curtis, t. 3, s. 1, iv-19-78.
17. F. J. Turner, oral comm., ii-7-78; E. R. Turner, oral comm., viii-7-78.
18. Irving, t. 1, s. 2, ii-7-78.
19. Irving, t. 1, s. 1, ii-7-78; I. McDougall, oral comm., viii-13-78; Richards, t. 1, s. 2; Sass, t. 1, s. 2.
20. F. J. Turner, t. 2, s. 1, ii-1-78.
21. Curtis, t. 3, s. 1, iv-19-78; oral comm., vi-30-78.
22. Irving, t. 1, s. 1, ii-7-78, quoted; corroborated by Richards, t. 1, s. 1, v-1-78.
23. Evernden, t. 4, s. 1, iii-14-78; M. Corbett, oral comm., iii-20-78.
24. Richards, t. 1, s. 1, v-1-78.
25. Evernden, t. 4, s. 1, iii-14-78; Evernden and Richards, 1962.
26. Evernden, t. 4, s. 1, iii-14-78.
27. Among the publications resulting from the pair's collaboration were Evernden and Curtis, 1961, 1965; Evernden, Curtis, Kistler, and Obradovich, 1961; Evernden and James, 1964; Curtis and Evernden, 1962; and Leakey et al., 1961.
28. "Résumé of Research on Potassium-Argon Dating," Document No. 11.
29. E. M. Irving, oral comm., ii-7-78; A. V. Cox, v-78; R. R. Doell, iv-78.
30. Creasey, t. 1, s. 1; P. C. Bateman, oral comm., ii-27-78.
31. Creasey, t. 1, s. 1.
32. P. C. Bateman, oral comm., ii-27-78.
33. S. S. Goldich, oral comm., iii-25-78; L. Pakiser, oral comm., iii-24-78.
34. S. S. Goldich, oral comm., iii-25-78.

35. S. S. Goldich, oral comm., iii-23-78.
36. Obradovich, t. 1, s. 1, iv-14-78.

Chapter 4

1. Brunhes and David, 1901, 1902, 1903, 1905; Brunhes, 1906.
2. Cox and Doell, 1960; Irving, 1964; McElhinny, 1973.
3. Needham, 1962.
4. Smith and Needham, 1967.
5. Smith, 1970.
6. Gilbert, 1958; Smith, 1970.
7. Gellibrand, 1634, *fide* McElhinny, 1973, p. 5.
8. Boyle, 1691, *fide* Koenigsberger, 1936, p. 225.
9. Folgheraiter (1899) gives earlier references in Italian journals relating to this work.
10. Brunhes and David, 1901, 1902, 1903, 1905; Brunhes, 1906.
11. Mercanton, 1926a, b, 1931, 1932.
12. Levinson-Lessing, 1927, *fide* Khramov, 1958.
13. Matuyama, 1929, p. 203–5.
14. Koenigsberger, 1936; he published related studies in 1930, 1934, 1935, and 1938.
15. Khramov, 1958, English translation of 1960, p. 3.
16. Thellier, 1938; Thellier and Thellier, 1951, 1952.
17. Nagata, 1953a, b, c.
18. Thellier and Rimbert, 1955.
19. Johnson et al., 1948.
20. Blackett, 1956, p. 5.
21. These studies were reviewed by Runcorn (1955b).
22. Nagata, 1953a.
23. Blackett, 1956.
24. Brunhes and David, 1901, 1902, 1903, 1905; Brunhes, 1906. Mercanton, 1926a; Matuyama, 1929a, b.
25. Irving, 1964; McElhinny, 1973.
26. R. L. Wilson, 1962b.
27. Gelletich, 1937.
28. Roche, 1950a, b, 1951, 1953, 1956, 1958, 1960.
29. Hospers, 1951.
30. Irving, 1954, 1957a.
31. In Clegg, 1956.
32. Einarsson and Sigurgeirsson, 1955, p. 892.
33. Einarsson, 1957a, p. 233.
34. Einarsson, 1957a, p. 238.
35. Khramov, 1955a, b, 1956, 1958.
36. Sigurgeirsson, 1957, p. 241.
37. Sigurgeirsson, 1957, p. 244–45.
38. Sigurgeirsson, 1957, p. 246.
39. Brynjolfsson, 1957, p. 254.
40. Irving and Runcorn, 1957; Hospers 1953, 1954a, b.
41. Irving and Runcorn, 1957, p. 99.

42. Irving and Runcorn, 1957, p. 99.
43. Blackett, 1956.
44. Graham, 1949, p. 153.
45. Balsley and Buddington, 1954, 1957a, b, 1958, 1960.
46. Nagata, Akimoto, and Uyeda, 1953b; see also Nagata, 1952.
47. Nagata, Akimoto, and Uyeda, 1953a; Nagata, Akimoto, Uyeda, Momose, and Asami, 1954; Uyeda, 1955, 1956, 1957, 1958, and 1962.
48. Uyeda, 1958.
49. Uyeda, 1958, p. 120.
50. Uyeda, 1958, p. 121.
51. Uyeda, 1958, p. 46.
52. Personal comm. of Saito to Uyeda (Uyeda, 1958, p. 48, 123).
53. Kawai et al., 1954.
54. Balsley, Buddington, and Fahey, 1952.
55. Balsley and Buddington, 1954, p. 177.
56. Balsley and Buddington, 1957b, p. 321–22.
57. Kawai et al., 1954, p. 864.
58. Kawai et al., 1954, p. 865.
59. Kawai et al., 1954, p. 867.
60. Kawai et al., 1954, p. 868.
61. Verhoogen, 1956.
62. See also Ishikawa and Syono, 1963.
63. Balsley and Buddington, 1957a.
64. Balsley and Buddington, 1957a, p. 561.
65. Balsley and Buddington, 1957a, p. 561. See Blackett, 1956.
66. Bullard, 1968.
67. Ade-Hall, 1964; R. L. Wilson, 1963; Carmichael, 1961; and others.
68. R. L. Wilson, 1965, p. 381.
69. Bullard, 1968, p. 517.
70. Verhoogen, 1956.
71. Verhoogen, 1956, p. 208.
72. Rutten, 1941.
73. Thiadens, 1970; Eschman, 1974.
74. Rutten, 1969.
75. Eschman, 1974, p. 176.
76. Hospers, 1954a, b; Einarsson, 1957a, b, c.
77. Rutten and Wensink, 1959.
78. Roche, 1950a, b.
79. Rutten and den Boer, 1954, p. 108.
80. Rutten, 1959, p. 373.
81. Rutten, 1960, p. 163.
82. Rutten, 1960, p. 164.
83. Rutten and Wensink, 1959.
84. Rutten, 1960, p. 163.
85. Alvarez, 1975.
86. Blanc, 1958.
87. Blanc, 1958, p. 397–98; translation by Walter Alvarez, iv-78. The term absolute age, commonly used as a synonym for isotopic or radiometric age, is to be avoided since it implies an exactness in dating that is almost never possible.
88. J. F. Evernden, oral comm., vi-15-79.
89. G. H. Curtis, oral comm., iv-12-78.
90. Records examined by R. F. Drake (oral comm., viii-8-78).

91. Evernden, Savage, Curtis, and James, 1964, p. 175, shows that KA 345, run on i-20-59, was a rerun of the earlier KA 334.
92. Rutten, 1959, p. 373.
93. Rutten, 1959, p. 373.
94. Rutten, 1959, p. 374.
95. Blanc, 1957, p. 101–03.
96. Rutten, 1959, p. 374.
97. Marinelli and Mittempherger, 1966.
98. Rutten and Wensink, 1960, p. 63.
99. Rutten, 1958; Rutten and Wensink, 1959.
100. Rutten and Wensink, 1960, p. 62.
101. Rutten and Wensink, 1960, p. 62.
102. Rutten, 1959, p. 374.
103. Evernden and Curtis, 1965; Evernden, Savage, Curtis, and James, 1964.
104. Evernden and Curtis, 1965, p. 360.
105. J. F. Evernden, oral comm., i-79.
106. Blanc, 1957, p. 95.
107. G. H. Curtis, oral comm., iv-19-78; J. F. Evernden, oral comm., iii-14-78.
108. Cox, Doell, and Dalrymple, 1963a, p. 1050.
109. G. B. Dalrymple, oral comm., ii-79.
110. A. V. Cox, oral comm., iv-6-74.

Chapter 5

1. Doell, 1955a.
2. Doell, t. 1, s. 1, iv-17-78.
3. Doell, t. 1, s. 2, iv-17-78.
4. Verhoogen, i-21-69 in Document No. 13.
5. Doell, t. 2, s. 1, iv-17-78.
6. Doell, t. 2, s. 1, iv-17-78.
7. Doell, 1955a.
8. Doell, t. 2, s. 2, iv-17-78.
9. Doell, t. 1, s. 2, iv-17-78.
10. Ruth Doell, t. 1, s. 1.
11. Doell, t. 2, s. 2, iv-17-78.
12. Letter from Doell to Verhoogen of x-15-55 is Document No. 14A.
13. Letter from Doell to Verhoogen of xii-10-55 is Document No. 14C.
14. Ruth Doell, oral comm., ix-16-78.
15. Doell, t. 3, s. 1, iv-17-78.
16. Cox and Doell, 1960, the major review paper on paleomagnetism in the Geological Society of America *Bulletin*.
17. Doell, t. 3, s. 1, iv-17-78.
18. Cox and Doell, 1960, p. 645.
19. Cox, t. 2, s. 1, v-4-78.
20. Doell, t. 3, s. 1, iv-17-78.
21. Doell, t. 3, s. 1, iv-17-78.
22. Emphasis added. Documents No. 17A–17D, correspondence between Balsley and Cox (vi-56 to iv-57), detail the plans and expectations that led to Cox's employment, verified by Document No. 17E.
23. Pakiser commented on his desire to have Doell join the Geophysics Branch of the Survey at that date (oral comm., iii-24-78).

24. Cox, t. 1, s. 1, v-2-78.
25. Standard Form 85, Civil Service Commission, lists Cox's employers from x-44 to ix-50 plus places of residence (Document No. 18A).
26. Cox, t. 1, s. 2, v-2-78.
27. Cox, t. 1, s. 1, v-2-78.
28. Wahrhaftig, t. 1, s. 1, ii-28-78.
29. Wahrhaftig, t. 1, s. 1, ii-28-78.
30. Cox, t. 1, s. 1, v-2-78.
31. Cox, t. 1, s. 1, v-2-78.
32. Employment papers for Cox transmitted by Clyde Wahrhaftig to the Director, 11th U.S. Civil Service region, vi-5-51 and to P. Lewis Killeen, Chief, Alaskan Section of U.S. Geological Survey, Washington, D.C., in Document No. 18C.
33. Verified in Document No. 17F.
34. Cox, t. 1, s. 1, v-2-78.
35. Wahrhaftig, t. 1, s. 2, ii-28-78.
36. Wahrhaftig, t. 1, s. 2, ii-28-78.
37. Wahrhaftig and Cox, 1959.
38. Cox, t. 1, s. 2, v-2-78.
39. Cox, t. 1, s. 2, v-2-78; Hopkins, t. 1, s. 1, iii-2-78; C. Wahrhaftig, oral comm., ii-28-78.
40. Cox, t. 1, s. 2, v-2-78.
41. Cox, t. 1, s. 2, v-2-78.
42. Cox, t. 1, s. 2, v-2-78.
43. Cox, t. 2, s. 1, v-2-78.
44. Verhoogen, t. 1, s. 1, viii-11-77.
45. Letter from Cox to Doell of i-13-58 is Document No. 19; Cox, t. 2, s. 1, v-2-78; Christensen, t. 1, s. 1.
46. Cox, t. 2, s. 1, v-2-78. On Du Bois's work, see Du Bois, 1957. Deutsch's findings were published later in 1958 and in Deutsch et al., 1958, 1959.
47. Document No. 19A details Cox's research plans at that time and confirms Verhoogen's obvious influence in their formulation.
48. Snavely et al., 1948, 1949.
49. Snavely, t. 1, s. 1.
50. Cox, t. 2, s. 1, v-2-78.
51. Snavely, t. 1, s. 1.
52. Cox, t. 1, s. 1, 2, v-2-78.
53. Hospers, 1953, 1954a, b.
54. Malde et al., 1962, 1963.
55. Cox, t. 2, s. 2, v-4-78.
56. Kuno, 1959.
57. Malde, t. 1, s. 1.
58. Malde, t. 1, s. 1.
59. Cox, t. 2, s. 2, v-4-78.
60. Cox, t. 3, s. 1, v-4-78.
61. Cox, t. 3, s. 2, v-4-78.
62. Cox's letter of ix-30-58 is Document No. 16; Doell, t. 3, s. 2, iv-17-78.
63. Letter of xi-7-58 from Balsley to Cox is Document No. 16-1.
64. Balsley, 1942, 1945.
65. Balsley, 1946; Hawkes and Balsley, 1946; Jensen and Balsley, 1946; Keller et al., 1947; Balsley et al., 1952.
66. Balsley, t. 1, s. 1; Balsley and Buddington, 1954, 1957a, b, 1958, 1960.
67. Carnegie Institute *Yearbook* No. 45, vii-1-45 to vi-30-46.
68. Cox, oral comm., v-23-78.
69. Doell, t. 3, s. 2, iv-17-78.
70. Doell, t. 3, s. 2, iv-17-78.
71. Cox, t. 5, s. 1, v-23-78.
72. Document No. 19 details equipment orders made through the Survey.
73. Cox, t. 5, s. 1, v-28-78.
74. Doell, t. 3, s. 2, iv-17-78.
75. Sherrill, t. 1, s. 1.
76. Doell, t. 4, s. 1, iv-26-78.
77. Monthly project report for iv-21 to v-20-59 is Document No. 20A.

78. Document No. 20B.
79. Doell, 1955a, b.
80. The report is Document No. 20C; for the article, see Doell and Cox, 1961a.
81. Letter of iv-19-64 from Doell to Wayne Hall, Assistant Chief Geologist for Experimental Geology, is Document No. 20D.
82. Letter of vi-25-63 from W. J. Robbins to Doell, Cox, and Dalrymple is Document No. 20E; letter of vi-25-63 from A. T. Waterman, Director, National Science Foundation, to Thomas B. Nolan, Director, U.S. Geological Survey, is Document No. 20F.
83. Cox, t. 3, s. 2, v-4-78 (emphasis added).
84. Dalrymple, t. 2, s. 1, iii-30-78.
85. Wahrhaftig, t. 3, s. 1, iii-12-78; Dalrymple, t. 2, s. 1, iii-30-78.
86. Dalrymple, t. 2, s. 1, iii-30-78.
87. Dalrymple, t. 1, s. 1, iii-30-78.
88. Dalrymple, t. 1, s. 2, iii-30-78.
89. Dalrymple, 1963.
90. Wahrhaftig, t. 3, s. 1, iii-4-78; Dalrymple, t. 2, s. 2, iii-30-78.
91. Dalrymple, t. 2, s. 2, iii-30-78.
92. Dalrymple, t. 1, s. 2, iii-30-78.
93. G. B. Dalrymple, oral comm., iii-30-78.
94. L. C. Pakiser, oral comm., iii-24-78; S. S. Goldich, oral comm., iii-23-78.
95. Document No. 20E.
96. Rock Magnetics Laboratory report for month ending vii-20-63 (Document No. 25) notes Dalrymple's arrival.
97. A number of Survey scientists made such remarks (unattributed by request).
98. Moxham, t. 1, s. 1.
99. Obradovich, t. 1, s. 2.
100. Letter of i-2-62 from R. M. Moxham to "Record" concerning U.S.-Japan magnetic studies is Document No. 26.
101. Moxham, t. 1, s. 1.
102. McDougall and Tarling, 1963b.
103. Lanphere, t. 1, s. 1, iii-7-78.
104. Dalrymple, t. 3, s. 1, iii-30-78; M. A. Lanphere, oral comm., iii-8-78.
105. Dalrymple, t. 2, s. 2, iii-30-78.
106. Document No. 28.
107. Document No. 28A (emphasis added).
108. Document No. 28B (emphasis added).
109. Document No. 28C-1.
110. Document No. 28C.
111. Letter of xi-1-63 from Doell through Moxham (chief of Branch of Theoretical Geophysics) to Wayne Hall (Asst. Chief Geologist for Experimental Geology) is Document No. 28D.
112. Letter of i-15-64 from Doell to Moxham is Document No. 28E.
113. The statement within the brackets was verified by Doell (oral comm., x-31-78).

114. Doell, t. 5, s. 2, iv-28-78.
115. Dalrymple, t. 3, s. 2, iii-30-78.
116. Document No. 28F.
117. Doell, t. 5, s. 2, iv-28-78.
118. Doell, t. 5, s. 2, iv-28-78.
119. Dalrymple, t. 4, s. 1, iii-31-78.
120. Monthly report of Rock Magnetics Project for month ending ii-20-64 is Document No. 20G.
121. Monthly report of Rock Magnetics Project for month ending viii-20-64 is Document No. 20I.
122. McDougall, t. 1, s. 1, viii-12-78.
123. McDougall, t. 1, s. 1, viii-12-78.
124. McDougall, t. 2, s. 1, viii-12-78.
125. McDougall, t. 2., s. 2, viii-12-78.
126. McDougall, t. 3, s. 1, viii-14-78; t. 4, s. 1, viii-14-78.
127. Hart et al., 1960; Hart, 1961a.
128. McDougall, t. 3, s. 2, viii-14-78.
129. McDougall, t. 4, s. 1, viii-14-78.
130. McDougall, t. 3, s. 2, viii-14-78.
131. On the rocks from Tasmania, see McDougall, 1961; on those from Antarctica and South Africa, see McDougall, 1963.
132. McDougall, t. 4, s. 1, viii-14-78.
133. Richards, t. 1, s. 2.
134. I. McDougall, oral comm., viii-15-78.
135. Tarling, t. 1, s. 1.
136. McDougall, t. 4, s. 2, viii-14-78.
137. Chamalaun and McDougall, 1966.
138. McDougall, t. 5, s. 1, viii-15-78.
139. McDougall, t. 3, s. 2, viii-14-78.
140. Tarling, t. 1, s. 1.
141. Tarling, t. 1, s. 1.
142. McDougall, t. 5, s. 1, viii-15-78 (emphasis added).
143. McDougall, t. 5, s. 2, viii-15-78.

Chapter 6

1. Document No. 29-1.
2. Berkeley Laboratory data card is Document No. 21 (noted earlier in Chapter 5).
3. Cox, t. 5, s. 2, v-25-78.
4. Cox, t. 5, s. 2, v-25-78.
5. Cox, t. 6, s. 1, v-25-78.
6. McDougall's letter of ii-4-63 is Document No. 29-1.
7. Grommé, t. 2, s. 1, iii-24-78.
8. Verhoogen, t. 2, s. 1, iii-29-78.
9. Cox, Doell, and Dalrymple, 1964a.
10. P. W. Birkland *fide* Cox, Doell, and Dalrymple, 1964a.
11. Doell, t. 7, s. 2, v-16-78.
12. Dalrymple, t. 3, s. 2, iii-30-78.
13. Cox, t. 5, s. 2, v-25-78.
14. Cox, t. 6, s. 1, v-25-78.

15. Doell, t. 7, s. 2, v-16-78.
16. G. B. Dalrymple, written comm., xii-21-78.
17. G. B. Dalrymple, oral comm., x-29-78.
18. C. S. Grommé, written comm., viii-19-78.
19. Document No. 31.
20. Letter is Document No. 31A.
21. Evernden, Curtis, and Kistler, 1957; Evernden, 1959; Evernden, Savage, Curtis, and James, 1964.
22. Gilbert, 1938; Bateman, 1956; Rinehart and Ross, 1956, 1957; Putnam, 1960.
23. Hopkins, t. 1, s. 1, iii-2-78.
24. Dalrymple, 1964.
25. Doell, Dalrymple, and Cox, 1966, p. 531 (emphasis added).
26. Doell and Dalrymple, 1966 (emphasis added).
27. Cox, t. 6, s. 1, v-28-78.
28. Doell, t. 7, s. 2, v-16-78.

Chapter 7

1. Clegg et al., 1954a, b.
2. Wegener, 1912a, b, 1915.
3. Matthews, t. 1, s. 1.
4. Matthews, t. 1, s. 1.
5. Matthews, t. 1, s. 1.
6. Matthews, t. 1. s. 1.
7. Vine, t. 1, s. 1.
8. Vine, t. 1, s. 2.
9. Vine, t. 1, s. 1.
10. Vine, t. 1, s. 1.
11. Vine, t. 1, s. 2.
12. Vine, t. 2, s. 1.
13. Matthews, t. 1, s. 1.
14. Vine, t. 1, s. 2.
15. Vine, t. 1, s. 2.
16. Cox, Doell, and Dalrymple, 1963a.
17. Vine, t. 1, s. 2.
18. Vine, t. 2, s. 1.
19. Revelle and Maxwell, 1952; Bullard, Maxwell, and Revelle, 1956; Maxwell and Revelle, 1956; Maxwell, t. 1, s. 1, viii-17-76.
20. Wayland, 1930; Willis, 1936.
21. Matthews, t. 1, s. 2.
22. The paper was published on xi-21-58.
23. Girdler, 1958, p. 102.
24. Girdler, 1958, p. 92.
25. Girdler, 1958, p. 92.
26. Girdler, 1958, p. 100 and 101.
27. Drake et al., 1959, p. 21.
28. Ewing et al., 1959, p. 25 (emphasis added).
29. Girdler, 1963, p. 43.
30. Girdler, 1963, p. 45.
31. Hill, 1960.
32. Vine, t. 1, s. 2; Matthews, t. 1, s. 1.
33. Bullard and Mason, 1963, p. 192.
34. Bullard and Mason, 1963, p. 192.

35. Ade-Hall, 1964, p. 86.
36. Vine, 1966, p. 1406.
37. Raff, 1963, p. 955.
38. Raff, 1963, p. 956.
39. Raff, t. 1, s. 2.
40. Raff, t. 1, s. 2.
41. Raff, t. 1, s. 2.
42. Raff, t. 1, s. 2.
43. Mason, 1958; Mason and Raff, 1961; Raff and Mason, 1961.
44. Vacquier, 1959; Vacquier et al., 1960; Raff, 1962.
45. Vacquier, 1962b, 143–44 (emphasis added).
46. Vacquier, t. 1, s. 1.
47. Morley, t. 1, s. 2, x-15-79.
48. Dietz, 1961a, b.
49. Morley, t. 1, s. 1, x-15-79.
50. Vine, t. 2, s. 1.
51. Vine, t. 2, s. 2.
52. Vacquier and Von Herzen, 1964.
53. Vacquier, 1965, p. 81.
54. Vine, t. 2, s. 1.
55. Matthews, t. 1, s. 2. The publications were Wilson, 1965a, b; and Vine and Wilson, 1965.
56. Vine and Wilson, 1965.
57. Vine, t. 2, s. 1 (emphasis added).
58. Vine and Wilson, 1965; J. T. Wilson, 1965b.
59. Dalrymple, Doell, and Cox, 1966.
60. Vine, t. 2, s. 2 (emphasis added).
61. Opdyke, Glass, Hays, and Foster, "Paleomagnetic Study of Antarctic Deep-Sea Cores," *Science*, x-21-66.
62. Vine, t. 2, s. 2.

Chapter 8

1. B. C. Heezen, oral comm., viii-23-76.
2. Herron, t. 1, s. 1.
3. Heezen, t. 2, s. 1.
4. Heirtzler and Le Pichon, 1965, p. 4019.
5. Talwani et al., 1965, p. 1109.
6. J. T. Wilson, 1965, p. 345.
7. Talwani et al., 1965, p. 1114.
8. Heirtzler, t. 1, s. 1, viii-17-76.
9. Heirtzler et al., 1966, p. 442.
10. Heirtzler et al., 1966, p. 435.
11. Wensink, 1964a.
12. Heirtzler et al., 1966, p. 442.
13. Matthews, 1965.
14. Bodvarrson and Walker, 1964, p. 285.
15. Bodvarsson and Walker, 1964, p. 288–89.
16. Bodvarsson and Walker, 1964, p. 297.
17. Bodvarsson and Walker, 1964, p. 298.
18. Heirtzler et al., 1966, p. 440.
19. Irving and Opdyke, 1965.
20. Foster, t. 2, s. 2, viii-13-79.
21. Opdyke, t. 3, s. 1, v-11-79.
22. Ryan, t. 1, s. 1; Glass, t. 1, s. 1; Broecker, t. 1, s. 1; Pitman, t. 5, s. 1, viii-16-79.
23. Ninkovich, t. 1, s. 1; Pitman, t. 5, s. 2, viii-15-79.
24. Ninkovich, t. 1, s. 1; J. H. Foster, oral comm., viii-15-79.

25. Broecker, t. 1, s. 1.
26. Foster, t. 3, s. 1, viii-13-79.
27. Foster, t. 3, s. 2, viii-13-79.
28. Foster, t. 3, s. 2, viii-13-79.
29. Opdyke et al., 1966.
30. Ninkovich et al., 1966.
31. Opdyke, t. 1, s. 2; t. 2, s. 1; t. 2, s. 2; v-11-79 (emphasis added).
32. Herron, t. 1, s. 1.
33. N. D. Opdyke, oral comm., v-11-79.
34. Herron, t. 1, s. 1.
35. Pitman, t. 1, s. 2, iv-24-79.
36. Pitman, t. 1, s. 1, iv-24-79.
37. Pitman, t. 2, s. 2, iv-24-79.
38. Vine, t. 2, s. 2.
39. Pitman, t. 3, s. 2, viii-6-79.
40. Cox, t. 6, s. 1, v-25-78.
41. Opdyke, t. 3, s. 1, v-11-79.
42. Vine, 1966, p. 1415.
43. Pitman, t. 2, s. 2, iv-24-79.
44. Pitman and Hays, 1968; Dickson et al., 1968; Le Pichon and Heirtzler, 1968; Heirtzler et al., 1968; etc.
45. Oral comm.: G. B. Dalrymple, vii-18-78; C. S. Grommé, vii-19-78; written comm., A. V. Cox, i-4-80.
46. J. T. Wilson, 1970.

Epilogue

1. Opdyke, t. 2, s. 2, v-11-79.
2. Morley, t. 1, s. 1, iv-24-79.
3. Dalrymple, t. 4, s. 2, iii-31-78; and others.
4. Vetleson Prize Award Dinner, Ewing, t. 1, s. 1.
5. Vetleson Prize Award Dinner, Kusch, t. 1, s. 1.
6. Vetleson Prize Award Dinner, Kusch, t. 1, s. 2.
7. Cox, 1972, p. 12.

Appendix C

1. Bullard et al., 1965.
2. Wilson, t. 1, s. 1, x-29-80.
3. Wilson, t. 1, s. 1 and 2, x-29-80.
4. Wilson, t. 1, s. 2, x-29-80.

References

Ade-Hall, J. M. 1964. The magnetic properties of some submarine oceanic lavas, *Roy. Astron. Soc. Geophys. Jour.*, 9: 85–91.

Ade-Hall, J. M., and R. L. Wilson. 1963. Petrology and the natural remanence of the Mull lavas, *Nature*, 198: 659–60.

Aldrich, L. T., and A. O. Nier. 1948a. Argon 40 in potassium minerals, *Phys. Rev.*, 74: 876.

———. 1948b. The occurrence of He^3 in natural sources of helium, *Phys. Rev.*, 74: 1590–94.

Allan, D. W. 1958. Reversals of the Earth's magnetic field, *Nature*, 182: 469–71.

Allan, T. D. 1964. A preliminary magnetic survey in the Red Sea and the Gulf of Aden, *Boll. Geofis, Teor. Appl.*, 6: 199–214.

———. 1969. Review of marine geomagnetism, *Earth Sci. Rev.*, 5: 217–54.

Allen, C. C., and R. E. Folinsbee. 1944. Scheelite veins related to porphyry intrusives, Hollinger Mine, *Econ. Geol.*, 39, no. 5: 340–48.

Allen, G. E. 1909. Magnetism of basalt, *Phil. Mag.*, Ser. 6, 17: 572–81.

Almond, M., J. A. Clegg, and J. C. Jaeger. 1956. Remanent magnetism of some dolerites, basalts, and volcanic tuffs from Tasmania, *Phil. Mag.*, Ser. 8, 1: 771–82.

Alpert, D. 1953. New developments in the production and measurement of ultra-high vacuum, *Jour. Appl. Phys.*, 24: 860.

Alvarez, W. 1972. The Treia Valley north of Rome: Volcanic stratigraphy, topographic evolution, and geological influences on human settlement, *Geol. Rom.*, 11: 153–76.

———. 1975. The Pleistocene volcanoes north of Rome, in C. H. Squyres, ed., *Geology of Italy*, vol. 2, pp. 355–77. Tripoli, Libya: Earth Science Society of the Libyan Arab Republic.

Amirkhanov, K. I., E. N. Bartnitskii, S. B. Brandt, and G. V. Voitkevich. 1959. On the migration of argon and helium in several rocks and minerals, *Akad. Nauk SSSR Doklady*, Geochem. Ser., 126: 160–62.

Anderson, E. M. 1951. *The dynamics of faulting*. 2d ed. Edinburgh: Oliver and Boyd.

Aston, F. W. 1920. The constitution of atmospheric neon, *Phil. Mag.*, Ser. 6, 39: 449–55.

———. 1921. The mass spectra of the alkali metals, *Phil. Mag.*, Ser. 6, 42: 436–41.

Avery, D. E. 1963. *Geomagnetic and bathymetric profiles across the North Atlantic Ocean* (U.S. Naval Oceanog. Office Tech. Rept. TR-161).

Baadsgaard, H., and R. E. Folinsbee. 1959. Potassium-argon age of biotites from Cordilleran granites, *Roy. Soc. Canada Proc. Programme of Papers*, pp. 20–21 (abstract).

Baadsgaard, H., R. E. Folinsbee, and J. Lipson. 1959. Caledonian and Acadian granites of the northern Yukon, *Can. Oil & Gas Indust.*, 12, no. 12: 50.

———. 1961. Potassium-argon dates of biotites from Cordilleran granites, *Geol. Soc. Amer. Bull.*, 72: 689–702.

Baadsgaard, H., J. Lipson, and R. E. Folinsbee. 1961. The leakage of radiogenic argon from sanidine, *Geochim. Cosmochim. Acta*, 25: 247–57.

Backus, G. E. 1964. Magnetic anomalies over oceanic ridges, *Nature*, 201: 591–2.

Balestrini, S. J. 1954. Mass-spectrometric study of deuteron-induced reactions in iodine, *Phys. Rev.*, 95: 6: 1502–8.

Balsley, J. R. 1942. *Chromite deposits of Red Bluff Bay and vicinity, Baranof Island, Alaska* (U.S. Geol. Survey Bull. 936–G).

———. 1945. *Vanadium-bearing magnetite-ilmenite deposits near Lake Sanford, Essex County, New York* (U.S. Geol. Survey Bull. 940–D).

———. 1946. *The airborne magnetometer* (U.S. Geol. Survey Geophys. Inv. Prelim. Rept. 3).

Balsley, J. R., and A. F. Buddington. 1954. Correlation of reverse remanent magnetism and negative anomalies with certain minerals, *Jour. Geomag. Geoelec.*, 6, no. 4: 176–81.

———. 1957a. Puzzles in the interpretation of paleomagnetism, *India Geol. Survey Records*, 86: 553–80.

———. 1957b. Remanent magnetism of the Russel belt of gneisses, northwest Adirondack Mountains, New York, *Adv. Phys.*, 6: 317–22.

———. 1958. Iron-titanium oxide minerals, rocks and aeromagnetic anomalies of the Adirondack area, New York, *Econ. Geol.*, 53: 777–805.

———. 1960. Magnetic susceptibility anisotropy and fabric of some Adirondack granites and orthogneisses, *Amer. Jour. Sci.*, 258A: 6–20.

Balsley, J. R., A. F. Buddington, and J. Fahey. 1952. Titaniferous hematite and ilmenohematite correlated with inverse polarization in rocks of the northwest Adirondacks, N.Y., *Amer. Geophys. Union Trans.*, 33: 320 (abstract).

Balsley, J. R., H. L. James, and K. L. Weir. 1949. *Aeromagnetic survey of parts of Baraga, Iron, and Houghton Counties, Michigan, with preliminary geologic interpretation* (U.S. Geol. Survey Geophys. Inv. Prelim. Map).

Barth, T. F. W. 1952. The differentiation of a composite aplite from the Pribilof Islands, Alaska, *Amer. Jour. Sci.* (Bowen Volume) 250, no. 1: 27–36.

Barzun, J., and A. F. Graff. 1977. *The Modern Researcher*. 3d ed. New York: Harcourt Brace Jovanovich.

Bassett, W. A., and P. F. Kerr. 1961. Potassium-argon ages of volcanics near Grants, New Mexico, *Jour. Geophys. Res.*, 66: 2512 (abstract).

Bateman, P. C. 1956. *Economic geology of the Bishop tungsten district, California* (Calif. Div. Mines Spec. Rept. 47).

Becker, G. F. 1908. Relations of radioactivity to cosmogony and geology. *Geol. Soc. Amer. Bull.*, 19: 113–46.

Becquerel, H. 1896. Sur les radiations invisibles émises par phosphorescence; Sur les radiations invisibles émises par les corps phosphorescents; Sur les radiations invisibles émises par les sels d'uranium, *Acad. Sci. Paris Compt. Rend.*, 122: 420–21, 501–3, 689–94.

Benioff, H. 1949. Seismic evidence for the fault origin of oceanic deeps, *Geol. Soc. Amer. Bull.*, 60: 1837–56.

———. 1951. Earthquakes and rock creep characteristics of rocks and the origin of aftershocks, *Seismol. Soc. Amer. Bull.*, 41: 31–62.

———. 1954. Orogenesis and deep crustal structure; additional evidence from seismology, *Geol. Soc. Amer. Bull.*, 65: 385–400.

———. 1955. Seismic evidence for crustal structure and tectonic activity, in A. Poldervaart, ed., *Crust of the Earth—A Symposium* (Geol. Soc. Amer. Spec. Paper 62), pp. 61–75. Washington, D.C.: Geological Society of America.

Beveridge, A. J., and R. E. Folinsbee. 1956. Dating Cordilleran orogenies, *Roy. Soc. Canada Trans.*, Ser. 3, 50, no. 4: 19–43.

Blackett, P. M. S. 1952. A negative experiment relating a magnetism and the Earth's rotation, *Roy. Soc. London Phil. Trans.*, Ser. A, 245: 309–70.

———. 1956. *Lectures on Rock Magnetism*. Jerusalem: Weizmann Science Press of Israel.

———. 1961. Comparisons of ancient climate with the latitude deduced from rock magnetic measurements, *Roy. Soc. London Proc.*, Ser. A, 263: 1–30.

———. 1962. Palaeomagnetism and palaeoclimatology, *Jour. Geomag. Geoelec.*, 13: 127–32.

Blackett, P. M. S., E. C. Bullard, and S. K. Runcorn, eds. 1965. A symposium on continental drift, *Roy. Soc. London Phil. Trans.*, Ser. A, 258: 1–323.

Blackett, P. M. S., J. A. Clegg, and P. H. S. Stubbs. 1960. An analysis of rock magnetic data, *Roy. Soc. London Proc.*, Ser. A, 256: 291–322.

Blakely, R. J. 1979. Marine magnetic anomalies, *Rev. Geophys. Space Phys.*, 17: 204–14.

Blanc, A. C. 1956. Sur le pleistocène de la région de Rome; stratigraphie, palaeoéchologie, archéologie préhistorique, in *Actes du IVe Congrès International du Quaternaire, Rome-Pise, 1953*, vol. 2, pp. 1097–111.

———. 1957. On the Pleistocene sequence of Rome: Paleoecologic and archeologic correlations, *Quaternaria*, 4: 95–109.

———. 1958. Discorso tenuto da A. C. Blanc all'-Assemblea Generale di chiusura del V Congresso Internazionale delle Scienze Preistoriche e Protostoriche, in Amburgo, il 30 Agosto 1958, *Quaternaria*, 5: 399–400.

Blanc, A. C., E. Tongiorgi, and L. Trevisan. 1953. Le pliocène et le quaternaire aux alentours de Rome. *Guidebook, 4th Intern. Cong. Quaternary, Rome-Pisa.*

Blundell, D. J. 1957. A palaeomagnetic investigation of the Lundy dyke swarm, *Geol. Mag.*, 94: 187–93.

Bödvarsson, G., and G. P. L. Walker. 1964. Crustal drift in Iceland, *Roy. Astron. Soc. Geophys. Jour.*, 8, no. 3: 285–300 (manuscript received viii-21-63).

Boer, J. C. den. 1957. *Étude géologique et paléomagnétique des montagnes du Coirons, Ardèche, France.* Thesis, Rijksuniv., Utrecht. (*Geologica Ultraiectina*, no. 1.)

Boltwood, B. B. 1907. The disintegration products of uranium (Part 2 of On the ultimate disintegration products of the radio-active elements), *Amer. Jour. Sci.*, 4th Ser., 23, no. 134: 77–88.

Bowen, N. L., and O. F. Tuttle. 1949. The system MgO-SiO_2-H_2O, *Geol. Soc. Amer. Bull.*, 60: 439–60.

Bremaecker, J.-C. de. 1956. Premières données seismologiques sur le Graben de l'Afrique Central, *Acad. Roy. Sci. Coloniales, Brussels Bull.*, 2: 762.

———. 1959. Seismicity of the West African Rift Valley, *Jour. Geophys. Res.*, 64: 1961–66.

Bruckshaw, J. M., and E. I. Robertson. 1949. The magnetic properties of the tholeiite dykes of north England, *Roy. Astron. Soc. Geophys. Supp. Mon. Not.*, 5, no. 8: 308–20.

Brunhes, B. 1906. Recherches sur la direction d'aimantation des roches volcaniques, *Jour. Phys. Radium*, Ser. 4, 5: 705–24.

Brunhes, B., and P. David. 1901. Sur la direction d'aimantation dans des couches d'argile transformée en brique par des coulées de lave, *Acad. Sci. Paris Compt. Rend.*, 133: 155–57.

———. 1902. Sur la direction d'aimantation dans des couches d'argile transformée en brique par des coulées de lave, *Soc. Belge Geol. Brux. Bull.*, 16: 30–32.

———. 1903. Sur la direction de l'aimantation permanente dans diverses roches volcaniques, *Acad. Sci. Paris Compt. Rend.*, 137: 975–77.

———. 1905. Les travaux récents de magnétisme terrestre dans la France centrale; le présent et le passé magnétique des volcans d'Auvergne, *Soc. Belge d'Astron. Bull.*, 11: 270.

Bryniolfsson, A. 1957. Studies of remanent magnetism and viscous magnetism in the basalts of Iceland, *Adv. Phys.*, 6: 247–54.
Bullard, E. C. 1936. Gravity measurements in East Africa, *Roy. Soc. London Phil. Trans.*, Ser. A, 235: 445–531.
———. 1939. Heat flow in South Africa, *Roy. Soc. London Proc.*, Ser. A, 173: 474–502.
———. 1949. The magnetic field within the Earth, *Roy. Soc. London Proc.*, Ser. A, 197: 433–53.
———. 1952. Discussion of paper by R. Revelle and A. E. Maxwell entitled "Heat flow through the floor of the eastern North Pacific Ocean," *Nature*, 170: 200.
———. 1955. The stability of a homopolar dynamo, *Cambridge Phil. Soc. Proc.*, 51: 744–60.
———. 1964. Continental drift, *Geol. Soc. London Quart. Jour.*, 120: 1–34.
———. 1968. Reversals of the Earth's magnetic field (Bakerian lecture of 1967), *Roy. Soc. London Phil. Trans.*, Ser. A, 263: 481–524.
———. 1975a. The effect of World War II on the development of knowledge in the physical sciences, *Roy. Soc. London Proc.*, Ser. A, 342: 519–36.
———. 1975b. The emergence of plate tectonics; a personal view, *Ann. Rev. Earth Planet. Sci.*, 3: 1–30.
Bullard, E. C., J. E. Everett, and A. G. Smith. 1965. Fit of continents around Atlantic, in P. M. S. Blackett, E. C. Bullard, and S. K. Runcorn, eds., A symposium on continental drift, *Roy. Soc. London Phil. Trans.*, Ser. A, 258: 41–75.
Bullard, E. C., and H. Gellman. 1954. Homogeneous dynamos and terrestrial magnetism, *Roy. Soc. London Phil. Trans.*, Ser. A, 247: 213–78.
Bullard, E. C., and R. G. Mason. 1963. The magnetic field over the oceans, in M. N. Hill, ed., *The Earth Beneath the Sea* [and] *History* (vol. 3 of *The Sea: Ideas and Observations on Progress in the Study of the Seas*), pp. 175–217. New York: Wiley-Interscience.
Bullard, E. C., A. E. Maxwell and R. Revelle. 1956. Heat flow through the deep sea floor, *Adv. Geophys.*, 3: 153–81.
Byström-Asklund, A. M., H. Baadsgaard, and R. E. Folinsbee. 1961. K/Ar age of biotite, sanidine, and illite from Middle Ordovician bentonites at Kinnekulle, Sweden, *Geol. Foreningens Forhand*, 83, no. 504, 92–96.
Campbell, N. R., and A. Wood. 1906. The radioactivity of the alkali metals, *Proc. Cambridge Phil. Soc.*, 14: 15–21.
Cann, J. R., and F. J. Vine. 1966. An area of the crest of the Carlsberg Ridge; petrology and magnetic survey, *Roy. Soc. London Phil. Trans.*, Ser. A, 259: 198–217.
Carey, S. W. 1955. The orocline concept in geotectonics, part 1, *Roy. Soc. Tasmania Pap. Proc.*, 89: 255–88.
———. 1958. The tectonic approach to continental drift, in S. W. Carey, ed., *Continental Drift: A Symposium*, pp. 177–355. Hobart: University of Tasmania.

———. 1976. *The Expanding Earth*. Amsterdam: Elsevier.

Carmichael, C. M. 1959. Remanent magnetization of the Allard Lake ilmenites, *Nature*, 183: 1239–41.

———. 1961. The magnetic properties of ilmenite-haematite crystals, *Roy. Soc. London Proc.*, Ser. A, 263: 508–30.

Carr, D. R., and J. L. Kulp. 1957. Potassium-argon method of geochronology, *Geol. Soc. Amer. Bull.*, 68: 763–84.

Carter, R. N., and H. B. S. Cooke, K. M. Creer, T. Einarsson, J. D. Hays, C. E. Helsley, E. Irving, A. Khramov, J. Kukla, I. McDougall, M. W. McElhinny, H. Nakagawa, N. Opdyke, D. M. Terchesky, and N. D. Watkins. 1973. Magnetic polarity-reversal time scale, *Geotimes*, 18, no. 5: 21–22.

Chamalaun, F. H., and I. McDougall. 1966. Dating geomagnetic polarity epochs in Réunion, *Nature*, 210: 1212–14 (June 18, 1966).

Charlesworth, J. K. 1957. *Quaternary Era with Special References to its Glaciation*. London: E. Arnold.

Chevallier, R. 1925. L'aimantation des laves de l'Etna et l'orientation du champ terrestre en Sicile du XIIe au XVIIe siècle, *Ann. Physique*, Ser. 10, 4: 5–162.

Clegg, J. A. 1956. Rock magnetism, *Nature*, 178: 1085–87.

Clegg, J. A., M. Almond, and P. H. S. Stubbs. 1954a. The remanent magnetism of some sedimentary rocks in Britain, *Phil. Mag.*, Ser. 7, 45: 583–98.

———. 1954b. Some recent studies of the prehistory of the Earth's magnetic field, *Jour. Geomag. Geoelec.*, 6: 194–99.

Clegg, J. A., E. R. Deutsch, C. W. F. Everitt, and P. H. S. Stubbs. 1957. Some recent palaeomagnetic measurements made at Imperial College, London, *Adv. Phys.*, 6: 219–31.

Clegg, J. A., E. R. Deutsch, and D. H. Griffiths. 1956. Rock magnetism in India, *Phil. Mag.*, Ser. 8, 1, no. 5: 419–31.

Clegg, J. A., C. Radakrishnamurty, and P. W. Sahasrabudhe. 1958. Remanent magnetism of the Rajmahal traps of northeastern India, *Nature*, 181: 830–31.

Collinson, D. W., K. M. Creer, and S. K. Runcorn, eds. 1967. *Methods in Palaeomagnetism* (Proceedings of the NATO Advanced Study Institute on Palaeomagnetic Methods held at the University, Newcastle-upon-Tyne, April 1–10, 1964). Amsterdam: Elsevier.

Coode, A. M. 1965. A note on oceanic transcurrent faults, *Canadian Jour. Earth Sci.*, 2: 400–401 (August).

Cox, A. 1957. Remanent magnetization of lower to middle Eocene basalt flows from Oregon, *Nature*, 179: 685–86.

———. 1960. *Anomalous remanent magnetization of basalt*. U.S. Geol. Survey Bull. 1083–E.

———. 1962. Analysis of present geomagnetic field for comparison with paleomagnetic results, *Jour. Geomag. Geoelec.*, 13: 101–12.

———. 1964. Angular dispersion due to random magnetization, *Roy. Astron. Soc. Geophys. Jour.*, 8: 4: 345–55.

———. 1965. Review of *Paleomagnetism and Its Application to Geological and Geophysical Problems* by E. Irving, *Science*, 147: 494.

———. 1968. Lengths of geomagnetic polarity intervals, *Jour. Geophys. Res.*, 73: 3257–60.

———. 1969. Geomagnetic reversals, *Science*, 163: 237–245.

———. 1972. A perfect case of serendipity, *Mosaic*, 3, no. 2: 2–12.

———. 1973. *Plate tectonics and geomagnetic reversals*. San Francisco: W. H. Freeman.

Cox, A., and G. B. Dalrymple. 1967a. Geomagnetic polarity epochs—Nunivak Island, Alaska, *Earth Planet. Sci. Lett.*, 3: 173–77.

———. 1967b. Statistical analysis of geomagnetic reversal data and the precision of potassium-argon dating, *Jour. Geophys. Res.*, 72: 2603–14.

Cox, A., G. B. Dalrymple, and R. R. Doell. 1967. Reversals of the Earth's magnetic field, *Sci. Amer.*, 216: 44–54.

Cox, A., and R. R. Doell. 1960. Review of paleomagnetism, *Geol. Soc. Amer. Bull.*, 71: 645–768.

———. 1961a. Palaeomagnetic evidence relevant to a change in the Earth's radius, *Nature*, 189: 45–47.

———. 1961b. Palaeomagnetic evidence relevant to a change in the Earth's radius [reply to comment by S. W. Carey], *Nature*, 190: 36–37.

———. 1962. Magnetic properties of the basalt in Hole EM 7, Mohole project, *Jour. Geophys. Res.*, 67: 3997–4004.

———. 1964. Long period variations of the geomagnetic field, *Seismol. Soc. Amer. Bull.*, 54B, no. 6: 2243–70.

———. 1967. Measurement of high coercivity magnetic anisotropy, in D. W. Collinson, K. M. Creer, and S. K. Runcorn, eds., *Methods in Palaeomagnetism*, pp. 477–82. Amsterdam: Elsevier.

Cox, A., R. R. Doell, and G. B. Dalrymple. 1963a. Geomagnetic polarity epochs and Pleistocene geochronometry, *Nature*, 198: 1049–51 (June 15, 1963; polarity-reversal scale one).

———. 1963b. Radiometric dating of geomagnetic field reversals, *Science*, 140: 1021–23.

———. 1963c. Geomagnetic polarity epochs—Sierra Nevada II, *Science*, 142: 382–85 (Oct. 18, 1963; polarity-reversal scale four).

———. 1964a. Geomagnetic polarity epochs, *Science*, 143: 351–52 (Jan. 24, 1964).

———. 1964b. Reversals of the Earth's magnetic field, *Science*, 144: 1537–43 (June 26, 1964; polarity-reversal scale seven).

———. 1965. Quaternary paleomagnetic stratigraphy, in H. E. Wright, Jr., and D. G. Frey, eds., *The Quaternary of the United States*, pp. 817–30. Princeton, N.J.: Princeton University Press (polarity-reversal scale nine).

———. 1968. Radiometric time scale for geomagnetic reversals, *Geol. Soc. London Quart. Jour.*, 124, no. 493: 53–66.

Cox, A., D. M. Hopkins, and G. B. Dalrymple. 1966a. Palaeomagnetism and potassium-argon ages of some volcanic rocks from the Galapagos Islands, *Nature*, 209: 776–77.

———. 1966b. Geomagnetic polarity epochs—Pribilof Islands, Alaska, *Geol. Soc. Amer. Bull.*, 77: 883–910.

Cox, A., and W. Stauder, S.J. 1964. Geometrical properties of groups of fault plane solutions, *Seismol. Soc. Amer. Bull.*, 54, no. 1: 87–104.

Creer, K. M. 1958. Preliminary palaeomagnetic measurements from South America, *Ann. Geophys.*, 14: 373–90.

———. 1959. A.C. demagnetization of unstable Triassic Keuper marls from S W England, *Roy. Astron. Soc. Geophys. Jour.*, 2: 261–75.

———. 1970. A review of paleomagnetism, *Earth Sci. Rev.*, 6: 369–466.

———. 1971. Mesozoic palaeomagnetic reversal column, *Nature*, 233: 545–46.

Currie, R. G., C. S. Grommé, and J. Verhoogen. 1963. Remanent magnetization of some Upper Cretaceous granitic plutons in the Sierra Nevada, California, *Jour. Geophys. Res.*, 68: 2263–79.

Curtis, G. H. 1955. Importance of Novarupta during eruption of Mt. Katmai, Alaska, in 1912, *Geol. Soc. Amer. Bull.*, 66: 1547 (abstract).

———. 1966. The problem of contamination in obtaining accurate dates of young geologic rocks, in O. A. Schaeffer, and J. Zahringer, comp., *Potassium Argon Dating*, pp. 151–62. New York: Springer-Verlag.

———. 1968. The stratigraphy of the ejecta from the 1912 eruption of Mount Katmai and Novarupta, Alaska, in R. R. Coats, R. L. Hay, and C. A. Anderson, eds., *Studies in Volcanology: A Memoir in Honor of Howel Williams* (Geol. Soc. Amer. Mem. 116), pp. 153–210. Boulder, Colo.: Geological Society of America.

Curtis, G. H., and J. F. Evernden. 1962. Age of basalt underlying Bed I, Olduvai Gorge, *Nature*, 194: 610–12.

Curtis, G. H., J. F. Evernden, and J. Lipson. 1958. *Age determination of some granitic rocks in California by the potassium-argon method* (California Div. Mines Spec. Rept. 54).

Curtis, G. H., J. Lipson, and J. F. Evernden. 1956. Potassium-argon dating of Plio-Pleistocene intrusive rocks (California), *Nature*, 178: 1360.

Curtis, G. H., and J. H. Reynolds. 1958. Notes on the potassium-argon dating of sedimentary rocks, *Geol. Soc. Amer. Bull.*, 69: 151.

Curtis, G. H., D. E. Savage, and J. F. Evernden. 1961. Critical points in the Cenozoic, *N.Y. Acad. Sci. Ann.*, 91: 342.

Cuvier, Georges and Alexandre Brongniart. 1811. *Saessai sur la Géographie Minéralogique des Environs de Paris*. Paris: Editions d'Ocagne.

Dalrymple, G. B. 1963a. *Potassium-Argon Dates and the Cenozoic Chronology of the Sierra Nevada, California*. Ph.D. thesis, Univ. California, Berkeley.

———. 1963b. Potassium-argon dates of some Cenozoic volcanic rocks of the Sierra Nevada, California, *Geol. Soc. Amer. Bull.*, 74: 379–90.

———. 1964a. Argon retention in a granitic xenolith from a Pleistocene basalt, Sierra Nevada, California, *Nature*, 201: 282.

———. 1964b. Potassium-argon dates of three Pleistocene interglacial basalt flows from the Sierra Nevada, California, *Geol. Soc. Amer. Bull.*, 75: 753–58.

———. 1964c. Cenozoic chronology of the Sierra Nevada, California, *California Univ. Pubs. Geol. Sci.*, 47: 1–41.

———. 1967. Potassium-argon ages of recent rhyolites of the Mono and Inyo Craters, California, *Earth and Planetary Sci. Letters*, 3: 289–98.

———. 1972. Potassium-argon dating of geomagnetic reversals and North American glaciations, in W. W. Bishop and J. A. Miller, eds., *Calibration of Hominoid Evolution*, pp. 107–34. Edinburgh: Scottish Academic Press.

Dalrymple, G. B., A. Cox, and R. R. Doell. 1965. Potassium-argon age and paleomagnetism of the Bishop Tuff, California, *Geol. Soc. Amer. Bull.*, 76: 665–73 (June 1965; polarity-reversal scale eight).

Dalrymple, G. B., and R. R. Doell. 1966. Portable core drill and thin-section laboratory for field use, in *Geological Survey Research, 1966* (U.S. Geol. Survey Prof. Paper 550–B), pp. B182–85.

Dalrymple, G. B., R. R. Doell, and A. Cox. 1966. Recent developments in the geomagnetic polarity [epoch time scale], in *Abstracts for 1965* (Geol. Soc. Amer. Spec. Paper No. 87), p. 41 (abstract). Washington, D.C.: Geological Society of America.

Dalrymple, G. B., and M. A. Lanphere. 1969. *Postassium-Argon Dating: Principles, Techniques, and Applications to Geochronology*. San Francisco: W. H. Freeman.

Damon, P. E., and J. L. Kulp. 1958. Excess helium and argon in beryl and other minerals, *Amer. Min.*, 43: 433–59.

David, P. 1904. Sur la stabilité de la direction d'aimantation dans quelques roches volcaniques, *Acad. Sci. Paris Compt. Rend.*, 138: 41–42.

Delesse, A. 1849. Sur le magnétisme polaire dans les minéraux et dans les roches, *Ann. Chim. Phys.*, 25: 194.

Deutsch, E. R. 1958. Recent palaeomagnetic evidence for the northward movement of India, *Alberta Soc. Petrol. Geol. Jour.*, 6: 155–62.

Deutsch, E. R., C. Radakrishnamurty, and P. W. Sahasrabudhe. 1958. The remanent magnetism of some lavas in the Deccan Traps, *Phil. Mag.*, Ser. 8, 3: 170–84.

———. 1959. Palaeomagnetism of the Deccan Traps, *Ann. Geophys.*, 15: 39–59.

Dewey, J. F., and F. M. Bird. 1970. Mountain belts and the new global tectonics, *Jour. Geophys. Res.*, 75: 2625–47.

Dickinson, W. R., and T. Hatherton. 1967. Andesitic volcanism and seismicity around the Pacific, *Science*, 157: 801–03.

Dickson, G. O., W. C. Pitman III, and J. R. Heirtzler. 1968. Magnetic anomalies in the South Atlantic and ocean floor spreading, *Jour. Geophys. Res.*, 73: 2087–100.

Dietz, R. S. 1961a. Continent and ocean basin evolution by spreading of the sea floor, *Nature*, 190: 854–57 (June 3, 1961).

———. 1961b. Ocean basin evolution by sea floor spreading, in *Tenth Pa-*

cific Science Congress; Abstracts of Symposium Papers, pp. 357–58 [material in volume available to the editor as of June 1, 1961]. Honolulu: Pacific Science Association.

———. 1968. Reply, *Jour. Geophys. Res.*, 73: 6567.

Doell, R. R. 1955a. *Remanent Magnetism in Sediments*. Ph.D. thesis, Univ. California, Berkeley.

———. 1955b. Paleomagnetic study of rocks from the Grand Canyon of the Colorado River, *Nature*, 176: 1167.

———. 1956. Remanent magnetization of the Upper Miocene 'blue' sandstones of California, *Amer. Geophys. Union Trans.*, 37, no. 2: 156–67.

———. 1957. Crystallization magnetization, *Adv. Phys.*, 6: 327–32.

———. 1963. Seismic depth study of the Salmon Glacier, British Columbia, *Jour. Glaciology*, 4: 425–37.

———. 1970. United States-Japan cooperative science program, fourth seminar on paleomagnetism, Honolulu, Hawaii, Feb. 2–6, 1970, *Amer. Geophys. Union Trans.*, 51: 551–52.

Doell, R. R., and A. Cox. 1960. Paleomagnetism, polar wandering, and continental drift, in *Short Papers in the Geological Sciences* (U.S. Geol. Survey Prof. Paper 400–B), pp. B426–27.

———. 1961a. Paleomagnetism, in *Advances in Geophysics*, vol. 8, pp. 221–313. New York: Academic Press.

———. 1961b. Palaeomagnetism of Hawaiian lava flows, *Nature*, 192: 645–46.

———. 1962. Determination of the magnetic polarity of rock samples in the field, in *Short Papers in Geology, Hydrology and Topography*. (U.S. Geol. Survey Prof. Paper 450–D), pp. D105–8.

———. 1963. The accuracy of the paleomagnetic method as evaluated from historic Hawaiian lava flows, *Jour. Geophys. Res.*, 68: 1997–2009.

———. 1965. *Measurement of the remanent magnetization of igneous rocks* (U.S. Geol. Survey Bull. 1203–A).

———. 1967a. Analysis of alternating field demagnetization equipment, in D. W. Collinson, K. M. Creer, and S. K. Runcorn, eds., *Methods in Palaeomagnetism*, pp. 241–53. Amsterdam: Elsevier.

———. 1967b. Analysis of palaeomagnetic data, in D. W. Collinson, K. M. Creer, and S. K. Runcorn, eds., *Methods in Palaeomagnetism*, pp. 340–46. Amsterdam: Elsevier.

———. 1967c. Analysis of spinner magnetometer operation, in D. W. Collinson, K. M. Creer, and S. K. Runcorn, eds., *Methods in Palaeomagnetism*, pp. 196–206. Amsterdam: Elsevier.

———. 1967d. Measurement of natural remanent magnetization at the outcrop, in D. W. Collinson, K. M. Creer, and S. K. Runcorn, eds., *Methods in Palaeomagnetism*, pp. 159–62. Amsterdam: Elsevier.

———. 1967e. Palaeomagnetic sampling with a portable coring drill, in D. W. Collinson, K. M. Creer, and S. K. Runcorn, eds., *Methods in Palaeomagnetism*, pp. 21–25. Amsterdam: Elsevier.

———. 1967f. Polarity reversals of the geomagnetic field, in S. K. Runcorn, ed., *International Dictionary of Geophysics*, vol. 2, pp. 1207–9. London: Pergamon Press.

———. 1967g. Recording magnetic balance, in D. W. Collinson, K. M. Creer, and S. K. Runcorn, eds., *Methods in Palaeomagnetism*, pp. 440–44. Amsterdam: Elsevier.

———. 1967h. A spinner magnetometer for igneous rocks, in D. W. Collinson, K. M. Creer, and S. K. Runcorn, eds., *Methods in Palaeomagnetism*, pp. 136–41. Amsterdam: Elsevier.

———. 1972. The Pacific geomagnetic secular variation anomaly and the question of lateral uniformity in the lower mantle, in E. C. Robertson, J. F. Hays, and L. Knopoff, eds., *The Nature of the Solid Earth*, pp. 245–84. New York: McGraw-Hill.

Doell, R. R., A. Cox, R. W. Kistler, and G. B. Dalrymple. 1962. Radiometric ages of Pleistocene reversal zones from igneous rocks in California, *Amer. Geophys. Union Trans.*, 43: 438 (abstract).

Doell, R. R., and G. B. Dalrymple. 1966. Geomagnetic polarity epochs—a new polarity event and the age of the Brunhes–Matuyama boundary, *Science*, 152: 1060–61 (May 20, 1966; polarity-reversal scale eleven).

Doell, R. R., G. B. Dalrymple, and A. Cox. 1966. Geomagnetic polarity epochs—Sierra Nevada data 3, *Jour. Geophys. Res.*, 71: 531–41 (Jan. 15, 1966; polarity-reversal scale ten).

Doell, R. R., G. B. Dalrymple, R. L. Smith, and R. A. Bailey. 1968. Paleomagnetism, potassium-argon ages and geology of the rhyolitic and associated rocks of the Valles Caldera, New Mexico, in R. R. Coats, R. L. Hay, and C. A. Anderson, eds., *Studies in Volcanology—A Memoir in Honor of Howell Williams* (Geol. Soc. Amer. Mem. 116), pp. 211–48. Boulder, Colo.: Geological Society of America.

Domen, H. 1960. Some magnetic properties of the early Pleistocene basalts in the northern part of Yamaguchi prefecture, West Japan, *Yamaguchi Univ. Fac. Ed. Bull.*, 9: 31–42.

Drake, C. L., and R. W. Girdler. 1964. A geophysical study of the Red Sea, *Roy. Astron. Soc. Geophys. Jour.*, 8: 473–95.

Drake, C. L., R. W. Girdler, and M. Landisman. 1959. Geophysical measurements in the Red Sea, in M. Sears, ed., *Preprints of Abstracts of Papers to Be Presented at Afternoon Sessions* (1st Intern. Oceanog. Cong.), pp. 21–22 (abstract). Washington: American Association for the Advancement of Science.

Du Bois, P. M. 1957. Comparison of palaeomagnetic results for selected rocks of Great Britain and North America, *Adv. Phys.*, 6: 177–86.

Du Toit, A. L. 1927. *A Geological Comparison of South America with South Africa* (Publ. No. 381). Washington, D.C.: Carnegie Institution.

———. 1937. *Our Wandering Continents*. Edinburgh: Oliver and Boyd.

Eddington, A. S. 1926. *Internal Constitution of the Stars*. Cambridge: Cambridge University Press.

Egyed, L. 1956. The change of the Earth's dimensions determined from paleogeographical data, *Geofis. Pura Appl.*, 33: 42–48.

———. 1957. A new dynamic conception of the internal constitution of the Earth, *Geol. Rundsch.*, 46: 101–21.

———. 1960. Some remarks on continental drift, *Geofis. Pura Appl.*, 45: 115–16.

Einarsson, T., D. M. Hopkins, and R. R. Doell. 1967. The stratigraphy of Tjornes, northern Iceland, and the history of the Bering Land Bridge, in D. M. Hopkins, ed., *The Bering Land Bridge*, pp. 812–25. Stanford, Calif.: Stanford University Press.

Einarsson, Tr. 1957a. Magneto-geological mapping in Iceland with the use of a compass, *Adv. Phys.*, 6: 232–39.

———. 1957b. Der Paläomagnetismus der islandischen Basalte und seine stratigraphische Bedeutung, *Neues Jb. Geol. Paläontol. Mh.*, 4: 159–75.

———. 1957c. Über den Wert alter sediments fur paläomagnetische Zwecke, *Neues Jb. Geol. Paläontol. Mh.*, 5: 193–95.

Einarsson, Tr., and T. Sigurgeirsson. 1955. Rock magnetism in Iceland, *Nature*, 197: 892.

Elsasser, W. M. 1939. On the origin of the Earth's magnetic field, *Nature*, 143: 374.

———. 1946a. Induction effects in terrestrial magnetism, *Phys. Rev.*, 69: 106.

———. 1946b. Induction effects in terrestrial magnetism, II, *Phys. Rev.*, 70: 202.

———. 1947. Induction effects in terrestrial magnetism, III, *Phys. Rev.*, 72: 821.

———. 1955. Hydromagnetism, I; a review, *Amer. Jour. Phys.*, 23: 590–609.

Erickson, G. P., and J. L. Kulp. 1961. Potassium-argon measurements on the Palisades sill, New Jersey, *Geol. Soc. Amer. Bull.*, 72: 649–52 (April 1961).

Eschman, D. F. 1974. Memorial to Martin Gerard Rutten 1910–1970, *Geol. Soc. Amer. Memorials*, 3: 175–81.

Everitt, C. W. F. 1959. *Studies in the Magnetism of Baked and Igneous Rocks*. Ph.D. thesis, Univ. London.

———. 1960. Rock magnetism and the origin of the Midland Basalts, *Geophys. Jour.*, 3: 203–10.

Evernden, J. F. 1951. *Direction of Approach of Rayleigh Waves and Related Problems*. Ph.D. thesis, Univ. California, Berkeley.

———. 1953. Direction of approach of Rayleigh waves and related problems (part 1), *Seismol. Soc. Amer. Bull.*, 43: 335–74.

———. 1954. Direction of approach of Rayleigh waves and related problems (part 2), *Seismol. Soc. Amer. Bull.*, 44: 159–84.

———. 1959. Dating of Tertiary and Pleistocene rocks by the potassium-argon method, *Geol. Soc. London Proc.*, 1565: 17–19.

Evernden, J. F., and G. H. Curtis. 1961. The present status of potassium-argon dating of Tertiary and Quaternary rocks, in *Intern. Quaternary Assoc. Proc. 6th Cong., Warsaw*, pp. 643–52.

———. 1965. The potassium-argon dating of Late Cenozoic rocks in East Africa and Italy, *Curr. Anthropol.*, 6: 343–85.

Evernden, J. F., G. H. Curtis, and R. W. Kistler. 1957. Potassium-argon dating of Pleistocene volcanics, *Quaternaria*, 4: 13–17.

Evernden, J. F., G. H. Curtis, R. W. Kistler, and J. Obradovich. 1960. Argon diffusion in glauconite, microcline, sanidine, leucite and phlogopite, *Amer. Jour. Sci.*, 258: 583–604.

Evernden, J. F., G. H. Curtis, and J. Lipson. 1956. Potassium-argon dating of Plio-Pleistocene intrusive rocks (California), *Nature*, 178: 1360.

———. 1957. Potassium-argon dating of igneous rocks. *Amer. Assoc. Petrol. Geol. Bull.*, 41: 2120–3126.

Evernden, J. F., G. H. Curtis, J. Obradovich, and R. Kistler. 1961. On the evaluation of glauconite and illite for dating sedimentary rocks by the potassium-argon method, *Geochim. Cosmochim. Acta*, 23: 78–99.

Evernden, J. F., and R. K. S. Evernden. 1970. The Cenozoic time scale, in *Geol. Soc. Amer. Spec. Paper* No. 124, pp. 71–90.

Evernden, J. F., and G. T. James. 1964. Potassium-argon dates and the Tertiary floras of North America, *Amer. Jour. Sci.*, 262: 945–74.

Evernden, J. F., and R. W. Kistler. 1970. *Chronology of emplacement of Mesozoic batholithic complexes in California and western Nevada* (U.S. Geol. Survey Prof. Paper 623).

Evernden, J. F., and J. R. Richards. 1962. Potassium-argon ages in eastern Australia, *Geol. Soc. Austral. Jour.*, 9, no. 1: 1–50.

Evernden, J. F., D. E. Savage, G. H. Curtis, and G. T. James. 1964. Potassium-argon dates and the Cenozoic mammalian chronology of North America, *Amer. Jour. Sci.*, 262: 145–98 (Feb. 1964; polarity-reversal scale five).

Ewing, J., and M. Ewing. 1959. Seismic-refraction profiles in the Atlantic Ocean basins, in the Mediterranean Sea, on the Mid-Atlantic Ridge and in the Norwegian Sea, *Geol. Soc. Amer. Bull.*, 70: 291–318.

Ewing, M., and B. C. Heezen. 1956. Some problems of Antarctic submarine geology, *Amer. Geophys. Union Geophys. Mon.*, 1: 75–81.

Ewing, M., B. C. Heezen, and J. Hirshman. 1957. Mid-Atlantic Ridge seismic belts and magnetic anomalies, *Seismol. Assoc. Intern. Union Geodesy Geophys. Gen. Assembly, Toronto* (abstract).

Ewing, M., J. Hirshman, and B. C. Heezen. 1959. Magnetic anomalies of the mid-ocean ridge system, in M. Sears, ed., *Preprints of Abstracts of Papers to Be Presented at Afternoon Sessions* (1st Intern. Oceanog. Cong.), p. 24 (abstract). Washington, D.C.: American Association for the Advancement of Science.

Faure, G. 1977. *Principles of Isotope Geology*. New York: John Wiley.

Fechtig, H., and S. Kalbitzer. 1966. The diffusion of argon in potassium-bearing solids, in O. A. Schaeffer and J. Zahringer, eds., *Potassium Argon Dating*, pp. 68–107. Berlin: Springer-Verlag.

Folgerhaiter, G. 1899. Sur les variations seculaires de l'inclinaison magnétique dans antiquité, *Jour. Phys.*, Ser. 3, 8: 5–16.

Folinsbee, R. E. 1941a. The chemical composition of garnet associated with cordierite, *Amer. Min.*, 26: 50–53.

———. 1941b. Optic properties of cordierite in relation to alkalies in the cordierite-beryl structure, *Amer. Min.*, 26: 485–500.

———. 1947. *Lac de Gras, Northwest Territories* (Canada Geol. Survey Paper 47–5).

———. 1949a. Determination of reflectivity of the ore minerals, *Econ. Geol.*, 44: 425–36.

———. 1949b. *Lac de Gras, District of Mackenzie, Northwest Territories* (Canada Geol. Survey Map 977A).

———. 1950. *Preliminary map, Walmsley Lake, Northwest Territories, north half* (Canada Geol. Survey Paper 50–4).

———. 1952. *Walmsley Lake, District of Mackenzie, Northwest Territories* (Canada Geol. Survey Map 1013A).

———. 1954. Cardium sandstone of central Foothills (Alberta), *Alberta Soc. Petrol. Geol. New Bull.*, 2, no. 6: 5.

———. 1955. Archean monazite in beach concentrates, Yellowknife Geologic Province, Northwest Territories, Canada, *Roy. Soc. Canada Trans.*, 3d Ser., sec. 4, 49: 7–24.

———. 1961. Report of the Subcomittee on Mineralogy, Geochemistry and Petrology, *Canada Geol. Survey Nat. Advis. Comm. on Res. in Geol. Sci., 10th Ann. Rept.*, 1959–60.

———. 1965. Fall of Revelstoke Stony Meteorite, Canada, *Meteoritical Bull.*, no. 34.

Folinsbee, R. E., and H. Baadsgaard. 1958. An absolute age for the Exshaw shale, *Alberta Soc. Petrol. Geol. Guidebook, 8th Ann. Field Conf., Nordegg*, pp. 69–73.

Folinsbee, R. E., H. Baadsgaard, and G. L. Cumming. 1963. Dating of volcanic ash beds (bentonites) by the K-Ar method, in *Nuclear Geophysics; Proceedings of a Conference, Woods Hole, Mass., June 7–9, 1962* (Natl. Res. Coun. Publ. 1075 and also Nuclear Sci. Ser. Rept. No. 38), p. 70. Washington, D.C.: National Academy of Sciences.

Folinsbee, R. E., H. Baadsgaard, and J. Lipson. 1958. Potassium-argon age dating of sediments, *Amer. Assoc. Petrol. Geol., 43d Ann. Meeting Abs.*, p. 17 (abstract).

———. 1960. Potassium-argon time scale, in T. Sorgenfrei, ed., *International Geological Congress; Report of the Twenty-first Session, Norden*, Pt. 3, pp. 7–17.

———. 1961. Potassium-argon dates of Upper Cretaceous ash falls, Alberta, Canada, *New York Acad. Sci. Ann.*, 91, no. 2: 352–59.

Folinsbee, R. E., R. H. Krouse, and A. Sasaki. 1966. Sulphur isotopes and the Pine Point lead-zinc deposits, Northwest Territories, *Econ. Geol.*, 61: 1307–8 (abstract).

Folinsbee, R. E., J. Lipson, and J. H. Reynolds. 1955. Further applications of the potassium-argon method of age determination, in *Geological Society of America, Cordilleran Section, Official Program, Berkeley, April 29, 1955*, p. 13 (abstract). Berkeley, Calif.: Geological Society of America.

———. 1956. Potassium-argon dating, *Geochim. Cosmochim. Acta*, 10: 60–68.

Folinsbee, R. E., and J. C. Moore. 1950. *Preliminary map, Matthews Lake, Northwest Territories* (Canada Geol. Survey Paper 50–2).

Folinsbee, R. E., and W. D. Ritchie. 1957. Late Cretaceous geochronology, *Union Géodésique et Géophys. Intern., XIe Assem. Gen., Abstracts*, p. 19.

Folinsbee, R. E., W. D. Ritchie, and G. F. Stansberry. 1957. The Crowsnest volcanics and Cretaceous Geochronology, in *Alberta Society of Petroleum Geologists Guidebook, 7th Annual Field Conference, Waterton*, pp. 20–26. Alberta: Alberta Society of Petroleum Geologists.

Foster, J. H. 1966. A paleomagnetic spinner magnetometer using a fluxgate gradiometer, *Earth Planet. Sci. Lett.*, 1, no. 6: 463–66.

Frankel, H. 1979. Why drift theory was accepted with the confirmation of Harry Hess's concept of sea-floor spreading, in C. J. Schneer, ed., *Two Hundred Years of Geology in America* (Proc. New Hampshire Bicentennial Conf. on the History of Geology, 1976), pp. 337–53. Hanover, N.H.: University Press of New England.

Fuller, N. D., C. G. A. Harrison, and Y. R. Nayudu. 1966. Magnetic and petrologic studies of sediments found above basalt in experimental Mohole core EM7, *Amer. Assoc. Petrol. Geol. Bull.*, 50: 556–73.

Gauss, C. F. 1839. *Allgemeine Theorie des erdmagnetis Mus*. Leipzig. (Also in Gauss, Werke 5, Göttingen, 1877).

Gelletich, H. 1937. Über magnetitführende eruptive Gange und Gangsysteme im mittleven Teil des südlichen Transvaals, *Beitr. Angew. Geophys.*, 6: 337–406.

Gentner, W., K. Goebel, and R. Prag. 1954. Argonbestimmungen an Kaliummineralien, III; vergleichende Messungen nach der Kalium-Argon- und Ruan-Helium-Methode, *Geochim. Cosmochim. Acta*, 5: 124.

Gentner, W., R. Prag, and F. Smits. 1953. Argonbestimmungen an Kaliummineralien, II. Das Alter eines Kalilagers im unteren Oligozän, *Geochim. Cosmochim. Acta*, 4: 11.

Gerling, E. K., and T. G. Pavlova. 1951. Determination of the geological age of two stony meteorites by the argon method, *Akad. Nauk SSSR Doklady*, 77: 85.

Gerling, E. K., and N. E. Titov. 1949. Über den K-Zerfall des Kaliums (in Russian), *Akad. Nauk SSSR Izv.*, 2: 128.

Gerling, E. K., N. E. Titov, and G. M. Ermolin. 1949. Determination of the K-capture constant of K^{40}, *Akad. Nauk SSSR Doklady*, 68: 553.

Gilbert, C. M. 1938. Welded tuff in eastern California, *Geol. Soc. Amer. Bull.*, 49: 1829–62.

Gilbert, William. 1600. *De Magnete* Book 4, Chap. 3 *fide* Runcorn, S. K. 1964. Rock magnetism, *Science*, 129: 1002–11.

Gilliland, W. N. 1964. Extension of the theory of zonal rotation to explain global fracturing, *Nature*, 202: 1276.

Girdler, R. W. 1958. The relationship of the Red Sea to the East African rift system, *Geol. Soc. London Quart. Jour.*, 114: 79–105.

———. 1959–60. The case for an expanding Earth, *Jour. Durham Colleges Geol. Soc.* (reprint from President's Report), pp. 1–5.

———. 1963. Geophysical evidence on the nature of magmas and intrusions associated with rift valleys, *Bull. Volcanol.*, 26: 37–47.

———. 1964. Geophysical studies of rift valleys, in L. H. Ahrens, F. Press, and S. K. Runcorn, eds., *Physics and Chemistry of the Earth*, vol. 5, pp. 121–56. Oxford: Pergamon Press.

Girdler, R. W., and G. Peter. 1960. An example of the importance of natural remanent magnetization in the interpretation of magnetic anomalies, *Geophys. Prospecting*, 8: 474–83.

Glen, William. 1975. *Continental Drift and Plate Tectonics*. Columbus, Ohio: Charles E. Merrill.

Godard, L. 1950. Thesis. Paris *fide* Wilson, R. L., 1962b: 198, 202.

Goldich, S. S. 1961. *The Precambrian geology and geochronology of Minnesota*. Minnesota Geol. Survey Bull. 41.

Goldich, S. S., H. Baadsgaard, G. Edwards, and C. E. Weaver. 1959. Investigations in radio-activity-dating of sediments, *Amer. Assoc. Petrol. Geol. Bull.*, 53: 654–62.

Goldich, S. S., H. Baadsgaard, and A. O. Nier. 1957. Investigations in A^{40}/K^{40} dating, *Amer. Geophys. Union Trans.*, 38: 547–51.

Gorter, E. W., and J. A. Schulkes. 1953. Reversal of spontaneous magnetization as a function of temperature in LiFeCr spinels, *Phys. Rev.*, 90: 487–88.

Gough, D. I. 1956. A study of the palaeomagnetism of the Pilansberg dykes, *Roy. Astron. Soc. Geophys. Supp. Mon. Not.*, 7: 196–213.

———. 1967. The spinner magnetometer at Salisbury, in D. W. Collinson, K. M. Creer, and S. K. Runcorn, eds., *Methods in Palaeomagnetism*, pp. 119–30. Amsterdam: Elsevier.

Gough, D. I., and C. B. Van Niekerk. 1959. A study of palaeomagnetism of the Bushveld Gabbro, *Phil. Mag.*, Ser. 8, 4: 126–36.

Gouldner, W. G., and A. L. Beach. 1950. Vacuum fusion furnace for analysis of gases in metals, *Anal. Chem.*, 22: 366–71.

Graham, J. W. 1949. The stability and significance of magnetism in sedimentary rocks, *Jour. Geophys. Res.*, 54: 131–67.

———. 1952. Note on the significance of inverse magnetization of rocks, *Jour. Geophys. Res.*, 57: 429–31.

———. 1953. Changes of ferromagnetic minerals and their bearing on ferromagnetic properties of rocks, *Jour. Geophys. Res.*, 58: 243–60.

———. 1954a. Rock magnetism and the Earth's magnetic field during Paleozoic time, *Jour. Geophys. Res.*, 59: 215–22.
———. 1954b. Magnetic susceptibility anisotropy, an unexploited petrofabric element, *Geol. Soc. Amer. Bull.*, 65: 1257–58.
———. 1955. Evidence of polar shift since Triassic time, *Jour. Geophys. Res.*, 60: 329–47.
———. 1956. Palaeomagnetism and magnetostriction, *Jour. Geophys. Res.*, 61: 735–39.
———. 1957. The role of magnetostriction in rock magnetism, *Adv. Phys.*, 6: 362–63.
Graham, J. W., A. F. Buddington, and J. R. Balsley. 1957. Stressed-induced magnetizations of some rocks with analyzed magnetic minerals, *Jour. Geophys. Res.*, 62: 465–74.
Graham, K. W. T., and A. L. Hales. 1957. Palaeomagnetic measurements on Karroo dolerites, *Adv. Phys.*, 6: 149–61.
Gregory, J. W. 1921. *The Rift Valleys and Geology of East Africa.* London: Seeley.
Grenet, G. 1945. Quelques mesures d'aimantation permanente de roches du Massif Central et remarques sur les méthodes de détermination de la valeur du champ magnétique terrestre dans le passé, *Ann. Geophys.*, 1, no. 3: 256–63.
Griffith, D. H., and R. F. King, 1954. Natural magnetization of igneous and sedimentary rocks, *Nature*, 173: 1114.
Griggs, D. T. 1939. A theory of mountain building, *Amer. Jour. Sci.*, 237: 611–50.
Grommé, C. S. 1963. *Remanent magnetization of igneous rocks from the Franciscan and Lovejoy Formations, northern California.* Ph.D. thesis, Univ. California, Berkeley.
Grommé, C. S., and R. L. Hay. 1963. Magnetizations of basalt of Bed I, Olduvai Gorge, Tanganyika, *Nature* 200: 560–61 (Nov. 9, 1963; polarity-reversal scale three).
———. 1967. Geomagnetic polarity epochs—new data from Olduvai Gorge, Tanganyika, *Earth Planet. Sci. Lett.*, 2: 111–15.
———. 1971. Geomagnetic polarity epochs; age and duration of the Olduvai normal polarity event, *Earth Planet. Sci. Lett.*, 10: 179–85.
Haarmann, E. 1926. Tektogenese oder gefugebildung statt orogenese oder gebirgsbildung. *Deutsche Geologische Gesellschaft Zeitschrift*, 78: 105–7.
Hales, A. L. 1936. Convection currents in the Earth, *Roy. Astron. Soc. Geophys. Supp. Mon. Not.*, 3: 372–79.
Hallam, A. 1973. *A Revolution in the Earth Sciences from Continental Drift to Plate Tectonics.* Oxford: Clarendon Press.
Harper, C. T., ed. 1973. *Geochronology; Radiometric Dating of Rocks and Minerals.* Stroudsburg, Penn.: Dowden, Hutchinson, and Ross.
Harrison, C. G. A. 1966. The paleomagnetism of deep sea sediments, *Jour. Geophys. Res.*, 71: 3033–43.
———. 1968. Formation of magnetic anomaly patterns by dyke injection, *Jour. Geophys. Res.*, 73: 2137–42.

Harrison, C. G. A., and B. M. Funnell. 1964. Relationship of palaeomagnetic reversals and micropalaeontology in two late Cainozoic cores from the Pacific Ocean, *Nature*, 204: 566.

Hart, S. R. 1961. The use of hornblendes and pyroxenes for K-Ar dating, *Jour. Geophys. Res.*, 66: 2995–3001.

Hart, S. R., and R. T. Dodd. 1962. Excess radiogenic argon in pyroxenes, *Jour. Geophys. Res.*, 67: 2998–99.

Hart, S. R., H. W. Fairbairn, W. H. Pinson, and P. M. Hurley. 1960. Use of amphiboles and pyroxenes for K-Ar dating, *Geol. Soc. Amer. Bull.*, 71: 1882 (abstract).

Hawkes, H. E., and J. R. Balsley. 1946. *Magnetic exploration for iron ore in northern New York*. U.S. Geol. Survey Strategic Min. Inv. Prelim. Rept. 3-194.

Hay, R. L. 1963. Stratigraphy of Beds I through IV, Olduvai Gorge, Tanganyika, *Science*, 139: 829–31.

Hayden, R. J., J. H. Reynolds, and M. G. Inghram. 1949. Reactions induced by slow neutron irradiation of europium, *Phys. Rev.*, 75: 1500.

Hedberg, H. D. 1977. *Preliminary report in magnetostratigraphic classification*. Intern. Subcomm. Stratig. Classification Circ. 55.

Heezen, B. C. 1959. Géologie sous-marine et déplacements des continents, in *La topographie et la géologie des profondeurs océaniques*, pp. 295–304 (with English summary). Paris: France Centre Natl. Recherche Sci.

———. 1960. The rift in the ocean floor, *Sci. Amer.*, 203: 98–110 (Oct.).

Heezen, B. C., and M. Ewing. 1961. The mid-oceanic ridge and its extension through the Arctic basin, in G. O. Raasch, ed., *Geology of the Arctic*, vol. 7, pp. 622–42. Toronto: University of Toronto Press.

Heezen, B. C., M. Ewing, and E. T. Miller. 1953. Trans-Atlantic profile of total magnetic intensity and topography, Dakar to Barbados, *Deep-Sea Res.*, 1, no. 1: 25–33.

Heezen, B. C., M. Tharp, and W. M. Ewing. 1959. *The Floor of the Oceans. I, North Atlantic* (Geol. Soc. Amer. Spec. Paper No. 65). Washington, D.C.: Geological Society of America.

Heirtzler, J. R. 1961. *Vema cruise no. 16 magnetic measurements* (Columbia Univ. Lamont Geol. Observ. Tech. Rept. 2). New York: Columbia University.

Heirtzler, J. R., G. D. Dickson, E. M. Herron, W. C. Pitman III, and X. Le Pichon. 1968. Marine magnetic anomalies, geomagnetic field reversals and motions of the ocean floor and continents, *Jour. Geophys. Res.*, 73: 2119–36.

Heirtzler, J. R., and X. Le Pichon. 1965. Crustal structure of the mid-ocean ridges [pt.]3, Magnetic anomalies over the Mid-Atlantic Ridge, *Jour. Geophys. Res.*, 70: 4013–33 (Aug. 15, 1965).

Heirtzler, J. R., X. Le Pichon, and J. G. Baron. 1966. Magnetic anomalies over the Reykjanes Ridge, *Deep-Sea Res.*, 13: 427–43.

Hess, H. H. 1933. The problem of serpentinization and the origin of certain chrysotile asbestos talc and soapstone deposits, *Econ. Geol.*, 28: 634–57.

———. 1938a. A primary pendotite magma. *Amer. Jour. Sci.*, 35: 321–44.

———. 1938b. Gravity anomalies and island arc structures, with particular reference to the West Indies. *Amer. Phil. Soc. Proc.*, 79: 71–96.

———. 1954. Serpentines, orogeny, and epeirogeny, in A. Poldervaart, ed., *Crust of the Earth—A Symposium* (Geol. Soc. Amer. Spec. Paper No. 62), pp. 391–408. Washington, D.C.: Geological Society of America.

———. 1962. History of the ocean basins, in A. E. J. Engel, H. L. James, and B. F. Leonard, eds., *Petrologic Studies: A Volume in Honor of A. F. Buddington*, pp. 599–620. New York: Geological Society of America.

———. 1965. Mid-oceanic ridges and tectonics of the sea floor, in W. F. Whittard and R. Bradshaw, eds., *Submarine Geology and Geophysics*, pp. 317–32. London: Butterworth.

———. 1968. Reply, *Jour. Geophys. Res.*, 73: 6569.

Hibbard, F. 1961. Secondary magnetization and the paleomagnetism of some Pliocene rocks of Japan, *Jour. Geomag. Geoelec.*, 12: 222–26.

Hilde, T. W. C. 1964. Magnetic profiles across the Gulf of California, in T. H. van Andel and G. G. Shor, eds., *Marine Geology of the Gulf of California* (Amer. Assoc. Petrol. Geol. Mem. 3), pp. 122–25. Tulsa, Okla.: American Association of Petroleum Geologists.

Hill, M. N. 1960. A median valley of the Mid-Atlantic Ridge, *Deep-Sea Res.*, 6: 193–205.

Hillhouse, J. W. 1977. Paleomagnetism of the Triassic Nikolai Greenstone, McCarthy Quadrangle, Alaska, *Canadian Jour. Earth Sci.*, 14: 2578–92.

Holmes, A. 1911. The association of lead with uranium in rock-minerals, and its application to the measurement of geologic time, *Proc. Roy. Soc.*, Ser. A, 85: 245–56.

———. 1913. *The Age of the Earth*. London: Harper Brothers.

———. 1929. Radioactivity and Earth movements, *Geol. Soc. Glasgow Trans.*, 18: 559–606 (discussion, pp. 614–15).

Hospers, J. 1951. Remanent magnetism of rocks and the history of the geomagnetic field, *Nature*, 168: 1111–12.

———. 1953. Reversals of the main geomagnetic field, I, II, *Kon. Ned. Akad. Wetensch. Proc.*, Ser. B, 56: 467–91.

———. 1954a. Reversals of the main geomagnetic field, III, *Kon. Ned. Akad. Wetensch. Proc.*, Ser. B, 57: 112–21.

———. 1954b. Magnetic correlation in volcanic districts, *Geol. Mag.*, 91: 352–60.

———. 1955. Rock magnetism and polar wandering, *Jour. Geol.*, 63: 59–74.

Houtermans, F. G. 1966. History of the K/Ar method of geochronology, in O. A. Schaeffer, and J. Zahringer, comps., *Potassium Argon Dating*, pp. 1–6. New York: Springer-Verlag

Hudson, L. 1966. *Contrary Imaginations*. New York: Schocken Books.

Humboldt, Alexander von. 1797. Ueber die merkwürdige magnetische Polarität eine Gebirgskuppe von Serpentinstein, Gren's *Neues Journal der Physik*, 4: 136–40.

Hurley, P. M., and C. D. Goodman. 1941. Helium retention in common rock and minerals, *Geol. Soc. Amer. Bull.*, 52: 545–59.

———. 1943. Helium age measurements; I, Preliminary magnetite index, *Geol. Soc. Amer. Bull.*, 54: 305–23.

Inghram, M. G. 1953. Trace element determination by the mass spectrometer, *Jour. Phys. Chem.*, 57: 809.

———. 1954. Stable isotope dilution analysis, *Ann. Rev. Nucl. Sci.*, 4: 81.

Inghram, M. G., H. Brown, C. Patterson, and D. C. Hess, Jr. 1950. The branching ratio of K^{40} radioactive decay, *Phys. Rev.*, 80: 916.

Inghram, M. G., D. C. Hess, Jr., and J. H. Reynolds. 1949. On the relative yields of fission cesium isotopes, *Phys. Rev.*, 76: 1717.

Inghram, M. G., and J. H. Reynolds. 1949. On the double beta-process, *Phys. Rev.*, 76: 1265.

———. 1950. Double beta-decay of tellurium 130, *Phys. Rev.*, 78: 822.

Inglis, D. R. 1955. Theory of the earth's magnetism, *Rev. Mod. Physics* 27: 212.

Irving, E. M. 1954. *The Palaeomagnetism of the Torridonian Sandstone Series of North-western Scotland*. M.Sc. thesis, Univ. Cambridge.

———. 1956a. The magnetization of the Mesozoic dolerites of Tasmania, *Roy. Soc. Tasmania Pap. Proc.*, 90: 157–68.

———. 1956b. Palaeomagnetic and palaeoclimatological aspects of polar wandering, *Geofis. Pura Appl.*, 33: 23–41.

———. 1957a. The origin of the palaeomagnetism of the Torridonian Sandstone Series of Northwest Scotland, *Roy. Soc. London Phil. Trans.*, Ser. A, 250: 100–10.

———. 1957b. Directions of magnetization in the Carboniferous glacial varves of Australia, *Nature*, 180: 280–81.

———. 1957c. Rock magnetism; a new approach to some palaeogeographic problems, *Adv. Phys.*, 6: 194–218.

———. 1958a. Rock magnetism; a new approach to the problems of polar wandering and continental drift, in S. W. Carey, ed. *Continental Drift—A Symposium*, pp. 24–61. Hobart: University of Tasmania.

———. 1958b. Palaeogeographic reconstruction from palaeomagnetism, *Roy. Astron. Soc. Geophys. Jour.*, 1: 224–37.

———. 1959. Palaeomagnetic pole positions; a survey and analysis, *Roy. Astron. Soc. Geophys. Jour.*, 2: 51–79.

———. 1964. *Paleomagnetism and Its Application to Geological and Geophysical Problems*. New York: Wiley.

Irving, E. M., and N. D. Opdyke. 1965. The paleomagnetism of the Bloomsburg redbeds and its possible application to the tectonic history of the Appalachians, *Roy. Astron. Soc. Geophys. Jour.*, 9: 153–67.

Irving, E. M., J. K. Park, S. E. Haggerty, F. Aumento, and B. Loncarevic.

1970. Magnetism and opaque mineralogy of basalts from the Mid-Atlantic Ridge at 45°N, *Nature*, 228: 974–76.

Irving, E. M., and L. G. Parry. 1963. The magnetism of some Permian rocks from New South Wales, *Roy. Astron. Soc. Geophys. Jour.*, 7: 395–411.

Irving, E. M., and S. K. Runcorn. 1957. Analysis of the palaeomagnetism of the Torridonian Sandstone Series of Northwest Scotland I, *Roy. Soc. London Phil. Trans.*, Ser. A, 250: 83–99.

Isacks, B., J. Oliver, and L. R. Sykes. 1968. Seismology and the new global tectonics, *Jour. Geophys. Res.*, 73: 5855–99.

Ishikawa, Y., and Y. Syono. 1963. Order-disorder transformation and reverse thermoremanent magnetism in the $FeTiO_3$–Fe_2O_3 system, *Phys. Chem. Solids*, 24: 517.

Ising, G. 1943. On the magnetic properties of varved clay, I, Line of investigation; measurements on a varve series from Viby in southern Sweden, *Arkiv Mat. Astron. Fysik* (*K. Svenska Vetensk*) 29A, pt. 1, no. 5: 1–37.

Jacobs, J. A., R. D. Russel, and J. T. Wilson. 1974. *Physics and Geology*, 2d ed. New York: McGraw-Hill.

Jaeger, J. C. 1957a. The temperature in the neighborhood of a cooling intrusive sheet, *Amer. Jour. Sci.* 255: 306–18.

———. 1957b. The variation of density and magnetic properties in dolerite sills, *Roy. Soc. Tasmania Pap. Proc.*, 91: 143–44.

Jaeger, J. C., and R. Green. 1956. The use of the cooling history of thick intrusive sheets for the study of the secular variation of the Earth's magnetic field, *Geofis. Pura Appl.*, 35: 49–53.

———. 1958. A cross-section of a tholeiite sill, in *Dolerite, A Symposium*, pp. 26–37. Hobart: University of Tasmania.

Jaeger, J. C., and E. M. Irving. 1957. Paleomagnetism and the reconstructions of Gondwanaland, *Compt. Rend. 3rd Cong. Pacific Indian Ocean Sci. Tananarive*, pp. 233–42.

Jaeger, J. C., and G. A. Joplin. 1955. Rock magnetism and the differentiation of a dolerite sill, *Geol. Soc. Austral. Jour.*, 2: 1–19.

Jaeger, J. C., and R. F. Thyer. 1960. Report on progress in geophysics; geophysics in Australia, *Roy. Astron. Soc. Geophys. Jour.*, 3: 450–61.

Jaggar, T. A., Jr. 1901. The laccoliths of the Black Hills, *U.S. Geol. Survey Ann. Rept.* 21, no. 3: 163–303.

Jeffreys, H. 1970. *The Earth*. 5th ed. Cambridge: University Press.

Jensen, H., and J. R. Balsley. 1946. Controlling plane position in aerial magnetic surveying, *Eng. Min. Jour.*, 147, no. 8: 94–95, 153–54.

Johnson, E. A., T. Murphy, and O. W. Torreson. 1948. Pre-history of the Earth's magnetic field, *Terr. Magn. Atmos. Elec.*, 53: 349–72.

Kato, Y., A. Takagi, and I. Kato. 1954. Reverse remanent magnetism of dyke and basaltic andesite, *Jour. Geomag. Geoelec.*, 6: 206–07.

Kawai, N. 1954. Instability of natural remanent magnetism of rocks, *Jour. Geomag. Geoelec.*, 6: 208–09.

Kawai, N., S. Kume, and S. Sasajima. 1954. Magnetism of rocks and solid phase transformation in ferromagnetic minerals; I and II, *Japan Acad. Proc.*, 30: 588–93, 865–68.

Keen, M. J. 1963. Magnetic anomalies over the Mid-Atlantic Ridge, *Nature*, 197: 888–90.

Keevil, N. B., A. W. Jolliffe, and E. S. Larsen, Jr. 1942. Helium age investigations of diabase and granodiorites from Yellowknife, Northwest Territories, Canada, *Amer. Jour. Sci.*, 240: 831–46.

Keller, F., J. R. Balsley, and W. J. Dempey. 1947. Field operations and compilation procedure incidental to the preparation of isomagnetic maps, *Photogram. Eng.*, 13: 644–47.

Kelvin, W. Thompson, Lord. 1899. The age of the earth as an abode fitted for life, *Phil. Mag.*, Ser. 5, 47: 66–90.

Khramov, A. N. 1955a. Issledovanie vozmozhnosti stratigraficheskoi korreliatsii osadochnykh tolshch magnitometricheskin metodon [a study of the possibility of stratigraphic correlation of sedimentary layers by the magnetometric method], *Vsesoyuz. Neft. Nauch.-Issled. Geol.-Razv. U., Avtoreferatz Nauchnykh Tr.* 16: 104–6.

———. 1955b. Iznchenie ostatochnoi namagnichennosti osadkov v sviazi i problemoi stratigraficheskoi korreliatsii i raschleneniia nenykh tolshch, *Akad. Nauk SSSR Doklady*: 100: 551–54.

———. 1956. O vozmozhnosti stratigraficheskoi korrelyatsii i raschleneniya osadochnykh tolschch po vektoru ostatochnoi namagnichennosti. *Vsesoyuz. Neft. Nauch.-Issled. Geol.-Razv. Inst. Tr.* (geol. Sbornik 2) 95: 198–208.

———. 1957. Paleomagnetism; the basis of a new method of correlation and subdivision of sedimentary strata, *Akad. Nauk SSSR Doklady* geol. ser. 114: 849–52 (in Russian), pp. 129–32 (English transl., 1958).

———. 1958. *Paleomagnetism and Stratigraphic Correlation*. Leningrad: Gostoptechizdat. English translation by A. J. Lajkine. 1960. Canberra: Geophysics Department, Australian National University.

Khramov, A. N., and G. N. Petrova. 1972. Paleomagnetism with special reference to research in the U.S.S.R., in A. R. Ritsema, ed., *The Upper Mantle: Tectonophysics*, vol. 13, pp. 325–40.

Kirsten, T. 1966. Determination of radiogenic argon, in O. A. Schaeffer, and J. Zahringer, comp., *Potassium Argon Dating*, pp. 7–39. Berlin: Springer-Verlag.

Kleinpell, R. M. 1938. *Miocene Stratigraphy of California*. Tulsa, Okla.: American Association of Petroleum Geologists.

Klemperer, O. 1935. On the radioactivity of potassium and rubidium, *Roy. Soc. London Proc.*, Ser. A, 148: 638–48.

Knopf, A. 1956. Argon-potassium determination of the age of the Boulder Bathylith, Montana, *Amer. Jour. Sci.*, 254: 744–45.

———. 1957. The Boulder Bathylith of Montana, *Amer. Jour. Sci.*, 255: 81–103.

Knopoff, L. 1964. The convection current hypothesis, *Rev. Geophys.*, 2: 89–122.

Koenigsberger, J. G. 1930. Grossenverhaltnis von remanenten zu induziertem Magnetismus in Gesteinen; Grosse und Richtung des remanenten zu induzierten Magnetismus. *Z. Geophys.*, 6: 190–207.

———. 1934. Magnetische Eigenschaften der ferromagnetischen Mineralien in den Gesteinen, *Beitr. Angew. Geophys.*, 4: 385–406.

———. 1935. Die Abhangigkeit der naturlichen remanenten Magnetisierung bei Eruptivgesteinen von deren Alter und Zusammensetzung, *Beitr. Angew. Geophys.*, 5: 193–246.

———. 1936. Residual magnetism and the measurement of geologic time, *S.-A. Rept., Internat. Geol. Cong. 16th Wash., D.C.* 7S: 225–31.

———. 1938. Natural residual magnetism of eruptive rocks, parts I and II, *Terr. Mag. Atmos. Elec.*, 43: 119–27, 299–320.

Krummenacher, D., and J. F. Evernden. 1960. Determination d'âge isotopique faites sur quelques roches des Alpes par la méthode potassium-argon, *Bull. Suisse Min. Pétr.*, 40: 267–77.

Krummenacher, D., J. F. Evernden, and M. Vuagnat. 1960. Sur l'âge absolu de la péridotite micacée de Finero (zone d'Ivrée), *Archives Sci.*, 13: 369–73.

Kuhn, T. S. 1962. *The Structure of Scientific Revolutions*. Chicago: University of Chicago Press. Second edition, 1970.

———. 1963. The essential tension; traditions and innovation in scientific research, in C. W. Taylor, and F. Barron, eds., *Scientific Creativity, Its Recognition and Development*, pp. 341–54. New York: Wiley.

Kuno, H. 1959. Origin of Cenozoic petrographic provinces of Japan and surrounding areas, *Bull. Volcanol.*, 20: 37–76.

Larmor, J. 1919. How could a rotating body such as the sun become a magnet?, *Brit. Assoc. Adv. Sci. Rept.*, 87: 159–60.

Laughton, A. S., M. N. Hill, and T. D. Allan. 1960. Geophysical investigations of a seamount 150 miles north of Madeira, *Deep-Sea Res.*, 7: 117–41.

Leakey, L. S. B., J. F. Evernden, and G. H. Curtis. 1961. The age of Bed I, Olduvai Gorge, Tanganyika, *Nature*, 191: 478.

Lear, J. 1967. Canada's unappreciated role as scientific innovator, *Sat. Rev.*, 50: 45–50 (Sept. 2, 1967).

Le Pichon, X. 1968. Sea floor spreading and continental drift, *Jour. Geophys. Res.*, 73: 3661–97.

Le Pichon, X., and J. R. Heirtzler. 1968. Magnetic anomalies in the Indian Ocean and sea floor spreading, *Jour. Geophys. Res.*, 73: 2101–17.

Lin'kova, T. I. 1965. [Some results of paleomagnetic study of Arctic Ocean floor sediments], in *Natoyascheye i Proshloye Magnetnozo Zemli*, pp. 279–81. Moscow: Nauka. Translated by E. R. Hope. 1966. Canada Dir. Sci. Inf. Serv. Publ. T 463R

Lipson, J. J. 1956. K-A dating of sediments, *Geochim. Cosmochim. Acta*, 10: 149–51.

———. 1958. Potassium-argon dating of sedimentary rocks, *Geol. Soc. Amer. Bull.*, 69: 137.

Lipson, J. J., and J. H. Reynolds. 1954. Performance data for a high sensitivity mass spectrometer, *Phys. Rev.*, 98: 283.

Lipson, J. J., J. H. Reynolds, and R. E. Folinsbee. 1956. Potassium-argon dating, *Geochim. Cosmochim. Acta*, 10: 60.

Lyell, Charles. 1830–35. *Principles of Geology*. London: J. Murray.

McClelland, D. C. 1970. On the dynamics of creative physical scientists, in L. Hudson, ed., *The Ecology of Human Intelligence*. Harmondsworth, Eng.: Penguin.

McDougall, I. 1961. Determination of the age of a basic igneous intrusion by the potassium-argon method, *Nature*, 190: 1184–86 (June 24, 1961).

———. 1963a. Potassium-argon ages from western Oahu, Hawaii, *Nature*, 197: 344–45 (Jan. 26, 1963).

———. 1963b. Potassium-argon age measurements on dolerites from Antarctica and South Africa, *Jour. Geophys. Res.*, 68: 1535–45 (Mar. 1, 1963).

———. 1963c. Potassium-argon ages of some rocks from Viti Levu, Fiji, *Nature*, 198: 677 (May 18, 1963).

———. 1964. Potassium-argon ages from lavas of the Hawaiian Islands, *Geol. Soc. Amer. Bull.*, 75: 107.

———. 1977. *The present status of the geomagnetic polarity time scale* (Austral. Natl. Univ. Res. Sch. Earth Sci. Publ. no. 1288). Canberra: Australian National University, Research School of Earth Sciences.

McDougall, I., W. Compston, and D. D. Hawkes. 1963. Leakage of radiogenic argon and strontium from minerals in Proterozoic dolerites from British Guayana, *Nature*, 198: 564.

McDougall, I., and D. H. Green. 1964. Excess radiogenic argon in pyroxenes and isotopic ages on minerals from Norwegian eclogites, *Norsk. Geol. Tidsskr.*, 44: 183.

McDougall, I., and D. H. Tarling. 1963a. Dating of polarity zones in the Hawaiian Islands, *Nature*, 200: 54–56 (Oct. 5, 1963; polarity-reversal scale two).

———. 1963b. Dating of reversals of the Earth's magnetic field, *Nature*, 198: 1012–13.

———. 1964. Dating geomagnetic polarity zones, *Nature*, 202: 171–72 (Apr. 11, 1964; polarity-reversal scale six).

McDougall, I., and H. Wensink. 1966. Paleomagnetism and geochronology of the Pliocene-Pleistocene lavas in Iceland, *Earth Planet. Sci. Lett.*, 1: 232–36.

McElhinny, M. W. 1973. *Palaeomagnetism and Plate Tectonics*. Cambridge: University Press.

McElhinny, M. W., and P. J. Burek. 1971. Mesozoic palaeomagnetic stratigraphy, *Nature*, 232: 98–102.

McKenzie, D. P., and W. J. Morgan. 1969. The evolution of triple junctions, *Nature*, 224: 125–33.

McKenzie, D. P., and R. L. Parker. 1967. The North Pacific; an example of tectonics on a sphere, *Nature*, 216: 1276–80.

McNish, A. G., and E. A. Johnson. 1938. Magnetization of unmetamorphosed varves and marine sediments, *Terr. Magn. Atmos. Elec.*, 53: 349–60.

Malde, H. E., and H. A. Powers. 1962. Upper Cenozoic stratigraphy of western Snake River Plain, Idaho, *Geol. Soc. Amer. Bull.*, 73: 1197–1219.

Malde, H. E., H. A. Powers, and C. H. Marshall. 1963. *Reconnaissance geologic map of west-central Snake River Plain, Idaho.* U.S. Geol. Survey Misc. Geol. Inv. Map I–373.

Marble, J. P. 1954. *Report of the Committee on the Measurement of Geologic Time, 1952–53.* Natl. Acad. Sci., Natl. Res. Coun. Publ. 319.

———. 1957. *Report of the Committee on the Measurement of Geologic Time, 1954–55.* Natl. Acad. Sci., Natl. Res. Coun. Publ. 500.

Marinelli, G. 1967. Genèse des magmas du volcanisme plio-quaternaire des Apennins, *Geol. Rundsch.*, 57, no. 1: 127–41.

Marinelli, G., and M. Mittempergher. 1966. On the genesis of some magmas of typical Mediterranean (potassic) suite, *Bull. Volcanol.*, 29: 113–40.

Marvin, U. B. 1973. *Continental Drift; the Evolution of a Concept.* Washington, D.C.: Smithsonian Institution Press.

Mason, R. G. 1958. A magnetic survey off the west coast of the United States between latitudes 32° and 36°N longitudes 121° and 128°W, *Roy. Astron. Soc. Geophys. Jour.*, 1: 320–29.

———. 1967. Magnetic surveys, results of, in S. K. Runcorn, ed., *International Dictionary of Geophysics*, vol. 2, pp. 878–96. Oxford: Pergamon Press.

Mason, R. G., and A. D. Raff. 1961. A magnetic survey off the west coast of North America 32°N to 42°N, *Geol. Soc. Amer. Bull.*, 72: 1259–65.

Matthews, D. H. 1961. Lavas from an abyssal hill on the floor of the Atlantic Ocean, *Nature*, 190: 158–59.

———. 1967. Mid-ocean ridges, in S. K. Runcorn, ed., *International Dictionary of Geophysics*, vol. 2, pp. 979–91. Oxford: Pergamon Press.

Matthews, D. H., and J. Bath. 1967. Formation of magnetic anomaly pattern of Mid-Atlantic Ridge, *Roy. Astron. Soc. Geophys. Jour.*, 13: 349–57.

Matthews, J., and W. Gardner. 1963. *Analysis of an axially symmetric two-disc dynamo.* U.S. Naval Res. Lab. Rept. 5886.

Mattias, P. P., and U. Ventriglia. 1970. La regione vulcanica dei Monti Sabatini e Cimini, *Soc. Geol. Italia Mem.*, 9: 331–84.

Matuyama, M. 1929a. On the direction of magnetization of basalt in Japan, Tyosen and Manchuria, *Pacific Sci. Cong., 4th Proc.*, pp. 1–3.

———. 1929b. On the direction of magnetization of basalt in Japan, Tyosen and Manchuria, *Imp. Acad. Japan Proc.* 5: 203–5.

Maxwell, A. E., and R. Revelle. 1956. Heat flow through the Pacific Ocean Basin, *Publ. Bur. Cent. Seismol. Intern. Trav. Sci.*, 19: 395–405.

Melloni, M. 1853a. Ricerche intorno al magnetismo delle roccie, *R. Ac. delle Sci. Napoli*, 1: 121.

———. 1853b. Sur l'aimantation des roches volcaniques, *Acad. Sci. Paris Compt. Rend.*, 37: 299.

———. 1853c. Sur le magnétisme des roches, *Acad. Sci. Paris Compt. Rend.*, 37: 966.

Menard, H. W. 1955. Deformation of the northeastern Pacific basin and the west coast of North America, *Geol. Soc. Amer. Bull.*, 66: 1149–98.

———. 1958. Development of median elevations in the ocean basins, *Geol. Soc. Amer. Bull.*, 69: 1179–86.

———. 1964. *Marine Geology of the Pacific*. New York: McGraw-Hill.

———. 1966. Fracture zones and offsets of the East Pacific Rise, *Jour. Geophys. Res.*, 71: 682–85.

———. 1971. *Science; Growth and Change*. Cambridge, Mass.: Harvard University Press.

———. 1979. Very much like a spear, in C. J. Schneer, ed., *Two Hundred Years of Geology in America* (Proc. New Hampshire Bicentennial Conf. on the History of Geology, Univ. New Hampshire, 1976). Hanover, N.H.: University Press of New England.

Menard, H. W., and T. Atwater. 1968. Changes in the direction of sea floor spreading, *Nature*, 219: 463–67.

Menard, H. W., and V. Vacquier. 1958. Magnetic survey of part of the deep sea floor off the coast of California, *U.S. Office of Naval Res. Res. Rev.* (June 1958), pp. 1–5.

Mercanton, P. L. 1910. Physique du globe, *Acad. Sci. Paris Compt. Rend.*, 151: 1092–97.

———. 1926a. Inversion de l'inclinaison magnétique terrestre aux âges géologiques, *Terr. Magn. Atmos. Elec.*, 31: 187–90.

———. 1926b. Magnétisme terrestre—aimentation de basaltes groenlandais, *Acad. Sci. Paris Compt. Rend.*, 182: 859–60.

———. 1931. Inversion inclinaison magnétique aux âges géologiques, nouvelles observations, *Acad. Sci. Paris Compt. Rend.*, 192: 978–80.

———. 1932. Inversion inclinaison magnétique aux âges géologiques, *Acad. Sci. Paris Compt. Rend.*, 194: 1371.

———. 1933. Inversion de l'inclinaison magnétique aux âges géologiques; nouvelles observations, *Acad. Sci. Paris Compt. Rend.*, 196: 16.

Meyer, S., and E. Schweidler. 1927. *Radioaktivitat*. 2nd ed. Leipzig: B. G. Teubner.

Miller, J. A., and A. E. Mussett. 1963. Dating basic rocks by the potassium-argon method, the Whin sill, *Roy. Astron. Soc. Geophys. Jour.*, 7: 547.

Miller, J. A., K. Shibata, and M. Munro. 1962. The potassium-argon age of the lava of Killerton Park, near Exeter, *Roy. Astron. Soc. Geophys. Jour.*, 6: 394–96.

Mitroff, I. I. 1974. *The Subjective Side of Science, a Philosophical Inquiry into the Psychology of the Apollo Moon Scientists*. New York: American Elsevier.

Mohr, P. A. 1970. The Afar triple junction and sea-floor spreading, *Jour. Geophys. Res.*, 75: 7340–52.

Momose, K. 1958. Paleomagnetic researches for the Pliocene volcanic rocks in central Japan, *Jour. Geomag. Geoelec.*, 10: 12–19.

Montanari, C., and T. D. Allan. 1963. An improvement in the design of the towed nuclear spin magnetometer, *Tech. Rept.* SACLANT A.S.W. *Res. Centre*, 16: 1–2.

Morgan, W. J. 1968. Rises, trenches, great faults, and crustal blocks, *Jour. Geophys. Res.*, 73: 1959–82.

———. 1971. Convection plumes in the lower mantle, *Nature*, 230: 42–43.

Morley, L. W., and A. Larochelle. 1964. Paleomagnetism as a means of dating geological events, in F. F. Osborne, ed., *Geochronology in Canada* (Roy. Soc. Canada Spec. Pub. no. 8), pp. 39–51. Toronto: University of Toronto Press.

Mousuf, A. K. 1952. K^{40} radioactive decay; its branching ratio and its use in geological age determinations, *Phys. Rev.*, 88: 150.

Musset, A. E. 1969. Diffusion measurements and the potassium-argon method of dating, *Roy. Astron. Soc. Geophys. Jour.*, 13: 257–303.

Nagata, T. 1943. The natural remanent magnetism of volcanic rocks and its relation to geomagnetic phenomena, *Tokyo Imp. Univ. Earthquake Res. Inst.*, B. 21: 1–196.

———. 1950. Natural remanent magnetism of igneous rocks and its mode of development, *Nature*, no. 4189: 165 (Feb. 11, 1931).

———. 1952. Reverse thermo-remanent magnetism, *Nature*, 169: 704.

———. 1953a. *Rock Magnetism*. Tokyo: Maruzen.

———. 1953b. Self-reversal of thermoremanent magnetization of igneous rocks, *Nature*, 172: 850.

———. 1953c. Ferrimagnetism in nature, in *Proceedings of the International Conference of Theoretical Physics*, pp. 714–18. Tokyo: Science Council of Japan.

———. 1961. *Rock Magnetism*. 2nd ed. Tokyo: Maruzen.

Nagata, T., S. Akimoto, and S. Uyeda. 1951. Reverse thermoremanent magnetism, *Japan Acad. Proc.*, 27: 643–45.

———. 1952a. Reverse thermo-remanent magnetism (II), *Japan Acad. Proc.*, 28: 277–81.

———. 1953a. Self-reversal of thermo-remanent magnetism of igneous rocks (III), *Jour. Geomag. Geoelec.*, 5: 168–84.

———. 1953b. Origin of reverse thermo-remanent magnetism of igneous rocks, *Nature*, 172: 630.

Nagata, T., S. Akimoto, S. Uyeda, K. Momose, and E. Asami. 1954. Reverse magnetization of rocks and its connection with the geomagnetic field, *Jour. Geomag. Geoelec.*, 6: 182–93.

Nagata, T., S. Akimoto, S. Uyeda, Y. Shimizu, M. Ozima, and K. Kobayashi. 1957a. Palaeomagnetic study on a Quaternary volcanic region in Japan, *Adv. Phys.*, 6: 255–63.

Nagata, T., S. Akimoto, S. Uyeda, Y. Shimizu, M. Ozima, K. Kobayashi, and H. Kuno. 1957b. Palaeomagnetic studies on a Quaternary volcanic region in Japan, *Jour. Geomag. Geoelec.*, 9: 23–41.

Nagata, T., S. Uyeda, and S. Akimoto. 1952b. Self-reversal of thermo-remanent magnetism of igneous rocks (I), *Jour. Geomag. Geoelec.*, 4: 22–38.

Nagata, T., S. Uyeda, S. Akimoto, and N. Kawai. 1952c. Self-reversal of thermo-remanent magnetism of igneous rocks (II), *Jour. Geomag. Geoelec.*, 4: 102–07.

Nairn, A. E. M. 1956. Relevance of palaeomagnetic studies of Jurassic rocks to continental drift, *Nature*, 178: 935–36.

———. 1957a. Palaeomagnetic collections from Britain and South Africa illustrating two points in weathering, *Adv. Phys.*, 6: 162–68.

———. 1957b. Observations paléomagnétiques en France; roches permiennes, *Soc. Geol. France Bull.*, 7: 721–27.

———. 1960a. A palaeomagnetic survey of the Karroo System, *Overseas Geol. Min. Resources Gt. Brit.*, 7: 398–410.

———. 1960b. Paleomagnetic results from Europe, *Jour. Geol.*, 68: 285–306.

Nakamura, S., and S. Kikuchi. 1912. Permanent magnetism of volcanic bombs, *Physico-Math. Soc. Japan Proc.*, 6: 268–73.

National Research Council, Committee on Nuclear Science. 1954. *Proceedings of the* [1st] *Conference on Nuclear Processes in Geologic Settings, Williams Bay, Wisconsin, September 21–23, 1953*. Washington, D.C.: National Academy of Sciences, National Research Council.

———. 1956. *Nuclear Processes in Geologic Settings; Proceedings of the Second Conference, Pennsylvania State University, September 8–10, 1955*. Natl. Res. Coun. Publ. 400.

National Science Foundation. 1955. *Fifth Annal Report*, FY 1955. Washington, D.C.: National Science Foundation.

Needham, J. 1962. *Physics and Physical Technology, I. Physics* (vol. 4 of *Science and Civilization in China*). Cambridge: University Press.

Néel, L. E. F. 1949. Théorie du traînage magnétique des ferromagnétiques au grains fins avec applications aux terres cuites, *Ann. Geophys.*, 5: 99–136.

———. 1951. L'inversion de l'aimantation permanente des roches, *Ann. Geophys.*, 7: 90–102.

———. 1953. Thermoremanent magnetization of fine powders, *Rev. Mod. Phys.*, 25: 293–97.

———. 1955. Some theoretical aspects of rock magnetism, *Adv. Phys.*, 4: 191–243.

Newmann, F. H., and H. J. Walke. 1935. The radioactivity of potassium and rubidium, *Phil. Mag.*, Ser. 7, 19: 767.

Nicholls, G. D. 1955. The mineralogy of rock magnetism, *Adv. Phys.*, 4: 113–90.

Nier, A. O. 1935. Evidence for the existence of an isotope of potassium of mass 40, *Phys. Rev.*, 48: 283.

———. 1936. A mass spectrographic study of the isotopes of argon, potassium, rubidium, zinc and cadmium, *Phys. Rev.*, 50: 1041.

———. 1937. A mass-spectrographic study of the isotopes of Hg, Xe, Kr, Be, I, As, and Cs. *Phys. Rev.*, 52: 933.

———. 1947. A mass spectrometer for isotope and gas analysis, *Rev. Sci. Inst.*, 18: 398.

———. 1981. Some reminiscences of isotopes, geochronology, and mass spectrometry, *Ann. Rev. Earth Planet Sci.*, 9: 1–17.

Ninkovich, D., N. D. Opdyke, B. C. Heezen, and J. H. Foster. 1966. Paleomagnetic stratigraphy, rates of deposition and tephrachronology in north Pacific deep-sea sediments, *Earth Planet. Sci. Lett.*, 1: 476–92.

Obradovich, J. D. 1964. *Problems in the Use of Glauconite and Related Minerals for Radioactivity Dating*. Ph.D. thesis, Univ. California, Berkeley.

Oliver, J., and B. Isacks. 1967. Deep earthquake zones, anomalous structures in the upper mantle, and the lithosphere, *Jour. Geophys. Res.*, 72: 4259–75.

Opdyke, N. D. 1969. The Jaramillo event as detected in oceanic cores, in S. K. Runcorn, ed., *The Application of Modern Physics to Earth and Planetary Interiors*, pp. 549–52. New York: John Wiley.

———. 1972. Paleomagnetism of deep-sea cores, *Rev. Geophys. Space Phys.*, 10: 213–49.

Opdyke, N. D., B. P. Glass, J. D. Hays, and J. H. Foster. 1966. Paleomagnetic study of Antarctic deep-sea cores, *Science*, 154: 349–57 (Oct. 21, 1966).

Opdyke, N. D., and S. K. Runcorn. 1956. New evidence for reversal of the geomagnetic field near the Pliocene-Pleistocene boundary, *Science*, 123: 1126–27.

Oppel, Albert. 1856–58. Die Juraformation Englands, Frankreichs, und des sudwestlichen Deutschlands, *Wurtemberg Naturwis. Verein*, 12–14: 1–438, 439–694, 695–857. (Ph.D. at Tubingen).

Oriel, S. S., R. W. MacQueen, J. A. Wilson, and G. B. Dalrymple. 1976. Stratigraphic commission; note 44—application for addition to Code concerning magnetostratigraphic units, *Amer. Assoc. Petrol. Geol. Bull.*, 60: 273–77.

Packard, M., and R. Varian. 1954. Free nuclear induction in the Earth's magnetic field, *Phys. Rev.*, Ser. 2, 93: 941.

Pahl, M., J. Hiby, F. Smits, and W. Gentner. 1950. Mass spectrometric determination of argon from potash salts. *Z. Naturforsch.*, 5a: 404.

Pecherski, D. M. 1970. Paleomagnetism and paleomagnetic correlation of Mesozoic formations of north-east USSR, *Akad. Nauk SSSR, Sci. Works of North-East Complex Institute (SVKNEE), Magadan*, 37: 58–114.

Pekeris, C. L. 1935. Thermal convection in the interior of the earth, *Roy. Astron. Soc. Geophys. Supp. Mon. Not.*, 3: 346–67.

Pettersson, H. 1949. Exploring the bed of the ocean, *Nature*, 164: 468–70.

Péwé, T. L., *et al.* 1953. *Multiple glaciations in Alaska; a progress report* (U.S. Geol. Survey Circ. 289).

Pichler, H. 1970. *Italienische Vulkan-Gebiete* (*Sammlung geologischer Führer*, vol. 51). Berlin: Gebruder Borntraeger.

Pitman, W. C., III, and D. E. Hayes. 1968. Sea-floor spreading in the Gulf of Alaska, *Jour. Geophys. Res.*, 73: 6571–80.

Pitman, W. C., III, and J. P. Heirtzler. 1966. Magnetic anomalies over the Pacific-Antarctic Ridge, *Science*, 154: 1164–71.

Popov, V. I. 1947. Paleomagnetism and the application of the magnetometric method, *Sov. Geol.*, 25: 21–28.

Press, F., and M. Ewing. 1952. Magnetic anomalies over oceanic structures, *Amer. Geophys. Union Trans.*, 33: 349–55.

Putnam, W. C. 1960. Relation of the McGee glacial stage to the late Cenozoic history of the Sierra Nevada, *Geol. Soc. Amer. Bull.*, 71: 1950–51 (abstract).

Rabbitt, J. C., and M. C. Rabbitt. 1954. The U.S.G.S., 75 years of service to the nation, 1879–1954, *Science*, 119: 741–58.

Raff, A. D. 1962. Further magnetic measurements along the Murray Fault, *Jour. Geophys. Res.*, 67: 417–18.

———. 1963. Magnetic anomaly over Mohole drill EM7, *Jour. Geophys. Res.*, 68: 955–56.

Raff, A. D., and R. G. Mason. 1961. Magnetic survey off the west coast of North America, 40°N latitude to 52°N latitude, *Geol. Soc. Amer. Bull.*, 72: 1267–70.

Raitt, R. W. 1956. Seismic refraction studies of the Pacific Ocean basin, *Geol. Soc. Amer. Bull.*, 67: 1623–40.

Revelle, R., and A. E. Maxwell. 1952. Heat flow through the floor of the eastern North Pacific Ocean, *Nature*, 170: 199–202.

Reynolds, J. H. 1950. A mass-spectrometric investigation of branching in copper 64, bromine 80, bromine 82 and iodine 128, *Phys. Rev.*, 79: 789.

———. 1952. The surface ionization of lanthanum, *Phys. Rev.*, 85: 770 (abstract).

———. 1953. The isotopic constitution of silicon, germanium, and hafnium, *Phys. Rev.*, 90: 1047.

———. 1954. A high sensitivity mass spectrometer, *Phys. Rev.*, 98: 283.

———. 1956a. High sensitivity mass spectrometer for rare gas analysis, *Rev. Sci. Inst.*, 27: 928.

———. 1956b. K-Ar dating, in National Research Council Committee on Nuclear Science. *Nuclear Processes in Geologic Settings; Proceedings of the Second Conference, Pennsylvania State University, September 8–10, 1955* (Natl. Res. Coun. Publ. 400), p. 135.

———. 1957. Comparative study of argon content and argon diffusion in mica and feldspar, *Geochim. Cosmochim. Acta*, 12: 177.

———. 1960a. Determination of the age of the elements, *Phys. Rev. Letters*, 4: 8.

———. 1960b. Isotopic composition of primordial xenon, *Phys. Rev. Letters*, 4: 351.

———. 1960c. Isotopic composition of xenon from enstatite chondrites, *Z. Naturforsch.*, 15a: 1112.

———. 1960d. I-Xe dating of meteorites, *Jour. Geophys. Res.*, 65: 3843.

———. 1960e. Rare gases in tektites, *Geochim. Cosmochim. Acta*, 20: 101.

———. 1960f. The age of the elements in the solar system, *Sci. Amer.*, 203: 171.

———. 1967. Isotopic abundance anomalies in the solar system, *Ann. Rev. Nuclear Sci.*, 17: 253–316.

Reynolds, J. H., E. C. Alexander, Jr., P. K. Davis, and B. Srinivasan. 1974. Potassium-argon dating and xenon from extinct radioactivity in breccia 14318; Implications for early lunar history, *Geochim. Cosmochim. Acta*, 38: 401–17.

Reynolds, J. H., and J. Lipson. 1954. A multicircuit control for ultra-high vacuum gauges, *Rev. Sci. Inst.*, 24: 1029.

———. 1957. Rare gases from the Nuevo Laredo stone meteorite, *Geochim. Cosmochim. Acta*, 12: 330.

Reynolds, J. H., and J. Verhoogen. 1953. Natural variations in the isotopic constitution of silicon, *Geochim. Cosmochim. Acta*, 3: 224–34.

Reynolds, J. H., and T. J. Ypsilantis. 1953. Techniques in isotopic abundance measurements on elements of group IV, *Phys. Rev.*, 90: 378 (abstract).

Rikitake, T. 1958. Oscillations of a system of disk dynamos, *Cambridge Phil. Soc. Proc.*, 54: 89–105.

Rinehart, C. D., and D. C. Ross. 1956. *Economic geology of the Casa Diablo Mountain quadrangle, California* (California Div. Mines Spec. Rept. 48).

———. 1957. *Geology of the Casa Diablo Mountain quadrangle, California* (U.S. Geol. Survey Geol. Quadrangle Map GQ–99).

Rittenberg, D. 1942. Some applications of mass spectrometric analysis to chemistry, *Journal Appl. Phys.*, 13: 561.

Roche, A. 1950a. Sur les caractères magnétiques du système éruptif de Gergovie, *Acad. Sci. Paris Compt. Rend.*, 230: 113–15.

———. 1950b. Anomalies magnétiques accompagnant des massifs de pépérites de la Limagne d'Auvergne, *Acad. Sci. Paris, Compt. Rend.*, 230: 1603–4.

———. 1951. Sur les inversions de l'aimantation rémanente des roches volcaniques dans les monts d'Auvergne, *Acad. Sci. Paris Compt. Rend.*, 233: 1132–34.

———. 1953. Sur l'origine des inversions de l'aimantation constantées dans les roches d'Auvergne, *Acad. Sci. Paris Compt. Rend.*, 236: 107–9.

———. 1954. Exposé sommaire des études relatives à l'aimantation de matériaux volcaniques, *Jour. Geomag. Geoelec.*, 6: 169–71.

———. 1956. Sur la date de la dernière inversion du champ magnétique terrestre, *Acad. Sci. Paris Compt. Rend.*, 243: 812–14.

———. 1957. Sur l'aimantation des roches volcaniques de l'Estérel, *Acad. Sci. Paris Compt. Rend.*, 244: 2952–54.

———. 1958. Sur les variations de direction du champ magnétique terrestre au cours du Quaternaire, *Acad. Sci. Paris Compt. Rend.*, 246: 3364–66.
———. 1959. Paléomagnétisme déplacements des poles et dérives des continents, *Rev. l'Indust. Min.*, 41: 1–10.
———. 1960. Sur l'aimantation de laves miocènes d'Auvergne, *Acad. Sci. Paris Compt. Rend.*, 250: 377–79.
Roche, A., and L. Cattala. 1959. Remanent magnetism of the Cretaceous basalts of Madagascar, *Nature*, 183: 1049–50.
Roche, A., L. Cattala, and J. Boulanger. 1958. Sur l'aimantation de basaltes de Madagascar, *Acad. Sci. Paris Compt. Rend.*, 246: 2922–24.
Roche, A., and B. Leprêtre. 1955. Sur l'aimantation de roches volcaniques de l'Ahaggar, *Acad. Sci. Paris Compt. Rend.*, 240: 2002–4.
Roche, A., H. Saucier, and J. Lacaze. 1962. Étude paléomagnétique des roches volcaniques permiennes de la région Nideck-Donon, *Bull. Serv. Carte Géol. Alsace-Lorraine*, 15: 59–68.
Roquet, J. 1954. Sur les rémanences des oxydes de fer et leur interêt en géomagnétisme (first and second parts), *Ann. Geophys.*, 10: 226–47, 282–325.
Rothé, J. P. 1954. La zone séismique médiane Indo-Atlantique, *Roy. Soc. London Proc.*, Ser. A, 222: 387.
Runcorn, S. K. 1955a. Palaeomagnetism of sediments from the Colorado Plateau, *Nature*, 176: 505–6.
———. 1955b. Rock magnetism—geophysical aspects, *Adv. Phys.*, 4: 244–91.
———. 1956a. Magnetization of rocks, *Handbuch der Physik*, 47: 470–97.
———. 1956b. Paleomagnetic comparisons between Europe and North America, *Geol. Assoc. Canada Proc.*, 8: 77–85.
———. 1956c. Paleomagnetic survey in Arizona and Utah; preliminary results, *Geol. Soc. Amer. Bull.*, 67: 301–16.
———. 1956d. Palaeomagnetism, polar wandering and continental drift, *Geol. Mijn.*, 18: 253–56.
———. 1957. The sampling of rocks for palaeomagnetic comparisons between the continents, *Adv. Phys.*, 6: 169–76.
———. 1959. Rock magnetism, *Science*, 129: 1002–11.
Russell, R. D., H. A. Shillibeer, R. M. Farquhar, and A. K. Mousuf. 1953. The branching ratio of potassium 40, *Phys. Rev.*, 91: 1223.
Rutherford, E. 1905. Present problems in radioactivity, *Pop. Sci. Monthly*, (May): 1–34.
Rutherford, E., and F. Soddy. 1902a. The cause and nature of radioactivity, Pts. I and II, *Phil. Mag.*, Ser. 6, 4: 370–96, 569–85.
———. 1902b. The radioactivity of thorium compounds, Pts. I and II, *Chem. Soc. London Jour.*, 81: 321–50, 837–60.
Rutten, M. G. 1941. A synopsis of the Orbitoididae, *Geol. Mijn.*, 3: 34–62.
———. 1958. Geological reconnaissance of the Esja-Hvalfjordur-Armanns-

fell area, southwestern Iceland, *K. Nederl. Geol.-Mijn. Gen. Verh.*, Geol. Ser., 17: 219–98.

———. 1959. Paleomagnetic reconnaissance of mid-Italian volcanoes, *Geol. Mijn.*, 21: 373–74.

———. 1960. Paleomagnetic dating of younger volcanic series, *Geol. Rundsch.*, 49: 161–67.

———. 1969. *The Geology of Western Europe*. Amsterdam: Elsevier.

Rutten, M. G., and J. C. den Boer. 1954. Inversion de l'aimantation dans les basaltes du Coiron (Ardèche), *Soc. Géol. France Compt. Rend. Som.*, 5: 106–7.

Rutten, M. G., R. O. v'Everdingen, and J. D. A. Zijderveld. 1957. Palaeomagnetism in the Permian of the Oslo graben (Norway) and of the Esterel (France), *Geol. Mijn.*, 19: 193–95.

Rutten, M. G., and J. Veldkamp. 1958. Paleomagnetic research at Utrecht University, *Ann. Geophys.*, 14: 519–21.

———. 1964. Paleomagnetic research in the Netherlands, *Geol. Mijn.*, 43: 183–95.

Rutten, M. G., and H. Wensink. 1959. Geology of the Hvalfjordur-Skorradalur area (southwestern Iceland), *Geol. Mijn.*, 21: 172–81.

———. 1960. Paleomagnetic dating, glaciations and the chronology of the Plio-Pleistocene in Iceland, in T. Sorgenfrei, ed., *International Geological Congress; Report of the Twenty-first Session, Norden*, Part 4, pp. 62–70.

Savage, D. E. 1975. Cenozoic—the Primate Episode, in F. S. Szalay, ed., *Approaches to Primate Paleobiology* (*Contributions to Primatology*, vol. 5), pp. 244–67. New York: S. Karger.

Savage, D. E., and J. F. Evernden. 1961. Critical points in the Cenozoic, *N.Y. Acad. Sci. Ann.*, 91: 342–51.

Schaeffer, O. A., and J. Zahringer, comps. 1966. *Potassium Argon Dating*. Berlin: Springer-Verlag.

Scheidegger, A. E., and J. T. Wilson. 1950. An investigation into possible methods of failure of the earth, *Proc. Geol. Assn. Canada*, 3: 167–190.

Schult, A. von. 1965. The effect of pressure on the Curie temperature of magnetite and some other ferrites, *Z. Geophys.*, 34: 505–11.

Shand, S. J. 1949. Rocks of the Mid-Atlantic Ridge, *Journ. Geol.*, 57: 89–92.

Shillibeer, H. A., and R. D. Russell. 1954. The potassium-argon method of geological age determination, *Can. Jour. Phys.*, 32: 681.

Shor, E. N. 1978. *Scripps Institution of Oceanography: Probing the Oceans, 1936 to 1976*. San Diego, Calif.: Tofua Press.

Sigurgeirsson, T. 1957. Direction of magnetization in Icelandic basalts, *Adv. Phys.*, 6: 240–46.

Slichter, L. B. 1950. The Rancho Sante Fe conference concerning the evolution of the Earth, *Nat. Acad. Sci. Proc.*, 36: 511–14.

Smith, P. J. 1968. Pre-Gilbertian conceptions of terrestrial magnetism, *Tectonophysics*, 6: 499–510.

———. 1970. *Petrus Peregrinus Epistola*—the beginning of experimental studies of magnetism in Europe, *Atlas (News Supp. to Earth Sci. Revs.)* vol. 6, pt. I, pp. A11–17.

Smith, P. J., and J. Needham. 1967. Magnetic declination in medieval China, *Nature*, 214: 1213–14.

Smith, R. L., R. A. Bailey, and C. S. Ross. 1961. Structural evolution of the Valles Caldera, New Mexico and its bearing on the emplacement of ring dykes, in *Geological Survey Research, 1961* (U.S. Geol. Survey Prof. Paper 424–D: D145–49).

Smith, William. 1815. *Memoir to the map and delineation of the strata of England and Wales with part of Scotland.* London: J. Cary.

———. 1820. *A new geological map of England and Wales with the inland navigations exhibiting the districts of coal and other sites of mineral tonnage.* London: J. Cary.

Smits, F., and W. Gentner. 1950. Argonbestimmungen an Kalium-Mineralien I. Bestimmungen an tertiaren Kalisalzen, *Geochim. Cosmochim. Acta*, 1: 22–27.

Smythe, W. R., and A. Hemmendinger. 1937. The radioactive isotope of potassium, *Phys. Rev.*, 51: 178.

Snavely, P. D., Jr., and E. M. Baldwin. 1948. Siletz River volcanic series, northwestern Oregon, *Amer. Assoc. Petrol. Geol. Bull.*, 32: 806–12.

Snavely, P. D., Jr., and H. E. Vokes. 1949. *Geology of the coastal area from Cape Kiwanda to Cape Foulweather, Oregon* (U.S. Geol. Survey Oil and Gas Inv. Prelim. Map 97).

Stearns, H. T. 1946. *Geology of the Hawaiian Islands* (Hawaii Terr. Div. Hydrography Bull. 8).

Steno, Nicolaus. 1669. *De Solido intra Solidum Naturaliter Contento Dissertationes Prodromus.*

Stevens, J. R., and H. A. Shillibeer. 1956. Loss of argon from minerals and rocks due to crumbling, *Proc. Geol. Assn. Canada*, 8: 71.

Stubbs, P. H. S. 1958. *Continental Drift and Polar Wandering: A Palaeomagnetic Study of British and European Triassic and of the British Old Red Sandstones.* Ph.D. thesis, Univ. London.

Suess, E. 1904–24. *The Face of the Earth* (*Das Antlitz der Erde*), transl. by H. B. C. Sollas. 5 vols. Oxford: Clarendon Press.

Suess, H. E. 1948. On the radioactivity of K^{40}, *Phys. Rev.*, 73: 1209.

Sullivan, Walter. 1974. *Continents in Motion—The New Earth Debate.* New York: McGraw-Hill.

Sutton, G. H., and E. Berg. 1959. Seismological studies of the western rift valley of Africa, *Amer. Geophys. Union Trans.*, 39: 474–81.

Sykes, L. R. 1966. Mechanism of earthquakes and nature of faulting on the mid-oceanic ridges, *Geol. Soc. Amer. Abstracts*, pp. 216–17 (GSA Special Paper no. 101, submitted for the meeting in San Francisco, Nov. 14–16, 1966).

———. 1967. Mechanism of earthquakes and nature of faulting on the mid-oceanic ridges, *Jour. Geophys. Res.*, 72: 2131–53.

———. 1968. Seismological evidence for transform faults, sea floor spreading and continental drift, in R. A. Phinney, ed., *The History of the Earth's Crust, a Symposium* (Contributions to a Conference held at the Goddard Institute for Space Studies, Nov. 10–11, 1966), pp. 120–50. Princeton, N.J.: Princeton University Press.

Takeuchi, H., and Y. Shimazu. 1952. On a self-exciting process in magneto-hydrodynamics, *Jour. Phys. Earth*, Tokyo, 1, 1: 57; 1953 *Jour. Geophys. Res.*, 58: 497.

Takeuchi, H., S. Uyeda, and H. Kanamori. 1967. *Debate About the Earth.* San Francisco: W. H. Freeman.

Talwani, M., X. Le Pichon, and J. R. Heirtzler. 1965. East Pacific Rise; the magnetic pattern and the fracture zones, *Science*, 150: 1109–15 (Nov. 26, 1965).

Tarling, D. H. 1962. Tentative correlation of Samoan and Hawaiian Islands using "reversals" of magnetization, *Nature*, 196: 882–83.

Thellier, E. 1936. Aimantation des briques et inclinaison du champ magnétique terrestre, *Paris Univ. Inst. Phys. du Globe Ann.*, 14: 65–70.

———. 1937a. Sur l'aimantation dite permanent des basaltes, *Acad. Sci. Paris Compt. Rend.*, 204: 876–78.

———. 1937b. Recherche de l'intensité du champ magnétique terrestre dans le passé; premier résultats, *Paris Univ. Inst. Phys. du Globe Ann.*, 15: 179–84.

———. 1937c. Aimantation des terres cuites; application à la recherche de l'intensité du champ magnétique terrestre dans le passé, *Acad. Sci. Paris Compt. Rend.*, 204: 184–86.

———. 1938. *Sur l'aimantation des terres cuites et ses applications géophysiques.* Thesis, Univ. Paris.

———. 1951. Propriétés magnétiques des terres cuites et des roches, *Jour. Phys. Radium*, 12: 205–18.

Thellier, E., and F. Rimbert. 1954. Sur l'analyse d'aimantations fossiles par action de champs magnétiques alternatifs, *Acad. Sci. Paris Compt. Rend.*, 239: 1399–401.

———. 1955. Sur l'utilisation en paléomagnétisme, de la désaimantation par champs alternatifs, *Acad. Sci. Paris Compt. Rend.*, 240: 1404–6.

Thellier, E., and O. Thellier. 1941. Sur les variations thermiques de l'aimantation thermorémanente du terres cuites, *Acad. Sci. Paris Compt. Rend.*, 213: 59–61.

———. 1942. Sur l'intensité du champ magnétique terrestre, en France, trois siècles avant les première mésures directes, *Acad. Sci. Paris Compt. Rend.*, 214: 382–84.

———. 1949. Sur les propriétés magnétiques des roches éruptives pyrenéennes, *Acad. Sci. Paris Compt. Rend.*, 228: 1958–60.

———. 1951. Sur la direction du champ magnétique terrestre, retrouvée sur des parois de fours des époques punique et romaine, à Carthage, *Acad. Sci. Paris Compt. Rend.*, 233: 1476–78.

———. 1952. Sur la direction du champ magnétique terrestre, dans la région de Trèves, vers 380 après J.-C., *Acad. Sci. Paris Compt. Rend.*, 234: 1464–66.

———. 1959. Sur l'intensité du champ magnétique terrestre dans le passé historique et géologique, *Ann. Geophys.*, 15: 285–376.

Thiadens, A. A. 1970. In memoriam, Prof. Dr. M. G. Rutten, *Geol. Mijn.*, 49: 433–38.

Thompson, F. C., and S. Rowlands. 1943. Dual decay of potassium, *Nature*, 152: 103.

Thompson, J. J. 1905. On the emission of negative corpuscles by the alkali metals, *Phil. Mag.*, Ser. 6, 10: 584.

Turner, F. J. 1948. Mineralogical and structural evolution of metamorphic rocks, *Geol. Soc. Amer. Mem.*, 30: 130–32.

Uyeda, S. 1955. Magnetic interaction between ferromagnetic materials contained in rocks, *Jour. Geomag. Geoelec.*, 7: 9–36.

———. 1956. Magnetic interaction between ferromagnetic minerals contained in rocks (II), *Jour. Geomag. Geoelec.*, 8: 39–70.

———. 1957. Thermo-remanent magnetism and coercive force of the ilmenite-hematite series, *Jour. Geomag. Geoelec.* 9: 61–78.

———. 1958. Thermo-remanent magnetism as a medium of paleomagnetism, with special reference to reverse thermo-remanent magnetism, *Jap. Jour. Geophys.*, 2: 1–123.

———. 1962. Thermoremanent magnetism and reverse thermoremanent magnetism, in T. Nagata, ed., *Proceedings of the Benedum Earth Magnetism Symposium*, pp. 87–106. Pittsburgh: University of Pittsburgh Press.

———. 1978. *The New View of the Earth.* San Francisco: W. H. Freeman.

Uyeda, S., and K. Horai. 1964. Terrestrial heat flow in Japan, *Jour. Geophys. Res.*, 69: 2121–41.

Vacquier, V. 1959. Measurement of horizontal displacement along faults in the ocean floor, *Nature*, 183: 452–53.

———. 1962a. A machine method for computing the magnitude and the direction of the magnetization of a uniformly magnetized body from its shape and a magnetic survey, in T. Nagata, ed., *Proceedings of the Benedum Earth Magnetism Symposium*, pp. 123–37. Pittsburgh: University of Pittsburgh Press.

———. 1962b. Magnetic evidence for horizontal displacements in the floor of the ocean, in S. K. Runcorn, ed., *Continental Drift*, pp. 135–44. New York: Academic Press.

———. 1965. Transcurrent faulting in the ocean floor, in P. M. S. Blackett, E. Bullard, and S. K. Runcorn, eds., Symposium on continental drift, *Roy. Soc. London Phil. Trans.*, Ser. A, 258: 77–81.

Vacquier, V., and R. P. von Herzen. 1964. Evidence for connection between heat flow and the Mid-Atlantic Ridge magnetic anomaly, *Jour. Geophys. Res.*, 69: 1093–101.

Vacquier, V., A. D. Raff, and R. E. Warren. 1960. Horizontal displacements in the floor of the northeastern Pacific Ocean, *Geol. Soc. Amer. Bull.*, 72: 1251–58.

Valiev, V. V. 1960a. [Paleomagnetic subdivision of the Margazar section of the Cenozoic continental molasse beds], *Akad. Nauk SSSR Izv.*, Geol. Ser., 7: 974–76 (in Russian). English translation, 1961.

———. 1960b. The position of the poles during the Tertiary period as determined from the remanent magnetization of rocks in northern Ferghana, *Akad. Nauk SSSR Izv.*, Geophys. Ser., 26: 1213–15.

Vening Meinesz, F. A. 1930. Maritime gravity surveys in the Netherlands East Indies, tentative interpretation of the results, *Ned. Akad. Wetensch. Proc.*, Ser. B, 33: 566–77.

———. 1954. The origin of continents and oceans, *Geol. Mijn.*, 31: 373–84.

Verhoogen, J. 1954. Petrological evidence on temperature distribution in the mantle of the Earth, *Amer. Geophys. Union Trans.*, 35: 50–59.

———. 1956. Ionic ordering and self-reversal of magnetization in impure magnetites, *Jour. Geophys. Res.*, 61: 201–9.

———. 1959. The origin of thermo-remanent magnetization, *Jour. Geophys. Res.*, 64: 2441–49.

———. 1960. Temperatures within the earth, *Amer. Sci.*, 48: 134.

Vine, F. J. 1966 (December 16). Spreading of the ocean floor; new evidence, *Science*, 154: 1405–15.

Vine, F. J., and D. H. Matthews. 1963. Magnetic anomalies over ocean ridges, *Nature*, 199: 947–49.

Vine, F. J., and J. T. Wilson. 1965. Magnetic anomalies over a young ocean ridge off Vancouver Island, *Science*, 150: 485–89.

Visser, S. W. 1936. Some remarks on the deep-focus earthquakes in the international seismological summary, *Gerlands Beitr. Geophys.*, 48: 254–67.

Wahrhaftig, C., and A. Cox. 1959. Rock glaciers in the Alaska Range, *Geol. Soc. Amer. Bull.*, 70: 383–436.

Wasserburg, G. J. 1954. Argon 40 potassium dating, in H. Faul, ed. *Nuclear Geology*, p. 342. New York: Wiley.

Wasserburg, G. J., and R. J. Hayden. 1954. The branching ratio of K^{40}, *Phys. Rev.*, 93: 645.

———. 1955a. A^{40}-K^{40} dating, *Geochim. Cosmochim. Acta*, 7: 51–60.

———. 1955b. Age of meteorites by the A^{40}-K^{40} method, *Phys. Rev.*, 97: 86–87.

———. 1956. A^{40}-K^{40} dating, in National Research Council Committee on Nuclear Science. *Nuclear Processes in Geologic Settings; Proceedings of the Second Conference, Pennsylvania State University, September 8–10, 1955*. (Natl. Res. Coun. Publ. 400), pp. 131–34.

Watershoot van der Gracht, W. A. J. M. van. 1928. *Theory of Continental Drift: A Symposium*. Tulsa, Okla.: American Association of Petroleum Geologists.

Watkins, N. D. 1971. Geomagnetic polarity events and the problem of "The Reinforcement Syndrome," *Comments Earth Sci. Geophys.*, 2, no. 2: 36–43.

———. 1972. Review of the development of the geomagnetic polarity time scale and discussion of prospects for its finer definition, *Geol. Soc. Amer. Bull.*, 83: 551–74.

———. 1974. L. W. Morley and sea-floor spreading, *Geology*, 2: 170 (April; letter).

Watkins, N. D., and S. E. Haggerty. 1968. Oxidation and magnetic polarity in single Icelandic lavas and dykes, *Roy. Astron. Soc. Geophys. Jour.*, 15: 305–15.

Wayland, E. J. 1930. Rift valleys and Lake Victoria, *International Geological Congress; Compte Rendu of the XV Session*, Pretoria, vol. 2, pp. 323–53.

Wegener, A. 1912a. Die Entstehung der Kontinente, *Petermanns Geogr. Mitt.*, 58: 185–95, 253–56, 305–08.

———. 1912b. Die Entstehung der Kontinente, *Geol. Rundsch.*, 3: 276–92.

———. 1915. *Die Entstehung der Kontinente und Ozeane*. Braunschweiz: Vieweg.

Weiss, P. E. 1911. Molecular magnets, *Acad. Sci. Paris Compt. Rend.*, 152: 79–81.

Weizsäcker, C. F. von. 1937. Über die Möglichkeit eines dualen Beta-Zerfalls von Kalium, *Phys. Z.*, 38: 623.

Wensink, H. 1964a. Paleomagnetic stratigraphy of younger basalts and intercalated Plio-Pleistocene tillites in Iceland, *Geol. Rundsch.*, 54: 364–84.

———. 1964b. Secular variation of Earth magnetism in Plio-Pleistocene basalts of eastern Iceland, *Geol. Minj.*, 43: 403–13.

Wertenbaker, W. 1974. *The Floor of the Sea*. Boston: Little, Brown.

Wetherill, G. W. 1966. K-Ar dating of Precambrian rocks, in O. A. Schaeffer, and J. Zahringer, comp., *Potassium Argon Dating*, pp. 107–17. New York: Springer-Verlag.

Wetherill, G. W., L. T. Aldrich, and G. L. Davis. 1955. A^{40}/K^{40} ratios of feldspars and micas from the same rock, *Geochim. Cosmochim. Acta*, 8: 171–72.

Wetherill, G. W., G. J. Wasserburg, L. T. Aldrich, G. R. Tilton, and R. J. Hayden. 1956. Decay of constants of K^{40} as determined by the radiogenic argon content of potassium minerals, *Phys. Rev.*, 103: 987.

Williams, H. 1929. Geology of the Marysville Buttes, California, *Calif. Univ. Pubs. Dept. Geol. Sci. Bull.*, 18: 103–220.

Willis, B. 1936. *Studies in Comparative Seismology; East African Plateaus and Rift Valley* (Carnegie Inst. Publ. 470). Washington, D.C.; Carnegie Institution.

———. 1944. Continental drift, ein Märchen, *Amer. Jour. Sci.*, 242: 509–13.

Wilson, J. T. 1962a. Cabot fault; an Appalachian equivalent of the San Andreas and the Great Glen faults and some implications for continental displacement, *Nature*, 195: 135–138.

———. 1962b. Some further evidence in support of the Cabot fault, a great Paleozoic transcurrent fault zone in the Atlantic provinces and New England, *Trans. Roy. Soc. Canada*, Ser. 3, 56: 31–36.
———. 1962c. The effect of new orogenetic theories upon ideas of the tectonics of the Canadian Shield, in *The Tectonics of the Canadian Shield* (*Roy. Soc. of Canada, Spec. Pub.* No. 4), pp. 174–180.
———. 1963a. Evidence from islands on the spreading of the ocean floor, *Nature*, 197: 536–38.
———. 1963b. A possible origin of the Hawaiian Islands, *Canadian Jour. Phys.*, 41: 863–870.
———. 1965a. A new class of faults and their bearing on continental drift, *Nature*, 207: 343–47.
———. 1965b. Transform faults, oceanic ridges and magnetic anomalies southwest of Vancouver Island, *Science*, 150: 482–85.
———. 1965c. Submarine fracture zones, aseismic ridges and the International Council of Scientific Unions lines; proposed western margin of the East Pacific Ridge, *Nature*, 207: 907–10.
———. 1966. Did the Atlantic close and then reopen? *Nature*, 211: 676–81.
———. 1970. A new class of faults and their bearing on continental drift, in P. Cloud, ed., *Adventures in Earth History*, pp. 351–58. San Francisco: W. H. Freeman.
Wilson, R. L. 1961. Palaeomagnetism in Northern Ireland, part I and II, *Geophys. Jour.*, 5: 45–69.
———. 1962a. An instrument for measuring vector magnetization at high temperatures, *Roy. Astron. Soc. Geophys. Jour.*, 7: 125–30.
———. 1962b. The paleomagnetic history of a doubly-baked rock, *Roy. Astron. Soc. Geophys. Jour.*, 6: 397–99.
———. 1962c. The palaeomagnetism of baked contact rocks and reversals of the Earth's magnetic field, *Roy. Astron. Soc. Geophys. Jour.*, 7: 194–202.
———. 1963. The palaeomagnetism of some rhyolites from Northern Ireland, *Roy. Astron. Soc. Geophys. Jour.*, 8: 149–64.
———. 1964. Magnetic properties and normal and reversed natural magnetization in the Mull lavas, *Roy. Astron. Soc. Geophys. Jour.*, 8: 424–39.
———. 1965. Does the Earth's magnetism reverse its polarity? *New Scientist*, 27: 380–81.
———. 1966a. Further correlations between the petrology and the natural magnetic polarity of basalts, *Roy. Astron. Soc. Geophys. Jour.*, 10: 413–20.
———. 1966b. Polarity inversions of the Earth's magnetic field, in W. R. Hindmarsh *et al.*, eds., *Magnetism and the Cosmos*, pp. 79–84 (NATO Advanced Study Institute on Planetary and Stellar Magnetism held in 1965 at the University, Newcastle-upon-Tyne). New York: American Elsevier.
———. 1970a. Palaeomagnetic stratigraphy of Tertiary lavas from Northern Ireland, *Roy. Astron. Soc. Geophys. Jour.*, 20: 1–9.
———. 1970b. Permanent aspects of the Earth's non-dipole magnetic field over Upper Tertiary times, *Roy. Astron. Soc. Geophys. Jour.*, 19: 417–37.

———. 1971. Dipole offset—the time average palaeomagnetic field over the past 25 million years, *Roy. Astron. Geophys. Soc. Jour.*, 22: 491–504.

Wilson, R. L., and J. M. Ade-Hall. 1970. Palaeomagnetic indications of a permanent aspect of the non-dipole field, in S. K. Runcorn, ed., *Palaeogeophysics*, pp. 307–12. London: Academic Press.

Wilson, R. L., and S. E. Haggerty. 1966. Reversals of the Earth's magnetic field, *Endeavour*, 25: 104–9.

Wilson, R. L., S. E. Haggerty, and N. D. Watkins. 1968. Variation of palaeomagnetic stability and other parameters in a vertical traverse of a single Icelandic lava, *Roy. Astron. Soc. Geophys. Jour.*, 16: 79–96.

Wilson, R. L., and N. D. Watkins. 1967. Correlation of magnetic polarity and petrological properties in Columbia Plateau basalts, *Roy. Astron. Soc. Geophys. Jour.*, 12: 405–24.

Winder, G. G. 1979. Peripatetic perambulation on publication, in C. J. Schneer, ed., *Two Hundred Years of Geology in America* (Proc. New Hampshire Bicentennial Conf. on the History of Geology, Univ. of New Hampshire, 1976), p. 32. Hanover, N.H.: University Press of New England.

Wiseman, J. D. H., and R. B. S. Sewell. 1937. The floor of the Arabian Sea, *Geol. Mag.*, 74: 219–30.

Wood, H. E., R. W. Cheney, J. Clark, E. H. Colbert, G. L. Jepsen, J. B. Reeside, and C. Stock. 1941. Nomenclature and correlation of the North American continental Tertiary, *Geol. Soc. Amer. Bull.*, 52: 1–48.

Wyllie, P. J. 1976. *The Way the Earth Works: An Introduction to the New Global Geology and Its Revolutionary Development*. New York: John Wiley and Sons.

Young, P. 1980. Earth's fountains of fire, *Mosaic*, 11: 1–9 (May/June).

Zartman, R. E., G. J. Wasserburg, and J. H. Reynolds. 1961. Helium, argon, and carbon in some natural gases, *Jour. Geophys. Res.*, 66: 277.

Zijl, J. S. V. van, K. W. T. Graham, and A. L. Hales. 1962. The paleomagnetism of the Stormberg lavas of South Africa (I and II), *Roy. Astron. Soc. Geophys. Jour.*, 7: 23–39, 169–82.

Table of Interviewees

Tapes, transcripts, running tables of contents, and other primary historical materials are on deposit in the History of Science and Technology Program, Bancroft Library, University of California at Berkeley. Duplicates of certain tapes and transcripts (marked by an asterisk) are on deposit also in the Center for History of Physics, American Institute of Physics, New York, N.Y.

Name and affiliation	Number of hours of interview: No tape record	Number of hours of interview: Tape record	Date(s) of interview
Aldrich, Lyman T. Carnegie Inst., Wash., D.C.	—	½	v-24-81
Balsley, James R. U.S.G.S., Reston, Va.	1	1	ii-7-78
Bateman, Paul C. U.S.G.S., Menlo Park, Calif.	2	—	ii-27-78
Broecker, Wallace S. Lamont-Doherty Geol. Observ., N.Y.	—	½	viii-30-79
Bullard, Edward C. (deceased) formerly at Univ. of Cambridge, U.K.	1	2	vii-2-79
Chamalaun, Francois H. Flinders Univ., Australia	correspondence only		i-18-78
Christensen, Mark Univ. of Calif., Berkeley	3	1	iii-20-78
Coe, Robert S. Univ. Calif., Santa Cruz	2	—	i-19-78
Corbett, Morley Univ. Calif., Berkeley	1	—	iii-20-78
Cox, Allan V.* Stanford Univ., Calif.	6	7	vii-28-76 to ix-5-78
Creasey, Cyrus S. U.S.G.S., Menlo Park, Calif.	3	1	iii-2-78

Name and affiliation	Number of hours of interview: No tape record	Number of hours of interview: Tape record	Date(s) of interview
Curtis, Garniss H.* Univ. Calif., Berkeley	10	6	iv-12-78 to v-10-78
Dalrymple, G. Brent* U.S.G.S., Menlo Park, Calif.	15	4	iii-30-78 to iii-31-78
Dewey, John F. State Univ. N.Y. at Albany	2	1	iii-13-81
Diment, William H. U.S.G.S., Reston, Va.	—	½	iii-14-78
Doell, Richard R.* Univ. Calif., Berkeley	10	7	x-25-77 to v-16-78
Doell, Ruth J. San Francisco State Univ., Calif.	7	1	iv-1-78
Durham, J. Wyatt Univ. Calif., Berkeley	1	1	vi-23-78
Elsasser, Walter Johns Hopkins Univ., Md.	3	1	viii-11-76
Evernden, Jack F.* U.S.G.S., Menlo Park, Calif.	10	5	xi-29-77 to iii-14-78
Folinsbee, Robert E. Univ. of Alberta, Canada	1	2	iii-20-78
Foster, John H.	3	5½	viii-13-79 to viii-20-79
Gilluly, James U.S.G.S., retired	—	¾	ii-1-78
Glass, Billy Univ. of Maryland	—	½	viii-30-79
Goldich, Samuel S. U.S.G.S., Denver, Colo.	6	—	iii-23-78 to iii-25-78
Grommé, Sherman C. U.S.G.S., Menlo Park, Calif.	4	2	xi-29-77 to iii-29-78
Hamilton, Warren B. U.S.G.S., Denver, Colo.	½	—	viii-1-79
Harrison, Christopher	2	1	i-21-81
Hay, Richard L. Univ. Calif., Berkeley	1	—	i-12-81
Heezen, Bruce C. (deceased) formerly at Lamont-Doherty Geol. Observ., N.Y.	4	6	viii-23-76
Heirtzler, James R. Woods Hole Oceanographic Inst., Mass.	1	2	viii-16-76 to viii-17-76
Herron, Ellen M. Lamont-Doherty Geol. Observ., N.Y.	—	½	viii-1-79
Hoare, Joseph U.S.G.S., Menlo Park, Calif.	½	1	iii-3-78
Hopkins, David M. U.S.G.S., Menlo Park, Calif.	6	2	iii-2-78 to iii-13-78
Irving, Edward M. Carleton Univ., Ottawa, Canada	½	1	i-31-78 to xi-30-78
Kern, John Exxon Corp., Houston, Tex.	—	1	v-8-78

Name and affiliation	Number of hours of interview: No tape record	Number of hours of interview: Tape record	Date(s) of interview
Kistler, Ronald W. U.S.G.S., Menlo Park, Calif.	1	1½	ii-10-78 to iv-14-78
Kleinpell, Robert M. Univ. Calif., Berkeley	1	1½	vi-21-78
Lachenbruch, Arthur U.S.G.S., Menlo Park, Calif.	½	—	iii-17-78
Lajoie, Kenneth R. U.S.G.S., Menlo Park, Calif.	2	—	
Langseth, Marcus G. Lamont-Doherty Geol. Observ., N.Y.	—	1½	xii-28-79
Lanphere, Marvin A. U.S.G.S., Menlo Park, Calif.	—	2	iii-7-78 to iii-8-78
Larochelle, Andre Div. Energy and Mines, Ottawa, Canada	—	½	viii-31-79
Lillard, Major U.S.G.S., retired	1	½	iii-6-78
Lipson, Joseph Nat. Sci. Found., Washington, D.C.	2	2	ii-1-78 to ii-4-78
Malde, Harold E. U.S.G.S., Denver, Colo.	2	1	iii-24-78
Matthews, Drummond H. Univ. of Cambridge, U.K.	¼	1	v-28-79
Maxwell, Arthur E. Woods Hole Oceanographic Inst., Mass.	1	1½	viii-17-76
McDougall, Ian Australian Nat. Univ., Canberra	12	5½	viii-12-78 to viii-15-78
McElhinny, Michael W. Australian Nat. Univ., Canberra	—	1	xi-29-78
Morgan, Jason Princeton Univ., N.J.	2	1	i-22-81
Morley, Lawrence W. Geol. Surv. of Canada	—	1	iv-24-79 to x-15-79
Moxham, Robert (deceased) formerly of U.S.G.S.	—	¼	iii-6-78
Nier, Alfred O. Univ. Minn., Minneapolis	—	½	vi-24-81
Ninkovich, Dragoslav Lamont-Doherty Geol. Observ., N.Y.	—	½	viii-14-79
Obradovich, John U.S.G.S., Denver, Colo.	1	2	iv-14-78
Opdyke, Neil D. Lamont-Doherty Geol. Observ., N.Y.	1	3	v-2-79 to viii-20-79
Pakiser, Louis C. U.S.G.S., Denver, Colo.	1	½	iii-24-78
Pitman, Walter C. Lamont-Doherty Geol. Observ., N.Y.	1	4	iv-24-79 to viii-6-79
Raff, Arthur D. Scripps Inst. of Oceanog., Calif.	1	1½	vi-12-79
Reynolds, John H. Univ. Calif., Berkeley	3	6	i-30-78 to vi-29-78

Name and affiliation	Number of hours of interview: No tape record	Number of hours of interview: Tape record	Date(s) of interview
Richards, John R. Univ. of Toronto	½	1	v-1-78
Ryan, William B. F. Lamont-Doherty Geol. Observ., N.Y.	½	½	viii-18-78
Sass, John H. U.S.G.S., Menlo Park, Calif.	½	1	v-9-78
Savage, Donald E. Univ. Calif., Berkeley	1	1	iv-15-78
Sherrill, Nathaniel D. U.S.G.S., Menlo Park, Calif.	1	1	viii-25-78
Snavely, Parke D. U.S.G.S., Menlo Park, Calif.	1	1	v-30-78
Spiess, Fred N. Scripps Inst. of Oceanog., Calif.	1	—	ix-25-80
Tarling, Donald H. Australian Nat. Univ., Canberra	—	1	ii-1-78 to vi-21-78
Turner, Esme Berkeley, Calif.	—	½	viii-7-78
Turner, Francis J. Univ. Calif., Berkeley (emeritus)	2	1½	ii-1-78
Vacquier, Victor Scripps Inst. of Oceanog., Calif.	1	½	iv-23-79
Van Andel, Tjeerd H.G. Stanford Univ., Calif.	1	½	vi-9-78
Verhoogen, John* Univ. Calif., Berkeley	3	6	viii-11-77 to ix-20-77
Vetleson Prize Award Dinner (Allan V. Cox, Richard R. Doell, Maurice Ewing, Polykarp Kusch, S. Keith Runcorn; recorded by G. Brent Dalrymple)	—	1	iv-8-71
Vine, Fred J. Univ. East Anglia, U.K.	1	2½	v-18-78
Von Herzen, Richard Woods Hole Oceanog. Inst., Mass.	½	2	viii-19-76
Wahrhaftig, Clyde U.S.G.S., Menlo Park, Calif.	4	3	ii-28-78 to iii-12-78
Washburn, Sherwood L. Univ. Calif., Berkeley	½	—	vi-26-78
Wasserburg, Gerald J. Calif. Inst. Tech.	1	1¼	iii-31-78
Wetherill, G.W. Dept. Terr. Magn., Carnegie Inst., Washington, D.C.	—	1	v-30-78
Wilson, J. Tuzo Ontario Science Centre, Canada	1½	2	v-25-79 to x-29-80

Illustration Credits

Photographs

Listed below are sources for the photographs; these do not carry figure numbers in their legends, and accordingly are indicated here by page number. Most have not previously appeared in print. Where a credit does not indicate "By [someone]," the photographer is not known. Photographs not listed at all below are by the author.

p. 28 Courtesy of John Verhoogen
p. 31 Courtesy of John Reynolds
p. 38 Courtesy of Joseph Lipson
p. 39 Courtesy of John Reynolds
p. 40 Courtesy of John Reynolds
p. 42 Courtesy of Joachim Hampel
p. 43 Courtesy of Robert Folinsbee
p. 80 By Richard Doell; courtesy of Richard Doell
p. 82 By Richard Doell; courtesy of Richard Doell
p. 89 By Richard Doell; courtesy of Richard Doell
p. 95 Courtesy of S. Keith Runcorn
p. 96 Courtesy of John Verhoogen
p. 102 Courtesy of Allan Cox
p. 105 By F. J. Lowes; courtesy of F. J. Lowes
p. 115 By Roderic Wilson; courtesy of Roderic Wilson
p. 117 By David Hopkins; courtesy of David Hopkins
p. 125 Courtesy of Roderic Wilson
p. 127 By Richard Doell; courtesy of Richard Doell
p. 129 Courtesy of Allan Cox
p. 131 *Journal of the Royal Geological and Mining Society of the Netherlands*, 49 (1970), p. 433; used with permission.
p. 143 Courtesy of Richard Doell
p. 146 By Richard Doell; courtesy of Richard Doell
p. 148 Courtesy of Richard Doell
p. 149 Courtesy of Richard Doell
p. 157 Courtesy of Allan Cox
p. 158 By Imogen Cunningham; courtesy of Clyde Wahrhaftig
p. 164 Courtesy of Allan Cox
p. 165 Courtesy of Allan Cox
p. 175 Courtesy of James Balsley
p. 178 Courtesy of U.S. Geological Survey
p. 180 Courtesy of Major Lillard
p. 184 By Richard Doell; courtesy of Richard Doell
p. 187 Courtesy of Richard Doell

p. 188 Courtesy of Brent Dalrymple
p. 192 Courtesy of Brent Dalrymple
p. 205 Courtesy of Brent Dalrymple
p. 209 Courtesy of Allan Cox
p. 211 Courtesy of Ian McDougall
p. 235 By Richard Doell; courtesy of Richard Doell
p. 236 Courtesy of Richard Hay
p. 244 Courtesy of Allan Cox
p. 248 Courtesy of Sherman Grommé
p. 256 By William Laughlin; courtesy of William Laughlin
p. 261 Courtesy of Allan Cox
p. 274 By J. R. Cann; courtesy of Fred Vine
p. 276 Courtesy of Allan Cox
p. 280 By Richard Doell; courtesy of Richard Doell
p. 283 By Richard Doell; courtesy of Richard Doell
p. 290 Courtesy of Arthur Raff
p. 293 Courtesy of Arthur Raff
p. 296 Courtesy of Victor Vacquier
p. 298 Courtesy of Lawrence Morley
p. 305 Courtesy of J. Tuzo Wilson
p. 313 Courtesy of Allan Cox
p. 323 Courtesy of Neil Opdyke
p. 324 Courtesy of Christopher Harrison
p. 333 Courtesy of Walter Pitman
p. 357 Courtesy of Richard Doell
p. 361 Courtesy of Brent Dalrymple, Richard Doell, Sherman Grommé, and NASA
p. 362 Courtesy of Allan Cox and Richard Doell

Figures

Sources for the numbered figures (most of them reproductions from scientific journals) are given in the figure legends themselves. Additional information for the figures from *Science* is as follows: Fig. 6.6 appeared in vol. 142 (1963), pp. 382–85; Fig. 6.9 in vol. 144 (1964), pp. 1537–43; Fig. 6.15 in vol. 152 (1966), pp. 1060–61; Figs. 7.4–7.5 in vol. 150 (1965), pp. 485–89; Fig. 8.3 in vol. 154 (1966), pp. 1164–71; Figs. 8.4–8.9 in vol. 154 (1966), pp. 1405–15; and Fig. 8.12 in vol. 154 (1966), pp. 349–57. (Copyright by the American Association for the Advancement of Science in the year of first publication.)

Figs. 6.12 and 6.13 are from H. E. Wright, Jr., and David G. Frey, eds., *The Quaternary of the United States*. Copyright 1965 by Princeton University Press. Publication authorized by the U.S. Geological Survey, Menlo Park, California. Figs. 7 and 8, pp. 823 and 824, reprinted by permission of Princeton University Press.

Index